Earth System Analysis for Sustainability

Edited by

Hans Joachim Schellnhuber, Paul J. Crutzen,
William C. Clark, Martin Claussen, and
Hermann Held

Program Advisory Committee:

Hans Joachim Schellnhuber, Paul J. Crutzen, and
William C. Clark, Chairpersons
Martin Claussen, Hermann Held, Tim M. Lenton, and
Will Steffen

The MIT Press
Cambridge, Massachusetts
London, U.K.

in cooperation with Dahlem University Press

This book was set in Times New Roman by Dahlem Konferenzen.

Printed and bound in the United States of America.

Library of Congress Control Number: 2004110774

ISBN: 0-262-19513-5

Dahlem Workshop on Earth System Analysis for Sustainability (91st : 2003 : Berlin, Germany)

Contents

Dahlem Workshops

History

During the last half of the twentieth century, specialization in science greatly increased in response to advances achieved in technology and methodology. This trend, although positive in many respects, created barriers between disciplines, which could have inhibited progress if left unchecked. Understanding the concepts and methodologies of related disciplines became a necessity. Reuniting the disciplines to obtain a broader view of an issue became imperative, for problems rarely fall conveniently into the purview of a single scientific area. Interdisciplinary communication and innovative problem-solving within a conducive environment were perceived as integral yet lacking to this process.

In 1971, an initiative to create such an environment began within Germany's scientific community. In discussions between the *Deutsche Forschungsgemeinschaft* (German Science Foundation) and the *Stifterverband für die Deutsche Wissenschaft* (Association for the Promotion of Science Research in Germany), researchers were consulted to compare the needs of the scientific community with existing approaches. It became apparent that something new was required: an approach that began with state-of-the-art knowledge and proceeded onward to challenge the boundaries of current understanding; a form truly interdisciplinary in its problem-solving approach.

As a result, the *Stifterverband* established *Dahlem Konferenzen* (the Dahlem Workshops) in cooperation with the *Deutsche Forschungsgemeinschaft* in 1974. Silke Bernhard, formerly associated with the Schering Symposia, was

Figure adapted from *L'Atmosphère: Météorologie Populaire*, Camille Flammarion. Paris: Librairie Hachette et Cie., 1888.

engaged to lead the conference team and was instrumental in implementing this unique approach.

The Dahlem Workshops were named after a district of Berlin known for its strong historic connections to science. In the early 1900s, Dahlem was the seat of the Kaiser Wilhelm Institutes where, for example, Albert Einstein, Lise Meitner, Fritz Haber, and Otto Hahn conducted research. Today the district is home to several Max Planck Institutes, the *Freie Universität Berlin*, the *Wissenschaftskolleg*, and the Konrad Zuse Center.

In its formative years, the Dahlem Workshops evolved in response to the needs of science. They soon became firmly embedded within the international scientific community and were recognized as an indispensable tool for advancement in research. To secure its long-term institutional stability, *Dahlem Konferenzen* was integrated into the *Freie Universität Berlin* in 1990.

Aim

The aim of the Dahlem Workshops is to promote an international, interdisciplinary exchange of scientific information and ideas, to stimulate international cooperation in research, and to develop and test new models conducive to more effective communication between scientists.

Concept

The Dahlem Workshops were conceived to be more than just another a conference venue. Anchored within the philosophy of scientific enquiry, the Dahlem Workshops represent an independently driven quest for knowledge: one created, nurtured, and carefully guided by representatives of the scientific community itself. Each Dahlem Workshop is an interdisciplinary communication process aimed at expanding the boundaries of current knowledge. This dynamic process, which spans more than two years, gives researchers the opportunity to address problems that are of high-priority interest, in an effort to identify gaps in knowledge, to pose questions aimed at directing future inquiry, and to suggest innovative ways of approaching controversial issues. The overall goal is not necessarily to exact consensus but to search for new perspectives, for these will help direct the international research agenda.

Governance

Just as the efforts of different disciplines are needed to address problems faced by science, so too are the skills of a wide variety of experts required to ensure the successful continuance of the Dahlem Workshops.

The Scientific Advisory Board, composed of representatives from the international scientific community, is responsible for the scientific content and future directions of the Dahlem Workshops. The Board meets biannually to review and approve all workshop proposals as well as to discuss future directions. The Program Director implements the decisions of the Scientific Advisory Board and develops the future directions.

The *Präsidium* of the *Freie Universität Berlin* is responsible for the administration of the institute. The Executive Director represents the *Präsidium* in administrative matters. An International Advisory Board was created in 2003 to strengthen international financial support for the institute.

Workshop Topics

Workshop topics are problem-oriented, interdisciplinary by nature, of high-priority interest to the disciplines involved, and timely to the advancement of science. Scientists who submit workshop proposals, and chair the workshops, must be internationally recognized experts active in their field.

Program Advisory Committee

Once a proposal has been approved, a Program Advisory Committee is formed for each workshop. Composed of 6–7 scientists representing the various scientific disciplines involved, the committee meets approximately one year before the Dahlem Workshop to develop the scientific program of the meeting. The committee selects the invitees, determines the topics that will be covered by the pre-workshop papers, and assigns each participant a specific role. Participants are invited on the basis of their international scientific reputation alone. The integration of young German scientists is promoted through special invitations.

Dahlem Workshop Model

A Dahlem Workshop can best be envisioned as a week-long intellectual retreat. Participation is strictly limited to forty participants to optimize the interaction and communication process.

Participants work in four interdisciplinary discussion groups, each organized around one of four key questions. There are no lectures or formal presentations at a Dahlem Workshop. Instead, concentrated discussion — within and between groups — is the means by which maximum communication is achieved.

To enable such an exchange, participants must come prepared to the workshop. This is facilitated through a carefully coordinated pre-workshop dialog: Discussion themes are presented through "background papers," which review a particular aspect of the group's topic and introduce controversies as well as unresolved problem areas for discussion. These papers are circulated in advance, and everyone is requested to submit comments and questions, which are then compiled and distributed. By the time everyone arrives in Berlin, issues have been presented, questions have been raised, and the Dahlem Workshop is ready to start.

Discussion takes place in moderated sessions as well as through informal interactions. Cross-fertilization between groups is both stressed and encouraged. By the end of the week, through a collective effort directed by a rapporteur, each group has prepared a draft report of the ideas, opinions, and contentious issues raised by the group. Directions for future research are highlighted, as are problem areas still in need of resolution. On the final day, the results of the draft reports are discussed in a plenary session, to which colleagues from the Berlin–Brandenburg area are invited.

Dahlem Workshop Reports

After the workshop, attention is directed toward the necessity of communicating the results of the workshop to a wider audience. A two-tier review process guides the revision of the background papers, and discussion continues to finalize the group reports. The chapters are carefully edited to highlight the perspectives, controversies, gaps in knowledge, and proposed research directions.

The publication of the workshop results in book form completes the process of a Dahlem Workshop, as it turns over the insights gained to the broad scientific community for consideration and implementation. Each volume in the Dahlem Workshop Report series contains the revised background papers and group reports as well as an introduction to the workshop themes. The series is published in partnership with The MIT Press.

Julia Lupp, Program Director and Series Editor
Dahlem Konferenzen der Freien Universität Berlin
Thielallee 50, 14195 Berlin, Germany

List of Participants

Meinrat O. Andreae Abteilung Biogeochemie, Max-Planck-Institut für Chemie, Otto-Hahn-Institut, Postfach 3060, 55020 Mainz, Germany
Biosphere–atmosphere exchange; biomass burning; aerosol–climate interactions

Bert Bolin S. Åsvägen 51, 18452 Österskär, Sweden
Climate change in the Anthropocene — the interplay between the environmental system and the global socioeconomic system — building awareness and public opinion; key issues and thresholds

Victor Brovkin Climate System Department, Potsdam Institute for Climate Impact Research (PIK), P.O. Box 601203, 14412 Potsdam, Germany
Climate–vegetation interaction; terrestrial and marine biogeochemistry; Quaternary carbon cycle; feedbacks in the climate system

Ken G. Caldeira Climate and Carbon Cycle Modeling Group, Energy and Environment Directorate, Lawrence Livermore National Laboratory, 7000 East Ave., L-103, Livermore, CA 94550, U.S.A.
Climate; carbon cycle; energy systems; geochemical cycles; ocean biogeochemistry and physics; numerical simulation

William C. Clark John F. Kennedy School of Government, Harvard University, 79 John F. Kennedy Street, Cambridge, MA 02138, U.S.A.
Linkages between global environmental change and economic development, policy analysis for resource and environmental management, and understanding the use of incomplete scientific knowledge in decision-making

Martin Claussen Climate System Department, Potsdam Institute for Climate Impact Research (PIK), P.O. Box 601203, 14412 Potsdam, Germany
Climate system modeling; paleoclimate modeling

Peter M. Cox Hadley Centre for Climate Prediction and Research, Met Office, FitzRoy Road, Exeter, Devon EX1 3PB, U.K.
Earth system modeling; climate carbon cycle feedback; dynamic global vegetation modeling; land–atmosphere interaction

Paul J. Crutzen Abteilung Luftchemie, Max-Planck-Institut für Chemie, Otto-Hahn-Institut, Postfach 3060, 55020 Mainz, Germany
Global modeling of atmospheric chemical processes for troposphere, stratosphere, and lower mesosphere; interactions of atmospheric chemistry with climate

Ulrich Cubasch Institut für Meteorologie, Freie Universität Berlin, Carl-Heinrich-Becker Weg 6–10, 12165 Berlin, Germany
Climate modeling; climate change modeling; climate variability; climate monitoring

Ottmar Edenhofer Potsdam Institute for Climate Impact Research (PIK), P.O. Box 601203, 14412 Potsdam, Germany
Sustainability science; integrated assessment models; economic growth theory

Paul G. Falkowski   Institute of Marine and Coastal Sciences and Department of Geological Sciences, Rutgers, The State University of New Jersey, 71 Dudley Road, New Brunswick, NJ 08901, U.S.A.
Global biogeochemical cycles; photosynthesis; symbiosis; paleoecology

Siegfried A. Franck   Potsdam Institute for Climate Impact Research (PIK), P.O. Box 601203, 14412 Potsdam, Germany
Earth system analysis; geophysiology; astrobiology

Gilberto C. Gallopín   Division of Environment and Human Settlements, ECLAC, Casilla 179-D, Santiago, Chile
Ecological systems analysis; environmental impact assessment; sustainable development; impoverishment and global simulation models

Hermann Held   Potsdam Institute for Climate Impact Research (PIK), P.O. Box 601203, 14412 Potsdam, Germany
Uncertainty propagation in Earth system models; emergent phenomena in complex nonlinear systems; integrated assessment

Gerda Horneck   Institute of Aerospace Medicine, Radiation Biology, DLR, German Aerospace Center, Linder Höhe, 51147 Cologne, Germany
Astrobiology; environmental limits of life; search for signatures of life beyond Earth; human exploratory missions in the solar system

Alison Jolly   The Old Brewery House, 32 Southover High Street, Lewes, Sussex BN7 1HX, U.K. [Affiliation: School of Biological Sciences, University of Sussex]
Lemur behavior and survival (local and international influences on conservation in Madagascar); human social behavior (evolution and future evolution)

Ann P. Kinzig   School of Life Sciences, Life Sciences C. Wing, Arizona State University, Tempe, AZ 85287, U.S.A.
Urban ecology; human–environment interactions; resilience; landscape ecology and ecological legacies

Timothy M. Lenton   Centre for Ecology and Hydrology, Bush Estate, Penicuik, Midlothian EH26 0QB, U.K.
GAIA theory; Earth system science and modeling evolution; astrobiology; Earth history; Quaternary climate changes; human-induced global change; Earth future

Wolfgang Lucht   Potsdam Institute for Climate Impact Research (PIK), P.O. Box 601203, 14412 Potsdam, Germany
Global biogeochemistry; Earth observation; interaction of sociosphere and biosphere; cognitive elements in evolution and transitions; sustainability

Ronald B. Mitchell   Department of Political Science, 1284 University of Oregon, Eugene, OR 97403–1284, U.S.A.
International environmental law and politics; the interface of science and policy; international institutions and their influence on behavior

Nebojsa Nakicenovic International Insitute for Applied Systems Analysis (IIASA), Schlossplatz 1, 2361 Laxenburg, Austria
Long-term patterns of technological change and economic development; response to climate change, and the evolution of energy, mobility, and information technologies; diffusion of new technologies and their interactions with the environment; dynamics of technological and social change, mitigation of anthropogenic impacts on the environment and on response strategies to global change

Anthony J. Payne School of Geographical Sciences, University Road, University of Bristol, Bristol BS8 1SS, U.K.
Glaciology; Antarctic ice sheet; paleoceanography; numerical modeling

Elke Rabbow Institute of Aerospace Medicine, Radiation Biology, DLR, German Aerospace Center, Linder Höhe, 51147 Cologne, Germany
Exo/astrobiology; toxicology; biosensorics

Stefan Rahmstorf Potsdam Institute for Climate Impact Research (PIK), P.O. Box 601203, 14412 Potsdam, Germany
The role of the ocean and climate

Patricia Romero Lankao Department of Politics and Culture, Autonomous Metropolitan University of Mexico at Xochimilco, Calzada del Hueso 1100, Villa Quietud, Delegación Coyoacán, C.P. 04960, D.F. México
Carbon and regional pathways of development; water; water reform and sustainability; structural change, state reform, and sustainability

Hans Joachim Schellnhuber Tyndall Center for Climate Change Research, School of Environmental Sciences, University of East Anglia, Norwich NR4 7TJ, U.K., and Potsdam Institute for Climate Impact Research (PIK), P.O. Box 601203, 14412 Potsdam, Germany
Earth system analysis; nonlinear dynamics; integrated assessment of climate change; science of integration

Robert J. Scholes CSIR Environmentek, P.O. Box 395, Pretoria 0001, South Africa
Savanna ecology; global change; biodiversity; assessment of ecosystem services

Daniel P. Schrag Department of Earth and Planetary Sciences, Harvard University, 20 Oxford Street, Cambridge, MA 02138, U.S.A.
Isotope geochemistry; paleoclimatology; geochemical oceanography

Frank Sirocko Institute of Geoscience, University of Mainz, Becherweg 21, 55099 Mainz, Germany
Paleoclimate; Quaternary geoscience; neotectonics

S. Sreekesh TERI, Darbari Seth Block, Habitat Place, Lodhi Road, New Delhi 110 003, India
Regulatory and policy research: remote sensing, GIS, and land resource management

Will Steffen Executive Director, IGBP Secretariat, Royal Swedish Academy of Sciences, Box 50005, 104 05 Stockholm, Sweden
Earth system science; global carbon cycle

Eörs Szathmáry   Collegium Budapest, 2 Szentháromság, 1014 Budapest, Hungary
Evolutionary biology; origin of language; origin of the genetic code; astrobiology; evolution of sex

Liana Talaue-McManus   Division of Marine Affairs, University of Miami, Rosenstiel School of Marine and Atmospheric Science, 4600 Rickenbacker Causeway, Miami, FL 33149, U.S.A.
Human–environment interactions on the coastal zone; integrated coastal management; tropical plankton ecology

Sir Crispin Tickell   Ablington Old Barn, Ablington, Cirencester, Gloucestershire GL7 5NU, U.K.
Environmental issues, especially climate change; sustainability generally; Earth systems science; paleontology

Billie L. Turner II   Graduate School of Geography, Clark University, Worcester, MA 01602, U.S.A.
Geography; human–environment studies; land change; sustainability science

Andrew J. Watson   School of Environmental Sciences, University of East Anglia, Norwich NR4 7TJ, U.K.
Evolution of the atmosphere–ocean–biosphere system, especially long-term carbon cycle, Quaternary carbon cycle; astrobiology and freqency of life in the Universe

Frances Westall   Centre de Biophysique Moléculaire, CNRS, Rue de Charles Sadron, 45000 Orléans Cedex 02, France
Origin of life; first life on Earth and its geological context; search for extraterrestrial life

Oran R. Young   Donald Bren School of Environmental Science and Management, Donald Bren Hall 4518, University of California, Santa Barbara, CA 93106–5131, U.S.A.
Social institutions; institutional dimensions of environmental change; human-dominated ecosystems

Gyorgy A. Zavarzin   Institute of Microbiology, Russian Academy of Sciences, Prospekt 60–letija, Oktiabria 7 korp. 2, Moscow, Russia
Functional microbial diversity; microbial communities; environmental aspects of global change; bacterial paleobiology

Heike Zimmermann-Timm   Potsdam Institute for Climate Impact Research (PIK), P.O. Box 601203, 14412 Potsdam, Germany
Ecology; freshwater and marine biology; microbial food web; extreme environments

1

Science for Global Sustainability

Toward a New Paradigm

W. C. CLARK[1], P. J. CRUTZEN[2], and H. J. SCHELLNHUBER[3]

[1]J.F. Kennedy School of Government, Harvard University,
Cambridge, MA 02138, U.S.A.
[2]Max Planck Institute for Chemistry, 55020 Mainz, Germany
[3]Tyndall Center for Climate Change Research, School of Environmental Sciences,
University of East Anglia, Norwich NR4 7TJ, U.K. and
Potsdam Institute for Climate Impact Research (PIK), 14412 Potsdam, Germany

INTRODUCTION

This paper provides a context for the Dahlem Workshop on "Earth Systems Science and Sustainability." We begin by characterizing the contemporary epoch of Earth history in which humanity has emerged as a major — and uniquely self-reflexive — geological force. We turn next to the extraordinary revolution in our understanding of the Earth system that is now underway, pointing out how it has built on and qualitatively extended the approaches that have served science and society so well since the first Copernican revolution. We then discuss the novel challenges posed by the urgent need to harness science and other forms of knowledge in promoting a worldwide sustainability transition that enhances human prosperity while protecting the Earth's life-support systems and reducing hunger and poverty. Finally, we provide an overview of how the contributions to this Dahlem Workshop addressed the themes and challenges outlined in this introductory chapter.

THE ANTHROPOCENE

We live today in what may appropriately be called the "Anthropocene" — a new geologic epoch in which humankind has emerged as a globally significant — and potentially intelligent — force capable of reshaping the face of the planet (Crutzen 2002).

History of the Idea

Humans have doubtless been altering their local environments since arriving on the scene as a distinct species several hundred thousand years ago. Our debut as major actors on the global stage — actors comparable in influence to the classic roles played by erosion, volcanism, natural selection, and the like — is a much more recent phenomenon. This dates back at most several thousand years, but has accelerated greatly in scope and influence over the last several centuries (Ruddiman 2003).

Self-awareness by humans of our role as global transformers is younger still. Seminal contributions began to emerge in the nineteenth century, for example *Ansichten der Natur* by the German geographer Alexander von Humboldt (1808) and *The Earth as Modified by Human Action* by the American diplomat George Perkins Marsh (1864, 1965). By 1873, the Italian geologist Antonio Stoppani was describing humanity's activities as a "new telluric force, which in power and universality may be compared to the greater forces of Earth" (Stoppani 1873). The theme was subsequently developed and given a much wider audience by the Russian geochemist V.I. Vernadsky in a series of lectures on the "biosphere" given at the Sorbonne in the early 1920s (Vernadsky 1998/1926, 1945). The last half century witnessed an accelerating program of scientific studies (e.g., Thomas 1956; Steffen et al. 2004) that have broadened and deepened our understanding of what Turner et al. (1990) have convincingly characterized as an "Earth transformed by human action."

Along with humanity's growing awareness of its role in transforming the Earth has come a growing recognition that how we use this awareness will shape the Earth's future, and our own. Vernadsky (1998/1926) himself speculated on "...the direction in which the processes of evolution must proceed, namely toward increasing consciousness and thought, and forms having greater and greater influence on their surroundings." Along with his French colleagues, the scientist and mystic P. Teilhard de Chardin and philosopher Édouard Le Roy, Vernadsky (1998/1926) coined the term "noösphere" to suggest a biosphere in which not only human action, but human thought and reflection on the consequences of its actions, would come to play a determinative role. Rapidly expanding efforts to manage the impact of human activities on the global environment show that humanity is taking seriously the idea and implications of a noösphere (Brown 1954; Clark 1989; Caldwell and Weiland 1996; Mitchell 2003).

The most recent big idea to emerge in the history of the Anthropocene is that of "sustainability"— a normative concept regarding not merely what *is*, but also what *ought to be* the human use of the Earth (Kates 2001). The concern for using our understanding of human impacts on the Earth's environment to help guide our use of the Earth in "sustainable" directions can be traced back to early work on the conservation of renewable resources. Much of that discussion was cast in terms of a contest between environmental protection and human development. By the late 1970s, however, the inadequacies of this traditional competitive framing were becoming increasingly clear. The World Conservation Strategy, published by the International Union for the Conservation of Nature (IUCN) in 1980, reframed the modern sustainability debate by arguing explicitly that goals for protecting the Earth's lands and wildlife could not be realized except through strategies that also addressed the improvement of human well-being in conservation areas. This view was formulated for environmental protection, in general, and the Earth, as a whole, in the report of the World Commission on Environment and Development (WCED — the Brundtland Commission) on *Our Common Future*, released in 1987.

The Brundtland Commission argued for advancing a global program of sustainable development that "meets the needs of the present without compromising the ability of future generations to meet their own needs" (WCED 1987, p. 8). Its conceptualization of the sustainability challenge was adopted by many world leaders in Rio de Janeiro at the UN Conference on Environment and Development (UNCED) in 1992, and was diffused broadly within governmental, business, and academic communities over the next decade. U.N. Secretary-General Kofi Annan (2000) reflected a growing consensus when he wrote in his Millennium Report to the General Assembly that "freedom from want, freedom from fear, and the freedom of future generations to sustain their lives on this planet" are the three grand challenges facing the international community at the dawn of the twenty-first century. By the time of the World Summit on Sustainable Development, held in Johannesburg in 2002, achieving sustainability had become a "high table" goal in international affairs, and on many regional, national, and local political agendas.

The need for harnessing science and technology in support of efforts to achieve the goal of environmentally sustainable human development in the Anthropocene was generally recognized at the Johannesburg Summit. How this might be accomplished was not. Exploring the options and opportunities for promoting such efforts was the central objective of this Dahlem Workshop on "Earth Systems Analysis for Sustainability."

State of the Transformation

An up-to-date understanding of how human actions are in fact transforming the Anthropocene is the necessary foundation for any serious effort to harness science and technology for sustainability. The recent report of the world scientific community's decade-long research program on *Global Environmental Change and the Earth System* (Steffen et al. 2004) provides such a foundation.

Drawing from the works of hundreds of researchers, the "Global Change" study (Steffen et al. 2004; Chapter 3) concluded that perhaps 50% of the world's ice-free land surface has been transformed by human action; the land under cropping has doubled during the past century at the expense of forests, which declined by 20% over the same period. More than half of all accessible freshwater resources have come to be used by humankind. Fisheries remove more than 25% of the primary production of the oceans in the upwelling regions and 35% in the temperate continental shelf regions (Pauly and Christensen 1995).

More nitrogen is now fixed synthetically and applied as fertilizers in agriculture than is fixed naturally in all terrestrial ecosystems. Over-application of nitrogen fertilizers in agriculture and its concentration in domestic animal manure have led to eutrophication of surface waters and groundwater in many locations around the world. They also lead to the microbiological production of N_2O, a greenhouse gas and a source of NO in the stratosphere, where it is strongly involved in ozone chemistry.

Humanity's exploitation of fossil fuels that were generated over several hundred million years has resulted in a large pulse of air pollutants. The release of SO_2 to the atmosphere by coal and oil burning is at least two times larger than the sum of all natural emissions, which occur mainly as marine dimethylsulfide from the oceans. The oxidation of SO_2 to sulfuric acid has led to acidification of precipitation and lakes, causing forest damage and fish death in biologically sensitive regions, such as Scandinavia and the northeast section of North America. As a result of substantial reduction in SO_2 emissions, the situation in these regions has improved somewhat over the last decades. However, the problem has gotten worse in East Asia.

The release of NO into the atmosphere from fossil-fuel and biomass combustion is likewise larger than the natural inputs, adding to rainwater acidity and giving rise to photochemical ozone ("smog") formation in extensive regions of the world.

Humanity is also responsible for the presence of many toxic substances in the environment and even some, the chlorofluorocarbon gases ($CFCl_3$ and CF_2Cl_2), which are not toxic at all, have nevertheless led to the Antarctic springtime "ozone hole"; CFCs would have destroyed much more of the ozone layer if international regulatory measures had not been taken to end their production by 1996. However, due to the long residence times of CFCs, it will take at least another 4–5 decades before the ozone layer will have recovered. The discovery of maximum reduction in stratospheric ozone came as a total surprise. It was not predicted and happened in a section of the atmosphere, where ozone loss was thought to be impossible and the furthest away from the regions of CFC releases to the atmosphere.

Due to fossil-fuel burning, agricultural activities, deforestation, and intensive animal husbandry, several climatically important "greenhouse" gases have substantially increased in the atmosphere over the past two centuries: CO_2 by more than 30% and CH_4 by even more than 100%, contributing substantially to the observed global average temperature increase by about 0.6°C, which has been observed during the past century. According to a report by the Intergovernmental Panel of Climate Change (IPCC 2001, p. 10): "There is new and stronger evidence that most of the warming observed over the last 50 years is attributable to human activities."

There is no question that humanity has done quite well by its transformation of the planet. Supported by great technological and medical advancements as well as by access to plentiful natural resources, we have colonized most places on Earth and even set foot on the Moon. The transformations of the last century helped humanity to increase the amount of cropland by a factor of 2, the number of people living on the planet by a factor of 4, water use by a factor of more than 8, energy use by a factor of 16, and industrial output by a factor of more than 40 (McNeill 2000; see Chapter 14, Table A-1, this volume). The quality of human life also increased, with average life expectancy up more than 40% in the last 50 years, literacy up more than 20% in the last 35 years, and substantial improvements in the female/male ratio in primary education, the number of people living in democratic countries, and the increased commitment of the international community to protect civilians from internal conflict and to defend the rights of national minorities (Kates and Parris 2003). The uneven distribution of these increases, their tenuous character, and the continued suffering of peoples left, or falling, behind are stark reminders that much more remains to be done. However, the fact remains that humanity, on average, has done very well indeed through its continuing transformation of the Earth. The question is whether past trends of increasing prosperity can be broadened and sustained as the Anthropocene matures.

Prognosis for the Future

The prognosis for continued and sustainable improvements in human well-being on a transformed planet Earth is, at best, guarded. The U.S. National Academy of Sciences has concluded that over the next half century, human population can be expected to increase by perhaps 50%. Associated with such an increase, the demand for food production could well increase by 80%, for urban infrastructure by 100%, and for energy services by substantially more than 200% (NRC 1999, p. 70). The resulting intensification of pressures on an already stressed biosphere could be overwhelming.

For example, depending on the scenarios of future energy use and model uncertainties, the increasing emissions and resulting growth in atmospheric concentrations of CO_2 are estimated to cause a rise in global average temperature by 1.4°–5.8°C during the present century, accompanied by sea-level rise of 9–88 cm (and 0.5–10 m until the end of the current millennium). According to Hansen (2004), considering only the warming of the globe over the past 50 years plus the warming already "in the pipeline" — together more than one degree Celsius — the Earth will return halfway to temperature conditions of the last interglacial, the Eemian (120 to 130 thousand years ago), when global sea levels were 5–6 meters higher than at present. Greater warming is, however, expected if humanity cannot drastically curtail the emissions of CO_2 and other greenhouse gases. The impact of current human activities is projected to last over very long periods. According to Loutre and Berger (2000), because of past and future anthropogenic emissions of CO_2, climate may depart significantly from natural over the next 50,000 years.

After a careful examination of the environmental, social, and economic implications of these and other intensifying transformations, the Academy concluded that "current trends of population and habitation, wealth and consumption, technology and work, connectedness and diversity, and environmental change are likely to persist well into the (21st) century and could significantly undermine the prospects for sustainability. If they do persist, many human needs will not be met, life-support systems will be dangerously degraded, and the numbers of hungry and poor will increase" (NRC 1999, p. 101). Based on its view of plausible social and technical options for breaking these trends, however, the Academy noted that "a successful transition toward sustainability is possible over the next two generations. This transition could be achieved without miraculous technologies or drastic transformation of human societies....What will be required, however, are significant advances in basic knowledge, in the social capacity and technological capabilities to utilize it, and in the political will to turn this knowledge to action" (NRC 1999, p. 160).

Are such advances in knowledge and its application possible? Will the Anthropocene simply turn out to be a very short era in which humanity blindly careens forward, continuing to transform the Earth until the planet loses its capacity to support us? Or might humanity rise to the challenge posed by Vernadsky, becoming the reflective, thinking, and proactive agent that transforms the biosphere into a noösphere, and consciously striving to shape a niche for ourselves in a sustainable Anthropocene? The answers to such questions will hinge in no small part on future developments of the sciences of the Earth system, and of sustainability. We turn to the opportunities and challenges facing such developments in the next sections of this chapter.

EARTH SYSTEM SCIENCE

In its quest to become an intelligent agent in and of the Anthropocene, what are the prospects for humanity developing a robust scientific understanding of the complex Earth system of which it is such a rambunctious part?

The Second Copernican Revolution

In 1530, Nikolaus Copernicus published his book *De Revolutionibus Orbium Coelestium*, which set the stage for the development of modern science. Not only was the Earth finally put in its correct astrophysical context, but the first principles of "exact and objective" reasoning, ultimately triumphing in the Enlightenment, were also established: The perception of cosmic reality became dominated by the clockwork metaphor, assigning a regular trajectory governed by eternal physical laws to each particle in the Universe. The production of wisdom became dominated by the curiosity-driven mode, confronting the brightest minds with the ultimate riddles of creation in splendid isolation from sociopolitical interests — and from each other. Thus the great Copernican Revolution generated a paradigm of science, where the lonely scholar wrestles with Nature in order to snatch some of her secrets encoded in mathematical formulae of utter beauty.

In 2001, delegates from more than 100 countries participating in the four big international research programs on global environmental change endorsed the "Amsterdam Declaration," which formally established the "Earth System Science Partnership" (Moore et al. 2002) and set the stage for what one might call a second Copernican Revolution (Schellnhuber 1999). This novel revolution is deeply rooted in the original one, yet transcends it in several crucial ways:

1. The scientific eye is re-directed from outer space to our "living Earth" (Lovelock 2003), which operates as one single dynamical system far from thermodynamical equilibrium.

2. The scientific ambition is re-qualified by fully acknowledging the limits of cognition as highlighted by the notorious uncertainties associated with nonlinearity, complexity, and irreproducibility (Schellnhuber 2002); if the Earth system is a clockwork at all, then it is an organismic one that baffles our best anticipatory capacities.

3. The scientific ethos is re-balanced at last by accepting that knowledge generation is inextricably embedded in the cultural–historical context (Nowotny et al. 2001) — there is nothing wrong with being particularly curious about the items and issues that matter most for society and with recognizing that the coveted borderlines between observing subjects and scrutinized objects have often been mere constructions of a preposterous reductionism. Thus the research community becomes part of their own riddles, the research specimens become part of their own explanations, and *co-production* becomes the (post)normal way of coping with the cognitive "challenges of a changing Earth" (Steffen et al. 2002).

The very fact that the Amsterdam Declaration resulted from an intricate cooperative process — and not from one ingenious idea of a stand-alone intellectual giant — adequately reflects the co-productive mode that will be instrumental for the much-debated "new contract between science and society" (see discussion below). Even a superficial look at the current state and dynamics of our planet indicates that the sustainability of modern civilization is at risk without such a contract. Actually the threats associated with anthropogenic global warming have already sparked the creation of an unprecedented format for the dialogue between researchers and decision makers, and for the co-establishment of global assessment power, namely the IPCC. This panel is a genuine post-Copernican creation that provides a panoramic, yet fragmentary and fuzzy, view of the myriads of facets involved in the climate change problem. This view, in turn, provides the most credible basis for international adaptive management strategies which have to be implemented and revised in phase with the highly irregular advancement of our pertinent knowledge as based, *inter alia*, on the monitoring of our own ecological footprints.

A Hilbertian Program for Earth System Science

Although humanity has been capable of transforming the Earth through a deluge of entangled but uncoordinated actions, it was evidently intellectually unprepared to do so or to cope with the consequences at the level of whole-systems wisdom. Nobody had a grand plan for planetary refurbishment after World War II, and nobody anticipated the scientific challenges arising when this refurbishment actually happened without plan, rhyme, or reason. (Annan 2000).

We are learning by "doing global change." The post-Copernican process, as epitomized (and accelerated) by the Amsterdam Declaration, keeps on setting unprecedented research agendas in unprecedented ways. Of course, there are Copernican role models for this, like David Hilbert's monumental program for the advancement of mathematics in the twentieth century (Hilbert 1901). This program basically consisted of a rather eclectic list of 23 problems to be solved by the pertinent community that had gathered at the World Conference for Mathematics in Paris in 1900. Some of Hilbert's riddles still stand unbroken by scientific siege or even unassailed by strong intellectual forces, but their very formulation launched a collective campaign venturing toward the borders of formal reasoning. Recently, the international Earth system science community formulated their own Hilbertian Program (Steffen et al. 2004, p. 265; Schellnhuber and Sahagian 2002), which lists 23 crucial questions that need to be addressed for global sustainability and may well drive global change research toward, and beyond, the limits of conventional scholarship raised to the planetary level.

The Hilbertian program for the advancement of Earth system understanding in the (first decades of the) twenty-first century emerged quite unconventionally, namely from an extended email conference organized in 2001 by GAIM (Sahagian and Schellnhuber 2002) — the transdisciplinary think-tank of the International Geosphere–Biosphere Programme (IGBP). The list of questions (see Box 1.1) is arranged in four blocks emphasizing a predominantly analytical, methodological, normative, and strategic character, respectively, and strongly reflects the three "post-Copernican" features discussed earlier.

To illustrate this, let us choose and briefly explain one question from each of the blocks. We begin with Question 3, which asks about the "critical elements" in the Earth system, i.e., those components, areas, processes, patterns, or substances within the planetary machinery that behave like control knobs: their alteration triggers persistent (if not irreversible), large-scale (if not global) change. There is clearly an analogy to the human body, where the destruction of delicate organs or the suppression of trace hormones can bring about significant transmutation, if not *exitus*.

A biogeophysical subset of the Earth's critical elements is compiled in the map of Figure 1.1. Its entries are underpinned by research results of rather

---

Box 1.1   A Hilbertian program for Earth system science.

Analytical Questions:

1. What are the vital organs of the ecosphere in view of operation and evolution?
2. What are the major dynamical patterns, teleconnections, and feedback loops in the planetary machinery?
3. What are the critical elements (thresholds, bottlenecks, switches) in the Earth System?
4. What are the characteristic regimes and timescales of natural planetary variability?
5. What are the anthropogenic disturbance regimes and teleperturbations that matter at the Earth-system level?
6. Which are the vital ecosphere organs and critical planetary elements that can actually be transformed by human action?
7. Which are the most vulnerable regions under global change?
8. How are abrupt and extreme events processed through nature–society interactions?

Normative Questions:

15. What are the general criteria and principles for distinguishing non-sustainable and sustainable futures?
16. What is the carrying capacity of the Earth?
17. What are the accessible but intolerable domains in the coevolution space of nature and humanity?
18. What kind of nature do modern societies want?
19. What are the equity principles that should govern global environmental management?

Operational Questions:

9. What are the principles for constructing "macroscopes", i.e., representations of the Earth system that aggregate away the details while retaining all systems-order items?
10. What levels of complexity and resolution have to be achieved in Earth System modelling?
11. Is it possible to describe the Earth system as a composition of weakly coupled organs and regions, and to reconstruct the planetary machinery from these parts?
12. What might be the most effective global strategy for generating, processing and integrating relevant Earth system data sets?
13. What are the best techniques for analyzing and possibly predicting irregular events?
14. What are the most appropriate methodologies for integrating natural science and social science knowledge?

Strategic Questions:

20. What is the optimal mix of adaptation and mitigation measures to respond to global change?
21. What is the optimal decomposition of the planetary surface into nature reserves and managed areas?
22. What are the options and caveats for technological fixes like geoengineering and genetic modification?
23. What is the structure of an effective and efficient system of global environment and development institutions?

---

varying conclusiveness, and the collections are far from being complete. In fact, new suspects are identified by global change research almost every year, such as the Indian monsoon, which may be pushed onto a roller-coaster dynamics by the combined driving forces of anthropogenic global warming, anthropogenic regional air pollution, and anthropogenic local land surface transformation (Zickfeld 2003).

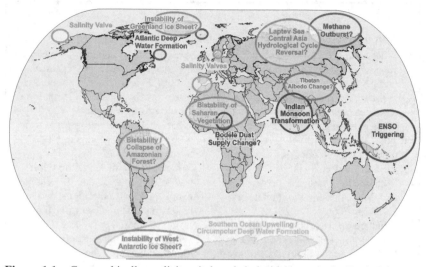

Figure 1.1   Geographically explicit switch and choke elements in the Earth system.

While the criticality analysis of the planetary ecosphere is making good progress and promises to support, in the not-too-distant future, global stewardship with a comprehensive list of neuralgic items that must be treated with utter caution, the complementary criticality analysis of the anthroposphere has not yet begun: What are the irreplaceable components of the global industrial metabolism? On which agricultural region will future world food production crucially depend? Are there institutions that can preserve/establish social cohesion and international equity throughout the globalization process? Which of the current megacities are bound to implode ultimately, and where will the new planetary centers of knowledge production lie? What technologies have the potential to transform radically humanity's interactions with its natural resources and its life-support systems? Genuine Earth system analysis for sustainability needs to address all of the questions, but there will be no quick answers.

The next illustration concerns Question 14, which asks about the best methodologies for integrating the Earth system knowledge produced by both the natural sciences and the social sciences. This question is part of a much wider, long-standing debate that attempts to bridge the "hard" and the "soft" disciplines. In recent years, two diametrically opposite schemes — perceived by many as battering rams rather than bridges — have been put forward: The first "integrating" strategy is the formalization of the social sciences along the lines of mathematical physics as epitomized by the invention of "econophysics" (see, e.g., the review articles in Bunde et al. 2002). This development reflects an epistemological attitude anticipated by David Hume in 1748 as follows: "The great advantage of the mathematical sciences above the moral consists in this, that the ideas of the former are always clear and determinate." A popular and sweeping state-of-the art account is given by Philip Ball (2004), who tries to demonstrate that socioeconomic behavior can be described, in principle, by Newtonian-like equations of motion.

Supporters of the second "integrating" strategy remain utterly unconvinced by this and suggest completely inverting the approach: Conduct particle science by engaging *the particles themselves* in the cognitive process! This proposition sounds much less bizarre if the "particles" addressed are actually human beings. Therefore, discursive interaction with the specimens to be explored is imperative. In the context of Earth system science, sustainability research and climate change assessment, this tenet implies that the "stakeholder dialogue" (O'Riordan et al. 1999) is the prime mode of holistic knowledge production. Whether this notion can be properly put into operation, or whether it ultimately tends to pervert the classy scientific theatre into a self-referential "big brother" show for the masses remains to be seen.

Between the two poles portrayed, however, there are many intermediate ways of reconciling — or constructively combining, at least — the natural science and the social science methodologies. For example, one can adopt a semi-quantitative, semi-discursive approach as employed in the syndromes analysis of global change dynamics (Schellnhuber et al. 1997). The main idea is to identify typical functional patterns of environment–society interactions from stakeholder-informed phenomenological inspection and to model the pattern dynamics by qualitative differential equations expressing robust place-based observations. Thus precision is realized only to the degree that it can be justified, not to the extent it can be handled.

This intermediate approach is illustrated in Figure 1.2, which displays results of a semi-quantitative analysis of the Sahel-Syndrome dynamics in Northeast Brazil (Seitz et al., submitted). In this analysis, the crucial processes governing the development of smallholder agriculture in the region are symbolically modeled on the basis of a massive body of newly generated empirical data. Note that such an analysis can only provide the *topology* — not the metrics — of the temporal succession of system states; however, this information may already suffice for designing intervention strategies for syndrome mitigation.

Semi-quantitative, yet fully formalized techniques, such as the ones employed in syndromes analysis, hold a huge potential for the adequate scientific description of complex systems characterized by strong nature–society interactions. There is no point in feigning exactness by treating, say, the atmospheric component of these systems with scrupulous precision while, for example, the lifestyle aspects are dealt with in cavalier vagueness. The playing field for the clash of disciplines needs a bit of leveling, at least.

From the block of normative questions, let us select Question 16, which asks about the carrying capacity of the Earth, that is, the maximum number of people (at a given lifestyle) that the planet can support. This very question has been posed and answered many times since Antoni van Leeuwenhoek, the great

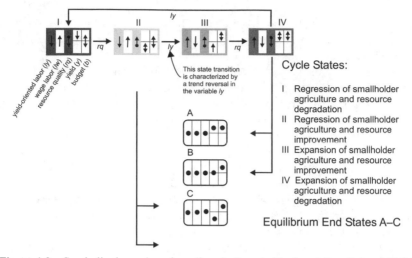

Figure 1.2 Symbolic dynamics of small agriculture in Northeast Brazil as a faithful projection to the five fundamental variables *ly* (yield-oriented labor), *lw* (wage-oriented labor), *rq* (resource quality), *y* (yield), and *b* (budget). The two sets of boxes represent cycle states (I–IV) and equilibrium end states (A–C), respectively, expressing magnitudes and trends of the fundamental variables by appropriate symbols. Dots indicate that the variable is constant over time, whereas trends are indicated by directed arrows. The crucial *ly–rq* trend combination is highlighted by color shading in the cycle state boxes. After Seitz et al., submitted.

Dutch scholar, provided the first serious estimate on April 25, 1679, in Delft: 13.4 billion. Unfortunately — and most interestingly — the sequence of subsequent estimates does not converge to a well-defined number, as Joel Cohen (1995a, b) has demonstrated (see Figure 1.3). In fact, the time series clearly exhibits the wild oscillations in the successive assessment numbers as well as the mostly increasing variability, which seems to peak in the science fiction-inspired decades after World War II. What is the explanation for this bewildering non-convergence of analysis? At least two factors have to be taken into account: First, there is the "supply side" of planetary carrying capacity, i.e., the totality of ecological services the Earth system can provide, including space, warmth, fresh air and fresh water, nutrition, shelter, and recreation (Millennium Ecosystem Assessment 2003). Our understanding of the structures and processes that determine the supply of these life-support items has dramatically grown, if not exploded, over the last thirty years, yet the number of riddles to be solved still appears to be almost infinite. For instance, nobody really knows the maximum sustainable protein yield of the world's oceans under *ceteris paribus* conditions, let alone under anthropogenic global warming. Without a full prognostic knowledge of the biogeophysical dynamics and the biogeochemical cycles involved, all of the figures put forward remain utterly elusive.

Second, there is the "demand side" of planetary carrying capacity, that is, the totality of ecological human needs that have to be satisfied according to

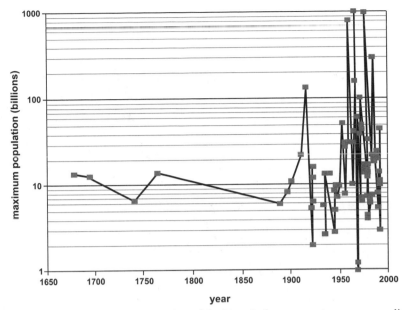

Figure 1.3   Time series representation of the historical assessment process regarding the human carrying capacity of the Earth. After Cohen (1995a, b).

judicious minimum standards. Who defines these standards? Dodging this awesome normative challenge by empirics does not help much, because one may calibrate the demand level either at the "American way of life" or at the subsistence needs of an Ethiopian farmer. To complicate matters, the ecological demand of X billion people is unlikely to equal the sum of X billion individual demands!

Thus, even the most unsophisticated approach to the Earth's carrying capacity boils down to calculating a ratio where both the numerator and the denominator are ill-defined. Still, Earth system science should provide a no-nonsense answer (or a sensible no-answer) to Question 16 to demonstrate its transdisciplinary worth.

To conclude our illustration of the Hilbertian program, let us consider Question 22, which asks about the options and caveats for top-down technologies supposed to fix global-scale sustainability problems. In recent years, there has been, for instance, a lot of discussion about the fantastically cheap and powerful geoengineering methods available to mitigate planetary warming; the world has also been told that genetic engineering of crops and entire ecosystems would turn the adaptation job almost into child's play. This seems to provide a dangerously overoptimistic picture biasing the mind-set of pragmatic decision makers, whereas solid opportunities may be dismissed precipitously by crucial parts of civil society in an intellectual backlash reaction. It is therefore the duty of the

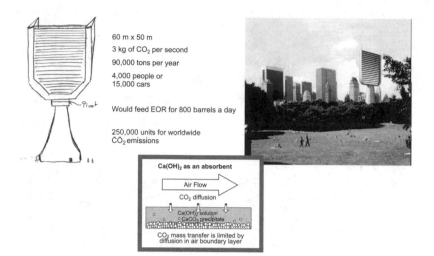

60 m x 50 m
3 kg of CO_2 per second
90,000 tons per year
4,000 people or
15,000 cars

Would feed EOR for 800 barrels a day

250,000 units for worldwide
CO_2 emissions

$Ca(OH)_2$ as an absorbent

Air Flow

CO_2 diffusion

$Ca(OH)_2$ solution
$CaCO_3$ precipitate

CO_2 mass transfer is limited by
diffusion in air boundary layer

Figure 1.4   A not entirely tongue-in-cheek vision of future CO_2 scrubbing, involving hundred thousands of macro-fans (courtesy of K. Lackner).

pertinent scientific community to explore carefully the possibility space of strategic schemes, including all its unconventional pockets, and to deliver a sober integrated assessment of the feasibility, effectiveness, efficiency, and acceptability of these schemes.

One remarkable attempt to meet this challenge was generated by the symposium on "Macro-engineering Options for Climate Change Management and Mitigation," organized by the Tyndall Centre and the Cambridge MIT Institute at the Isaac Newton Institute on January 7–9, 2004 (see http://www.tyndall.ac. uk/events/past_events/cmi.html). This event brought together experts from all relevant fields and institutions to discuss the profile of an optimal portfolio of macro-mitigation measures and potential macro-adaptation schemes (e.g., creating large-scale migration corridors for global warming-driven species and ecosystems) needed for coping with anthropogenic climate change.

One of the emerging ideas highlighted at the symposium was CO_2 capture from ambient air to address the problem of emissions from diffuse sources (for an illustration, see Figure 1.4). As Earth system and sustainability science advances further, the assessment of top-down options for global change management will have to be revised at an accelerating pace, not dissimilar to the situation concerning the carrying-capacity question sketched above.

SUSTAINABILITY SCIENCE

The last half century has seen a number of transitions in how society views the relationships among environment, development, and knowledge. Only very recently, however, has it become evident that the Anthropocene crisis forces hu manity to manage consciously a transition toward sustainable use of the Earth. Looking back over the last twenty years, few science-based ideas have risen from obscurity to take such a conspicuous position in international affairs as "sustainable development." Beginning shortly after the Rio Conference of 1992, however, it became increasingly clear that the enthusiasm with which much of the political world embraced sustainability ideas put environmental politics, negotiations, and agreements at center stage in the resulting debate, with science and technology relegated to the side wings if not thrown out altogether. At the same time, efforts to make progress in the implementation of sustainable development were increasingly being stalled by lack of technical knowledge rather than just weakness of political will: How could the decline in productivity for African agriculture be reversed while preserving biodiversity? How much greenhouse warming was too much? How could progress toward sustainability be reliably measured? The realization gradually began to sink in with the advocates of sustainability that all the negotiations in the world were not going to eaxct much progress on technical questions such as these.

In response to this mismatch of demand and supply, a number of efforts were initiated during the 1990s to reconsider how science might be better harnessed to achieve social goals of sustainable development in Vernadsky's would-be semi-intelligent "noösphere." The results of those efforts were synthesized as part of the international scientific community's input to the Johannesburg Summit (ICSU et al. 2002). One immediate outcome from this activity was the realization that the range of organized, disciplined, reflective activity needed for intelligently and effectively guiding a sustainability transition was much broader than what is conventionally subsumed under the term of "science." The Earth systems sciences noted above clearly have a role to play in promoting such a transition. So, however, does technology, innovation, and the tacit knowledge of practice. Even more broadly, there was clearly a need to mobilize the humanistic perspectives that would help us to understand where ideas about environment, development, and sustainability interacted with other dimensions of human thought about what we think we are and want to be. The term that has come closest to embracing this wide range of activities in English is "knowledge." Perhaps even more appropriate, within the backdrop of this Dahlem Workshop, is the German idea of *Wissenschaft*, embracing as it does the systematic pursuit of all knowledge, learning, and scholarship. Some of the key findings of this dialogue regarding what is needed from *Wissenschaft* — and the *Wissenschaftler* who pursue it — in a noösphere bent on sustainability are summarized below.

Changing Orientations

If *Wissenschaft* is to help advance sustainability, then a substantial part of our agenda needs to be driven by what society thinks it needs, not just by what scholars think is interesting (ICSU et al. 2002). This is not to advocate a return to sterile debates about the primacy of "basic" versus "applied," or "disciplinary" versus "interdisciplinary" research. Rather, it is to embrace the historical experience summarized by Donald Stokes (1997) in his book *Pasteur's Quadrant: Basic Science and Technological Innovation*, which argues that just as Pasteur created the field of microbiology in his pursuit of practical solutions to problems of great social importance, so it is possible today to do "cutting-edge research and development in the service of public objectives" (Branscomb et al. 2001). *Which* objectives is, of course, a matter of values — in this case values about what society actually means when it declares "sustainable development" to be a "high table" goal for the twenty-first century. Much debate has been expended in efforts to answer this question, and it is clear that different groups in society have reached different conclusions. Still, most of those debates share common concerns while differing largely in their emphasis on what is to be developed, what is to be sustained, what should be the relation of the developed to the sustained, and over how long a period the relationship should hold (see Figure 1.5).

At the international level, a broad consensus can be discerned that sustainable development should be development that, over the next two generations, promotes progress "to meet the needs of a much larger but stabilizing human population, to sustain the life-support systems of the planet, and to substantially

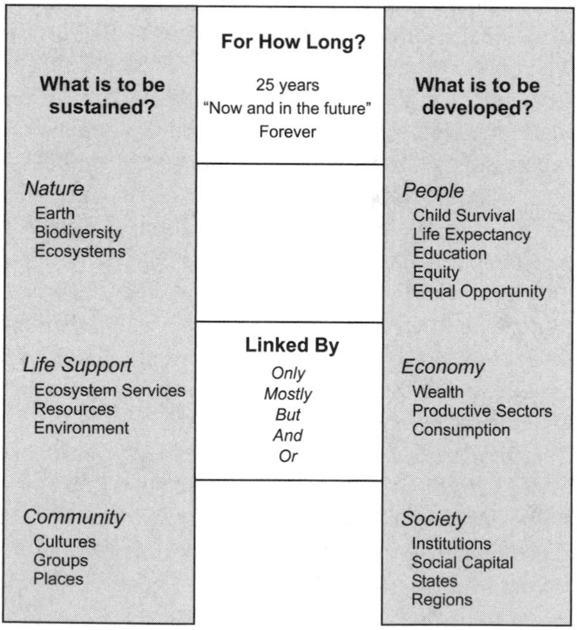

What is to be sustained?	For How Long? 25 years "Now and in the future" Forever	What is to be developed?
Nature Earth Biodiversity Ecosystems		*People* Child Survival Life Expectancy Education Equity Equal Opportunity
Life Support Ecosystem Services Resources Environment	**Linked By** Only Mostly But And Or	*Economy* Wealth Productive Sectors Consumption
Community Cultures Groups Places		*Society* Institutions Social Capital States Regions

Figure 1.5 Sustainable development: Common concerns, differing emphases. The U.S. National Research Council (NRC 1999, p. 24) produced this figure to summarize its exhaustive review of the different treatments of the sustainability concept present in the scholarly and political literature.

reduce hunger and poverty" (NRC 1999, p. 31). Clearly, science, technology, and *Wissenschaft* more generally have roles to play in devising instrumental means to help reach these goals. In addition, however, society needs knowledge to help it discover what it means by what it believes it values. The scholarly community has done a much better job of this for the "development" side of sustainability than for the "environment" side (Parris and Kates 2003a, b). For example, we have so far given the policy world little help in articulating what it would mean, in specific terms, to "sustain the life-support systems of the planet." The implications of this shortfall for future research agendas are addressed in Working Group 4 (see Kinzig et al., Chapter 20, this volume).

A second major conclusion from recent efforts to reassess the role of *Wissenschaft* in promoting sustainability concerns questions of scale. The international consensus on broad sustainability goals is helpful as a general frame for discussions. Experience makes it clear that both the ends and the means of sustainable development need to be tailored or tuned to the context of particular places. This is, in part, because the basic ecological, climatic, and social structures that define sustainability needs and opportunities vary from place to place. It is also partly because some of the greatest threats to sustainability derive from "multiple, cumulative, and interactive stresses" (NRC 1999, p. 8) that intersect in particular ways in particular places. The importance of such place-based calibration became clear in the course of the Green Revolution, where initial efforts to transfer new varieties directly from international agricultural research centers to the field had to give way to systems of research and innovation that linked international centers to local crop improvement efforts through intermediary systems of national and regional agricultural research universities (Bell et al. 1994). Modern efforts to promote sustainability need to balance the scientific community's long-standing regard for knowledge that is universally true, with an appreciation of the fruits of multi-scale, integrated research that connects local, regional, and global perspectives to produce understanding that is true for specific places (E. Miles, pers. comm.).

Increasingly, it has become widely accepted that development, in general, and sustainable development, in particular, is a knowledge-intensive activity (World Bank 1999; UNDP 2001). However, a final insight to emerge from the last decade's reconsideration of the role of science in achieving sustainability is a shift of emphasis from the importance of "knowing" to the centrality of "learning." In part, this shift follows the reconceptualization of goals noted above. If sustainable development is about progress "to meet the needs of a much larger but stabilizing human population, to sustain the life-support systems of the planet, and to reduce substantially hunger and poverty," then sustainability itself can be thought of less as a state or condition and more as a direction or bias for development activities. This puts "sustainability" in the same camp as other great goals of the last century, such as "freedom" and "justice" — goals that we think more about moving toward than we do about achieving. If achieving sustainable development in some ultimate sense may seem problematic, promoting a transition toward sustainability should not (NRC 1999). An even more important reason for the shift of emphasis in sustainability thinking from "knowing" toward "learning" is simply that we have so much to learn. Understanding sustainability is understanding a complex, dynamic system of nature–society interactions — a system made all the more unpredictable by both our interest in what goes on in particular places and by our active, reflective engagement in the system whose behavior we are trying to predict. Trying to discover or write blueprints for such turbulent, rapidly evolving systems will in many cases prove futile. More important is that we recognize the extent of our ignorance, accept the concomitant necessity to treat policies and other management interventions as experiments, and take measures that will increase our prospects for surviving, and learning from, the experiments we are forced to conduct on ourselves. Sustainable development thus becomes viewed as a process of adaptive management and social learning in which knowledge plays a central role (Cash et al. 2003; Steffen et al. 2004, Chapter 6.5).

Vulnerability Analysis: An Illustration of Sustainability Science

As an example of the kind of knowledge needed from the sciences of sustainability, let us start with what we are trying to sustain. What would it mean to "sustain the life-support systems of the planet?" Such questions are very much on managers' and policy makers' minds, as suggested by the language about preventing "dangerous anthropogenic interference with the climate system" inserted into the Framework Convention on Climate Change. When science has been able to characterize unambiguously what constitutes "dangerous interference" with our environmental life-support systems, society has been reasonably successful in adjusting its behavior to remain within safe limits (e.g., the European use of "critical load" estimates for managing sulfur emissions and the risk of acid rain). Conversely, so long as opponents of management have been able to declare — as did U.S. President George W. Bush in opposing the Kyoto Protocol — that "no one knows what that (dangerous) level is" (press conference, June 11, 2001), science-based management remains a ready excuse for inaction. The S&T community could therefore significantly improve "the prospects for humanity consciously managing a transition toward sustainability" by developing an understanding of the vulnerability and resilience of the Earth's life-support systems to "dangerous" disruption.

Early work on the "limits to growth," "Earth's carrying capacity," and "ecological footprints" addressed important issues but generally failed to develop a dynamic, causal understanding of how complex nature–society systems respond to stress. In contrast, more than a quarter century of serious scientific work on the resilience of ecological systems and the vulnerability of social systems has provided a solid foundation for such understanding. Recent efforts to

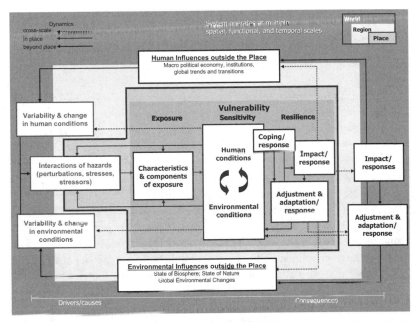

Figure 1.6 Vulnerability framework (Turner et al. 2003), summarizing the major elements of an emerging framework for analysis of vulnerability in coupled nature–society systems. It is explicitly scale-embedded and incorporates elements of exposure, sensitivity and resilience. Used with permission from the National Academy of Sciences, U.S.A.

synthesize those two historical strands of work have highlighted the importance of incorporating multiple stresses, teleconnections, explicit pathways of exposure, the possibility of threshold responses, explicit treatment of scale, and attention to the components of adaptive capacity in frameworks for the analysis of vulnerability and resilience (Turner et al. 2003; see Figure 1.6). These synthetic efforts have also drawn attention to the parallels between climatic and chemical "life-support systems" long discussed by Earth science researchers, the elements of "livelihood security" (e.g., access to and use of resources) stressed by development practitioners, and the newer emphasis by ecologists and resource economists on "ecosystem services." Needed now for management is problem-driven research that utilizes these conceptual vulnerability/resilience frameworks to illuminate the kinds, rates, and magnitudes of specific disturbances beyond which the "the ability of society to advance human well-being" can no longer be sustained.

Guidance Systems for Sustainability

We noted earlier the growing consensus that management systems for a sustainability transition need to be systems for adaptive management and social learning. The broad elements of such systems are reasonably straightforward: they require appropriate information, incentives, and institutions. What can *Wissenschaft* contribute to the development of such systems?

Information: Information is central to guidance, and guidance for a sustainability transition needs information on both where we want to go as well as how well we are doing at getting there. These are matters of setting goals and targets, defining indicators to track performance toward achieving them, and implementing the observational systems to measure the indicators. The scientific community has no monopoly on these tasks, but neither should it stand back and leave them wholly to others in the mistaken belief that to discuss the values inherent in the selection of goals and indicators is to lose scientific objectivity. Unfortunately, that is precisely what has happened for many of the sciences and scientists required to inform an intelligent social dialogue on goals and indicators of sustainability. Surprisingly, the situation is particularly grim with respect to the natural sciences. Whereas social scientists have been relatively successful at informing the debate on specific goals and indicators for the "meeting human needs" dimensions of sustainability, natural scientists have not contributed effectively to specifying goals and indicators for "protecting life-support systems." In a recent review of international efforts in this area, Parris (2003) found that only with respect to the global atmosphere was a reasonably integrated system of specific goals, targets, indicators, and monitoring in place. For the dimensions of "life-support systems" relating to ocean productivity, freshwater availability, land-use change, biodiversity, and toxic releases, no such system exists.

The call for problem-driven work on assessing vulnerability and resilience of the Earth's life-support systems noted above could provide the foundations for improvements in this area. However, even with improvements in the basic understanding of such key "life-support" concepts, a place-based strategy of goal-setting, indicator selection, and monitoring will still be needed for guiding the actions needed for a transition toward sustainability. More broadly still, the ultimate need is for a problem-driven, theoretically grounded, integrated approach for characterizing and measuring what we most value in coupled nature–society systems we inhabit. Partha Dasgupta (2001) has recently outlined one such an approach in his treatise on *Human Well-being and the Natural Environment*. The challenge for an emerging field of Earth systems analysis is to build on such frameworks, and to enrich them with our deepening understanding of how the biosphere — and noösphere — actually work.

Incentives: A second component of the guidance systems needed for managing a transition toward sustainability concerns getting the incentives right. How can people be induced to make production and consumption choices that are relatively less stressful to the environment than others that generate comparable increases in real well-being? How can perspectives incorporating long-term biosphere responses be appropriately factored into short-term social decision-making? More broadly, what is the right level and focus for the investments in the science, technology, and knowledge that are necessary for a transition toward sustainability? What sorts of inducements or feedback will best assure the provision of adequate constraint in individual human uses of nature's commons and of adequate investment in the "public goods" of ecosystem services?

When the economic costs and benefits of "sustainable" behaviors accrue to the same private parties, well-functioning markets perform admirably in aligning incentives and allocating society's investments. However, the real world is full of well-known features that undercut the efficacy of the market as an efficient allocator of investments in sustainability. These include distorting subsidies (e.g., to land clearing), mispriced environmental externalities (e.g., pollution), a strong public goods component of the social benefits provided by healthy environments (e.g., wetlands) and "clean" technologies (e.g., sewage treatment plants), the privatization of relevant data (e.g., genomics), and imperfect property rights regimes (e.g., fisheries). Although serious attention is being paid to each of these sources of "incentive failure," and a modest number of exciting new ideas have been broached in recent years (e.g., Sandler 1997), progress has been slow on both the theoretical and practical fronts. A concerted research and applications program on the incentives — market and otherwise — that could fulfill Vernadsky's vision of an intelligently reflective, self-guiding "noösphere" is still badly needed.

Institutions: "Institutions" include the norms, expectations, rules, and organizations through which societies figure out what to do and organize themselves to do it. "Sustainability" itself is a norm, and thus part of the emerging institutional structure of Vernadsky's self-reflective noösphere. So are the international treaties and related arrangements that society has developed and deployed over the last several decades in hopes of bringing some degree of rational governance to the interaction of society with nature. Our focus here, however, is on the kinds of institutional reforms and innovations that are needed to harness science and technology better to the tasks of enabling and guiding a transition toward sustainability.

Such institutions are not impossible to design, as illustrated by experience in efforts to employ science and technology to enhance agricultural productivity, combat disease, and protect the global atmosphere. Such successes, however, have been partial, reversible, and idiosyncratic, producing little in the way of consensus on what sorts of institutional designs are most likely to enhance the use of knowledge to create an effectively reflective "noösphere." Over the last couple of years, however, consultations throughout the international community have identified a number of specific shortcomings in the present institutional system, and have proposed some directions in which reform efforts might usefully head. Below we address some of the crucial issues identified:

- *Mobilizing the right knowledge*: In the dialogues leading up to the Johannesburg Summit, one of the most persistent complaints was that today's agenda of R&D for sustainability reflects the priorities of global programs, disciplines, and donors more than it does those of the local decision-makers on who the prospects for sustainable development are so dependent. Moreover, it tends to devote much more attention to the identification of problems than the production of solutions (ICSU et al. 2002). What kinds of institutions can best improve the chances that research conducted in the name of sustainable development will actually focus on the most pressing problems as defined by relevant decision makers in the field? Institutions that can meet these challenges need one foot in the politics of problem definition, responsive to issues of appropriate participation and representation, and the other in the world of S&T, responsive to issues of credibility and quality control. Few overtly "political" or "scientific" institutions seem to be able to perform such "boundary-spanning" functions effectively. A better record has been accumulated by organizations that explicitly cast themselves in a boundary-spanning role, responsible to the worlds of both knowledge and action, but not expected to conform fully to the norms of either (Cash et al. 2003). Examples include IIASA's "RAINS" effort to link S&T with efforts to develop sustainable energy policy in Europe; a number of groups involved in making ENSO forecasts useful to decision-makers (IIASA 1992); and local organizations for technology innovation, such as India's Honey-Bee network. At their best, such institutions have facilitated two-way communication between experts and decision-makers, and provided neutral "sites" for the co-production of useful knowledge by scientists and problem-solvers.

• *Integrating knowledges*: Today's S&T remains insufficiently inclusive and integrative to realize its full potential for helping with the complex, messy problems that need to be overcome in promoting a transition toward sustainability. What kinds of institutions can better integrate the "tacit" knowledge of practice (whether it be that of a cracking plant operator or a rice farmer) with the formal knowledge of laboratory science to produce practical insights on solutions to particular sustainability problems? How can the expertise of the private sector be integrated with that of the university and governmental S&T communities to produce the public good of sustainable development without unacceptably undermining the incentives of business? How can the traditionally "island empires" of research, observations, assessment, and applications be better integrated into problem-solving systems of S&T for sustainability? Examples of institutions that have successfully performed all of these desirable integrations are very few indeed. However, the community might usefully devote some attention to identifying effective models.

• *Balancing flexibility and stability*: The challenge of sustainable development is simultaneously long term and rapidly evolving. S&T programs designed to promote sustainability need themselves to be sustained long enough to make a difference, but not at the cost of being stuck fighting the last war, or failing to learn from experience. No single institutional approach to this tension is likely to work everywhere. Some critical attention to which sorts of approaches is likely to be more effective under which kinds of circumstances seems long overdue. One response to the challenge of balancing flexibility and stability in R&D agendas for sustainability has been to expand explicitly long-standing R&D organizations, such as the International Agricultural Research Centers, to make them more responsive to emerging sustainability needs. Another is to assemble ad hoc task forces commissioned to address particular problems, for example the World Commission on Dams or the Millennium Ecosystem Assessment. A third particularly promising approach has been to combine the two models above, retaining a small professional secretariat to facilitate learning and the building of trust relations, but convening ad hoc teams to address particular problems. The Interacademy Panel on International Issues is an example of one institution seeking to pursue this approach.

- *Infrastructure and capacity*: Most of the world lacks the physical infrastructure and human capacity to do as well as it might in harnessing S&T to sustainability. Additional investments, however, need to be strategically targeted if they are to improve the situation in an efficient manner. In particular, it seems clear that a balanced portfolio is needed that invests simultaneously in individuals, organizations, and networks. Furthermore, in those regions where basic education — the most fundamental source of R&D capacity — is underdeveloped, priority must be given to building the educational base and enhancing an appreciation for the methods and potential contribution of science. Finally, a critically important infrastructure need is for institutions that support cross-scale linkages among researchers and problem-solvers. These need to be structured to facilitate both "vertical" connections between the best research anywhere in the world and the specific circumstances of particular field applications. At the same time they will need to foster "horizontal" connections among regional research and application centers to promote learning from one another. Precedents are rare, and even more rarely are they widely known.

SCIENCE AND SOCIETY: A NEW CONTRACT FOR PLANETARY STEWARDSHIP?

A remarkably productive "social contract" between society and the science community evolved throughout the industrialized and industrializing world in the latter half of the twentieth century. In essence, the contract held that society would invest heavily in basic science on the presumption that such investments would eventually result in better economic growth and national security. The stunning accomplishments of the original contract notwithstanding, it has become increasingly clear that for achieving many important social goals — among them sustainable development — the original contract might be necessary but is certainly not sufficient. The result has been an increasing number of calls for "a new social contract for science," beginning with UNESCO's 1989 Conference on Science for the Twenty-first Century, focused by Jane Lubchenco's 1998 Presidential Address to the American Association for the Advancement of Science, and reaffirmed by the international science community in its preparations for the Johannesburg Summit on Sustainable Development (full citations in ICSU et al. 2002, p. 17).

The idea of the "new contract" has attracted a great deal of attention and has evolved in a number of parallel but not identical directions. Under most versions, however, the S&T community would agree to devote an increasing fraction of its overall efforts to R&D agendas reflecting society's goals for sustainable development. In return, society would undertake to invest an increasing fraction of its wealth to assure that science, technology, and *Wissenschaft* generally could be adequately mobilized to fulfill their role in guiding a transition toward sustainability, thus fulfilling their role in contributing to the informed governance of Vernadsky's noösphere. Moving from the intensely felt rhetoric of the "new contract" toward the practical reality of agendas for Earth system science and scientists — *Wissenschaft* and *Wissenschaftler* — is a suitably ambitious challenge for a Dahlem Workshop.

THE DAHLEM CONTRIBUTION

The observations and considerations of the previous sections may be summarized in just one statement:

> We are currently witnessing the emergence of a new scientific paradigm that is driven by unprecedented planetary-scale challenges, operationalized by transdisciplinary centennium-scale agendas, and delivered by multiple-scale co-production based on a new contract between science and society.

All crucial aspects of this statement were actually addressed at the 91st Dahlem Workshop, held in May 2003, by an exceptional collection of scholars from all corners of the international scientific community. This volume presents the information that supported the meeting, condensed into sixteen state-of-the-art papers, as well as the pertinent results distilled into the group reports. The intimidating intellectual challenges involved were tackled in Dahlem by four working groups with a division-of-labor strategy roughly orienting itself by the successive qualitative stages of coevolution in the Earth system.

A remarkable clash of scientific cultures was staged in Group 1 (Lenton et al., Chapter 6), where researchers mainly concerned with geosphere–biosphere interactions on planet Earth met with astrobiologists primarily interested in the existence and habitability of other planets inside and outside the solar system. The common themes were the general possibility of (intelligent) life in our Universe and the long-term, large-scale coevolution of dead and living matter through complex self-organization processes far from thermodynamic equilibrium. The group addressed a number of exciting issues, such as the evolutionary topology of the biosphere, the interactive development of environmental dynamics and information processing through the great planetary transitions, the terraforming potential provided by Mars, the probability for the emergence of intelligence, and the failure of the SETI project (thus far) to track down messages from extraterrestrial civilizations. The GAIA theory (Lovelock 2000) served as an integrating factor and unifying metaphor in the group's debates.

Group 2 (Watson et al., Chapter 10) moved the analytic focus to what geologists might call the "recent planetary past," i.e., the Quaternary. The main idea was to scrutinize the Earth system machinery in a state as similar as possible to the contemporary one — yet without human interference with the relevant biogeophysicochemical inventories and processes. Special emphasis was given to the stability and variability of the Quaternary mode of operation of our planet, an analysis clearly involving the identification and quantification of major feedback loops, phase thresholds, and other critical elements. Not unexpectedly, a certain "row" between stability optimists (led by the geologists) and stability pessimists (led by the climatologists) took place in this group, resulting in very specific demands for high-quality data for settling the case. There was also agreement on the indispensability of major progress in Earth system modeling for understanding the roller-coaster dynamics of the Quaternary as illustrated by quasi-periodic glaciation episodes.

Almost everything on Earth has changed with the advent of *Homo sapiens* and the establishment of the modern anthroposphere. Group 3 (Steffen et al., Chapter 16) made the heroic effort to describe how the human factor has already modified the Quaternary mode of operation of our planet, to identify potential anthropogenic phase transitions lurking around the corner, to specify the scientific advancements necessary for timely anticipation of dangerous Anthropocene dynamics, and to assess the prospects of large-scale technological fixes of the accelerating sustainability crisis all around us. An in-depth analysis of the notorious climate sensitivity conundrum and a thorough delineation of "Earth system geography in the Anthropocene" (intercomparing the role of the mid-latitudes to the tropics and the polar regions) were among the highlights in the group's deliberations.

The most difficult task of all, however, remained for Group 4 (Kinzig et al., Chapter 20), who were to transgress the borderline between purely analytical reasoning and solution-driven strategic thinking. In other words, the group tried to identify pathways toward global sustainability, to evaluate the conceivable management schemes for steering our planet clear of the Anthropocene crisis, and to imagine all the scientific, technological, socioeconomic, and institutional innovations necessary for implementing the right strategy. Not surprisingly, a number of heated debates ensued when issues such as adaptive management; participatory decision making; integrated systems of production, consumption, and distribution; capacity building for coping with environmental change; and up-scaling of successful local/regional institutional designs were addressed. The discussions culminated in the group's attempt to sketch the crucial features of a future science–policy dialogue that allows for the true co-production of sustainability wisdom and to derive the pertinent conclusions for the novel organization of science and technology in the twenty-first century.

Altogether, the four groups actually succeeded in covering much of the vast terrain encompassed by the extraordinarily ambitious conference theme. Of course, there remain huge gaps and blatant superficialities, but all participants were convinced that this Dahlem Workshop was a milestone event that truly advanced Earth system analysis for sustainability. We hope that some of the excitement and inspiration that we experienced at Dahlem is conveyed through the following documentation.

REFERENCES

Annan, K. 2000. We, the Peoples: The Role of the United Nations in the 21st Century. New York: United Nations. http://www.un.org/millennium/sg/report/full.htm.

Ball, P. 2004. Critical Mass: How One Thing Leads to Another. Oxford: Heinemann.

Bell, D.E., W.C. Clark, and V.W. Ruttan. 1994. Global research systems for sustainable development: Agriculture, health and environment. In: Agriculture, Environment and Health: Sustainable Development in the 21st Century, ed. V.W. Ruttan, 358–379. Minneapolis: Univ. of Minnesota Press.

Branscomb, L., G. Holton, and G. Sonnert. 2001. Science for Society: Cutting-edge basic research in the service of public objectives. http://www.cspo.org/products/reports/scienceforsociety.pdf

Brown, H. 1954. The Challenge of Man's Future: An Inquiry Concerning the Condition of Man during the Years that Lie Ahead. New York: Viking Press.

Bunde, A., J. Kropp, and H.J. Schellnhuber, eds. 2002. Theories of Disaster: Scaling Laws Governing Weather, Body, and Stock Market Dynamics. Heidelberg: Springer.

Caldwell, L.K., and P.S. Weiland. 1996. International Environmental Policy: From the Twentieth to the Twenty-first Century. Durham: Duke Univ. Press.

Cash, D.W., W.C. Clark, F. Alcock et al. 2003. Knowledge systems for sustainable development. *Proc. Natl. Acad. Sci. USA* **100**:8086–8091

Clark, W.C. 1989. Managing planet Earth. *Sci. Am.* **261(3)**:47–54.

Cohen, J.E. 1995a. How Many People Can the Earth Support? New York: W.W. Norton.

Cohen, J.E. 1995b. Population growth and Earth's human carrying capacity. *Science* **269**:341.

Crutzen, P.J. 2002. The Anthropocene: Geology of mankind. *Nature* **415**:23.

Dasgupta, P. 2001. Human Well-being and the Natural Environment. Oxford: Oxford Univ. Press.

Hansen, J. 2004. Defusing the global warming time bomb. *Sci. Am.* **290**:68–77.

Hilbert, D. 1901. Mathematische Probleme. *Archiv der Mathematik und Physik* **3**:44–63, 213–237.

Humboldt, A. von. 1808. Ansichten der Natur mit wissenschaftlichen Erläuterungen. Tübingen: J.G. Cotta.

ICSU, TWAS, and ISTS (Intl. Council for Science, Third World Academy of Science, and Initiative on Science and Technology for Sustainability). 2002. Science and Technology for Sustainable Development. Series on Science for Sustainable Development, vol. 9. Paris: Intl. Council for Science. http://www.icsu.org/Library/ WSSD-Rep/Vol1.pdf.

IIASA (Intl. Institute for Applied Systems Analysis). 1992. Science and sustainability: Selected papers on IIASA's 20th anniversary, p. 1. Laxenburg: IIASA.

IPCC (Intergovernmental Panel on Climate Change). 2001. Climate Change 2001: The Scientific Basis. Working Group I Contribution, Third Assessment Report, ed. J.T. Houghton, Y. Ding, D.J. Griggs et al. Cambridge: Cambridge Univ. Press.

IUCN (Intl. Union for the Conservation of Nature). 1980. World Conservation Strategy: Living Resource Conservation for Sustainable Development. Gland: IUCN.

Kates, R.W. 2001. Queries on the human use of the Earth. *Ann. Rev. Energy Environ.* **26**: 1–26.

Kates, R.W., and T.M. Parris. 2003. Long-term trends and a sustainability transition. *Proc. Natl. Acad. Sci. USA* **100**:8062–8067.

Loutre, M.F., and A. Berger. 2000. Future climatic changes: Are we entering an exceptionally long interglacial? *Clim. Change* **46**:61–90.

Lovelock, J.E. 2000. The Ages of Gaia. Oxford: Oxford Univ. Press.

Lovelock, J.E. 2003. Gaia: The living Earth. *Nature* **426**:769–770.

Marsh, G.P. 1864. Man and Nature: Or, Physical Geography as Modified by Human Action. New York: Scribner.

Marsh, G.P. 1965. The Earth as Modified by Human Action. Harvard: Harvard Univ. Press., Belknap Press.

McNeill, J.R. 2000. Something New under the Sun: An Environmental History of the Twentieth-century World. New York: W.W. Norton.

Millennium Ecosystem Assessment. 2003. Ecosystems and Human Well-being: A Framework for Assessment. Washington, D.C.: Island Press.

Mitchell, R.B. 2003. International environmental agreements: A survey of their features, formation, and effects. *Ann. Rev. Env. Resour.* **28**: 429–461.

Moore, B., III, A. Underdal, P. Lemke, and M. Loreau. 2002. The Amsterdam Declaration on global change. In: Challenges of a Changing Earth, ed. W. Steffen, J. Jäger, D.J. Carson, and C. Bradshaw. Heidelberg: Springer.

Nowotny, H., P. Scott, and M. Gibbons. 2001. Re-thinking Science: Knowledge Production in an Age of Uncertainty. Cambridge: Polity Press.

NRC (National Research Council). 1999. Our Common Journey: A Transition toward Sustainability. Washington, D.C.: Natl. Acad. Press.

O'Riordan, T., J. Burgess, and B. Szerszynski. 1999. Deliberative and Inclusionary Processes: A Report from Two Seminars. CSERGE Working Paper: PA 99-06. Centre for Social and Economic Research on the Global Environment, Univ. of East Anglia, Norwich, U.K.

Parris, T.M. 2003. Toward a sustainability transition. *Environment* **45**:13–22.

Parris, T.M., and R.W. Kates. 2003a. Characterizing a sustainability transition: Goals, targets, trends, and driving forces. *Proc. Natl. Acad. Sci. USA* **100**:8068–8073.

Parris, T.M., and R.W. Kates. 2003b. Characterizing and measuring sustainable development. *Ann. Rev. Energy Environ.* **28**:559–586.

Pauly, D., and V. Christensen. 1995. Primary production required to sustain global fisheries. *Nature* **374**:255–257.

Royal Society of Chemistry (Great Britain). 2003. Sustainability and Environmental Impact of Renewable Energy Sources. Issues in Environmental Science and Technology 19. Cambridge: Royal Society of Chemistry.

Ruddiman, W.F. 2003. The anthropogenic greenhouse era began thousands of years ago. *Clim. Change* **61**:261–293.

Sandler, T. 1997. Global Challenges: An Approach to Environmental, Political, and Economic Problems. Cambridge: Cambridge Univ. Press.

Sahagian, D., and H.J. Schellnhuber. 2002. GAIM in 2002 and beyond: A benchmark in the continuing evolution of global change research. *Global Change Newsl.* **50**:7–10.

Schellnhuber, H.J. 1999. "Earth system" analysis and the Second Copernican Revolution. *Nature* **402(2)**:C19–C23.

Schellnhuber, H.J. 2002. Coping with Earth system complexity and irregularity. In: Challenges of a Changing Earth, ed. W. Steffen, J. Jaeger, D.J. Carson, and C. Bradshaw, pp. 151–159. Berlin: Springer.

Schellnhuber, H.J., A. Block, M. Cassel-Gintz et al. 1997. Syndromes of global change. *GAIA* **6**:19–34.

Schellnhuber, H.J., and D. Sahagian. 2002. The twenty-three GAIM questions. *Global Change Newsl.* **49**:20–21.

Steffen, W.L. 2002. Challenges of a changing Earth. Proc. Global Change Open Science Conf., Amsterdam, July 10–12, 2001. Global Change, IGBP Series. Berlin: Springer.

Steffen, W., J. Jäger, D.J. Carson, and C. Bradshaw, eds. 2002.The Challenges of a Changing Earth. Heidelberg: Springer.

Steffen, W., A. Sanderson, P.D. Tyson et al., eds. 2004. Global Change and the Earth System: A Planet Under Pressure. The IGBP Book Series. Berlin: Springer.

Stokes, D. 1997. Pasteur's Quadrant: Basic Science and Technological Innovation. Washington, D.C.: Brookings Institution.

Stoppani, A. 1873. Corso di geologia. Vol. II. Milan: G. Bernardoni E.G. Brigola.

Thomas, W.L. 1956. Man's Role in Changing the Face of the EARTH. Chicago: Published for the Wenner-Gren Foundation for Anthropological Research and the National Science Foundation by the University of Chicago Press.

Turner, B.L., II, W.C. Clark, R.W. Kates et al., eds. 1990. The Earth as Transformed by Human Action: Global and Regional Changes in the Biosphere over the Past 300 Years. Cambridge: Cambridge Univ. Press.

Turner, B.L., II, R.E. Kasperson, P.A. Matson et al. 2003. A framework for vulnerability analysis in sustainability science. *Proc. Natl. Acad. Sci. USA* **100**:8074–8079.

UNDP (United Nations Development Programme). 2001. Making New Technologies Work for Human Development. Oxford: Oxford Univ. Press.

Vernadsky, V.I. 1945. The biosphere and the noösphere. *Am. Sci.* **33**:1–12.

Vernadsky, V.I. 1998/1926. The Biosphere (translated and annotated version from the original of 1926). New York: Copernicus, Springer.

WCED (World Commission on Environment and Development). 1987. Our Common Future. Oxford: Oxford Univ. Press.

World Bank. 1999. Knowledge for Development: The World Development Report for 1998/1999. Washington, D.C.: The World Bank.

Zickfeld, K. 2003. Modeling large-scale singular climate events for integrated assessment. Ph.D. Thesis, University of Potsdam.

2

What Does History Teach Us about the Major Transitions and Role of Disturbances in the Evolution of Life and of the Earth System?

T. M. LENTON[1], K. G. CALDEIRA[2], and E. SZATHMÁRY[3]

[1]Centre for Ecology and Hydrology, Penicuik, Midlothian, EH26 0QB, U.K.
[2]Lawrence Livermore National Laboratory, Livermore, CA 94550, U.S.A.
[3]Collegium Budapest, 1014 Budapest, Hungary and
Institute for Advanced Study Berlin, 14193 Berlin, Germany

ABSTRACT

This chapter explores the connections between the evolution of life and past changes of the Earth system. The concept of major transitions in evolution is extended beyond those associated with the storage and transmission of information to encompass those associated with the transformation of free energy and matter. A tentative synthesis of major transitions in the history of the Earth system is offered. Our review suggests that major transitions in the evolution of life are usually associated with major transitions in the global state of the environment, and that cause and effect are often difficult to disentangle. This is consistent with the notion that the Earth with abundant life is a tightly coupled feedback system ("Gaia"). It is unclear whether major transitions of life and of the planet are always associated, not least because there are a number of competing hypotheses to explain each major transition of the Earth system. External disturbances (e.g., asteroid impacts, massive volcanic eruptions) appear in some cases to have triggered significant transitions of the system between different (quasi-stable) states. However, the largest transitions in the state of the Earth system appear to have been internally generated with evolutionary innovation playing a leading role.

INTRODUCTION

In this chapter we review the major transitions in the evolution of life as well as in the history of the Earth system and explore the connections between them. In particular, are they always associated? If not, why not? Furthermore, what is the role of disturbances (e.g., mass extinctions) in the coupled evolution of life and the Earth system? All of this is essential background for understanding the potential impact of human activities and for addressing the challenge of Earth system analysis for sustainability.

Major transitions in the evolution of life have been discussed by a number of authors. In their book, *The Major Transitions in Evolution*, Maynard Smith and Szathmáry (1995) concentrated on those associated with the storage and transmission of information. Given our focus here on the planetary impact of life, evolutionary innovations in the transformation of free energy and matter are at least as important. Almost all of these occurred among bacteria and archaea. In some cases, most notably oxygenic photosynthesis, they prove to have been difficult to evolve, but had profound planetary consequences. The notion of major transitions of the Earth system is less familiar. Some have commented on the applicability of the notion of "punctuated equilibria" to describe Earth history (as well as the history of life). However, attempts to synthesize a sequence of major transitions of the Earth system are lacking. We offer our views as a tentative first step upon which further discussion can be based.

CONCEPTUAL FRAMEWORK: GAIA AND EVOLUTIONARY BIOLOGY

It is widely accepted that life and its environment at the surface of the Earth form a coupled system and that the living and nonliving components of this system coevolve, in some sense. How tightly the system is coupled, the extent to which it is self-regulating, the contribution of life to maintaining habitable conditions, the role of natural selection in planetary regulation, and in what sense the system can be said to evolve are among many key issues that are being actively debated. We term this the "Gaia debate" as most (if not all) of the issues were raised in the Gaia hypothesis (Lovelock and Margulis 1974), subsequent criticisms of it (Doolittle 1981; Dawkins 1983), its replacement, the Gaia theory (Lovelock 1988), and recent developments of that (Lenton 1998). Some progress toward a consensus or at least a clarification of differences is evident in recent exchanges in the journal *Climatic Change* (March 2002 and May 2003 issues). Here we summarize some of the points of agreement and disagreement.

Defining the System

"Gaia" has been used to refer both to a system and to propositions about the functioning of that system. At the outset it is worth asking: Are the concepts "Earth system" and "Gaia" synonymous? The Gaia system is defined as the thermodynamically open system at the Earth's surface comprising life (the biota), atmosphere, hydrosphere (ocean, ice, fresh water), dead organic matter, soils, sediments, and those parts of the lithosphere (crust) and mantle that interact with surface processes (including sedimentary rocks and rocks subject to weathering). The Earth's internal heat source is considered (like the Sun) to be "outside" of the influence of the system. In practice, the "surface" Earth system, as illustrated, for example, in the Bretherton diagram (NASA 1986), is the Gaia system minus some of the slowest processes. In contrast, geologists sometimes include the entire interior of the planet in what they also refer to as the "Earth system." Perhaps a useful compromise would be to distinguish an "interior" Earth system (driven by the internal heat source) that overlaps with the "surface" Earth system. A further distinction is that the surface Earth system is often taken to include states without life, including those before the origin of life, whereas "Gaia" refers to a system with abundant life. Thus Gaia may be considered to comprise (all states of) the surface Earth system with abundant life. Therefore, it has a shorter life span than the Earth system.

Self-regulation

There is agreement that the tendency of the Earth system with abundant life (Gaia) to return to a stable state following a perturbation is not particularly remarkable, since many abiotic systems possess this property. The fact that for at least 3.8 billion years the Earth has always tended to stabilize in a habitable state is more interesting, although some argue that this could still have been the case had life never emerged. There is some support for the view that the system with abundant life (Gaia) possesses negative, stabilizing feedbacks that make it more resistant or resilient to perturbation, compared to those that would exist in an abiotic state. An important issue is the extent to which these stabilizing feedbacks act to regulate planetary properties at specific set points (as would a thermostat in a home). Perhaps the most contentious and difficult point to address is: Why does the system possess these negative, and possibly regulatory, feedbacks? Is it just a chance occurrence ("lucky" Gaia), or is it because a planet with abundant life tends to develop such feedbacks ("probable" Gaia)? Astrobiology may ultimately provide information about the presence of life on extrasolar planets, which together with knowledge of the age distribution of the parent stars may help distinguish between "lucky" and "probable" Gaia. In the meantime, it is generally agreed that either case poses an important research agenda: to elucidate the feedbacks.

The Role of Natural Selection

The evolution of life involves natural selection acting on inherited variation in a population, whereas any "evolution" of the Earth system does not, as far as we know. It may be the case that there is a population of inhabited planets, and that a long-lived biosphere is selected for on those planets that develop strong negative feedbacks. However, there is no population of inhabited planets undergoing natural selection in the sense of inherited variation. This means that the opportunities for a selection process to generate or refine properties, such as regulation at the level of the planet, are limited. This in turn has led to a shift in focus toward what might be described as "bottom-up" explanations for the emergence of planetary self-regulation (Lenton 1998). The current focus is on the emergence of environmental altering properties at the individual level and the feedbacks that can result. Feedback between organisms and their nonliving environment can only emerge when direct benefit of a trait to the organisms carrying it outweighs the cost. Usually the benefit is not conferred (at least initially) by the changes in the environment. Thus the environmental consequences are often described as "by-products." In cases where the changes in the environment never (or only negligibly) alter the selective advantage of possessing the responsible trait, they can still affect the growth of carriers and noncarriers alike, and thus generate "feedback on growth." Such feedbacks on growth are thought to be widespread and to be the most important at the planetary scale. They include those caused by marine algal dimethyl sulfide emissions and biologic amplification of rock weathering. Alternatively (and it is thought more rarely), if changes in the environment significantly alter the forces of selection, "feedback on selection" can occur. An interesting example is nitrogen fixation, which is selected for when available nitrogen is scarce. This ultimately has the effect of increasing the amount of available nitrogen in the environment and, in turn, suppresses the selective advantage of being a nitrogen fixer. The resulting feedback is negative in sign, tending to damp changes in available nitrogen. Although changes in the amount of available nitrogen in the environment are initially an inadvertent by-product, they become a factor influencing the selection of the responsible trait. More generally, any evolutionary innovation that results in significant large-scale environmental change will likely apply new selective pressures on the innovative organisms.

From the Individual to the Global

Given that such feedbacks are continually emerging as products of evolution at the local level, and there is an innate tendency for life to spread (grow and replicate), it is not surprising that they can come to have a large-scale and in some cases global influence. However, we return to the key question: Is there any innate reason why regulatory feedbacks should predominate at the global scale and tend to maintain habitable conditions? Could not destabilizing effects of life on the environment assume global proportions? It has been argued that regulatory feedbacks are a more probable outcome at the planetary scale because (by definition) they are more persistent. Furthermore, if life is altering the environment toward uninhabitable conditions, this is a self-limiting process: feedback becomes negative as the limits of habitability for the responsible organisms are approached. However, there is the caveat that abiotic positive feedback in the system may exist with the potential to switch the system to an uninhabitable or barely habitable state, such as a "snowball Earth." Biotic feedback may inadvertently move the system into the basin of attraction for such a state before it becomes self-limited. Unstable systems tend to be dominated by positive feedbacks, whereas stable systems tend to be dominated by negative feedbacks. Furthermore, unstable systems are by their very nature transitory and short-lived, whereas stable systems tend to be persistent, almost by definition. Therefore, were we to observe a planet at a random point in time, we would likely observe the planet in a stable mode dominated by negative feedbacks. Because life plays such a large role in so many processes important to Earth's climate and geochemical cycles, it is perhaps not surprising that these negative feedbacks incorporate important biological components (Caldeira 1991b). The data that we discuss here support a view ("probable" Gaia) that regulatory feedback at the planetary scale may be the more probable outcome, but not so probable as to prevent global destabilizing feedback appearing at times and contributing to major transitions in the state of the system or occasional and temporary switches to barely habitable conditions. If one accepts this as a working hypothesis, it offers a further interesting possibility:

Sequential Selection

Selection operating on a series of systems over time, rather than on a population of systems coexisting at the same time, may help explain the emergence of planetary regulation (R. A. Betts, pers. comm.). Let us assume that regulatory and antiregulatory processes involving the biota are equally likely to arise as a by-product of evolution at the individual level. In the case where evolution produces an antiregulatory system, it has been suggested that the planet would be driven toward an uninhabitable state (e.g., "snowball Earth"). However, such a process would be self-limiting. As global conditions approached the limits of habitability, heterogeneity in surface conditions would provide some more favorable sites where extinction would be more gradual than elsewhere. A point would be reached where life was sparse but still in existence. Sparse life would cease to influence the global environment and the antiregulatory mechanism could be removed. There would be no *a priori* reason to assume that the antiregulatory population would be the last to suffer extinction; hence it is likely that the antiregulatory population would become extinct while other populations still remained on the planet. Thus, an antiregulatory system could destroy itself without the loss of all life. Unless the planet had been driven to a new stable state, the removal of the antiregulatory influence would allow the planet to return to a state more similar to that in which life began. Life could then begin to spread once more. Evolution would almost certainly take a different pathway to that previously followed, and it would be possible for the biota to evolve new properties of which a by-product is regulation; if not, the system would again destabilize and life would approach extinction, "re-setting" the system again and allowing evolution to explore yet another pathway. The process would continue, either until regulation emerged or until the planet became subject to an extreme external forcing which left no refugia for life. Therefore, even if regulation can only emerge as a fortuitous by-product of evolution by natural selection, the biota could have a number of opportunities to evolve properties that lead to regulation.

How We Describe Gaia

Much of the apparent disagreement over Gaia can be seen as a disagreement about how to talk about the system rather than a more fundamental discrepancy between the theoretical frameworks used to understand it. A consensus has been developing around the view that biological processes play key roles in negative feedbacks that act to stabilize Earth's climate and the chemistry of the atmosphere, ocean, and land surface. Disagreement arises when terms such as "homeostasis" and "regulation" are used, which appear to some to imply teleological causality. An example raised by the title of this article is: In what sense can we talk of the "evolution" of the Earth system? Readers who use "evolution" as shorthand for "evolution by natural selection" may prefer to think of the "development" rather than the "evolution" of the Earth system, although this should also be used with caution, since it is quite unlike the programmed development of organisms that arose as a result of variation and selection.

MAJOR TRANSITIONS IN THE EVOLUTION OF LIFE

The major transitions associated with the storage and transmission of information (Maynard Smith and Szathmáry 1995; Szathmáry and Maynard Smith 1995) and two associated with the transformation of free energy and materials (oxygenic photosynthesis and phagotrophy) are summarized in Table 2.1. Major transitions share a number of recurring features, including the emergence of higher-level units of evolution, the evolution of novel inheritance systems, and an increase in complexity:

- *Emergence of higher-level units of evolution*: Units of evolution have to multiply, possess heredity, and produce variation. If there are hereditary traits that affect survival and/or fecundity of the units, then in a population of such units, evolution by natural selection can take place (Maynard Smith 1986). Ecologically interacting units have given rise to evolutionary units at a higher level a number of times. For example, plastids and mitochondria, as well as the ancestor of the eukaryotic nucleocytoplasm, were once free-living cells; now they can reproduce only as parts of an integrated whole. Evolutionary innovations have spread that reduce the competition among the units of the lower level, since otherwise the higher-level units would be disrupted.

Table 2.1 Major transitions in the evolution of life.

Major Transitions	Age (Ga) [possible range][1]
In the storage and transmission of information[2]:	
Replicating molecules to populations of molecules in compartments; unlinked replicators to chromosomes; RNA as gene and enzyme to DNA and protein (genetic code)	3.0–3.5 [2.6–3.85]
Prokaryotes to eukaryotes; asexual clones to sexual populations	1.2–1.9 [0.8–2.7]
Protists to animals, plants, and fungi (cell differentiation)	0.57–1.2 [0.57–1.9]
Solitary individuals to colonies (nonreproductive castes)	0.092
Primate societies to human societies (language)	0.0001–0.0002
In the transformation of free energy and matter:	
Restricted to unrestricted free-energy capture (oxygenic photosynthesis)	2.6–2.8 [2.6–3.2]
Origin of phagotrophy	1.2–1.9

[1] See Chapter 6 (this volume)
[2] According to Maynard Smith and Szathmáry (1995)

- *Evolution of novel inheritance systems*: DNA is the most commonly known basis of inheritance, yet there is more to inheritance than just DNA. First, it is an undisputed observation that there must have been a long phase of not only chemical but also biological evolution before DNA appeared. Second, multicellular organisms (be they bacteria or eukaryotes) depend on a second (epigenetic) inheritance system: a liver cell and a white blood cell are genetically almost identical; the difference between them is caused by which genes are on (i.e., expressed) and which are off (silent). This regulated state of gene expression can be passed on from cell to cell, and the most spectacular accomplishments of this kind have been achieved by plants, fungi, and animals, which have a large number of cell types that can be maintained and propagated.
- *Increase in complexity*: Organismic complexity is hard to define, although all biologists have at least some intuitive feeling for the degree of complexity. At the morphological level, a convenient measure is the number of cell types (Bonner 1988), which is linked to epigenetic inheritance, hence gene regulation. Whereas the idea (Maynard Smith and Szathmáry 1995) that such complexity would be correlated with the number of coding genes (mostly for proteins, but also for rRNA and tRNA) is by now untenable (a nematode has more genes than a fruit fly), the *interaction density* (connectivity) among genes in regulatory networks seems to be a good genetic

correlate of organismic complexity (Szathmáry et al. 2001). Complexity can be increased by variation from within (duplication and divergence) or by symbiosis (e.g., the encapsulation of foreign replicators).

There is one final general point that we want to discuss in connection with the major transitions, namely, the problem of irreversibility. Clearly, some transitions have been reversed in certain lineages; for example, yeast is secondarily unicellular and there are solitary insects that obviously stem from social relatives. In other cases, there seems to be no way back: we do not know of an organism that has lost its genetic code and reverted back to some RNA-based metabolism. It is important to realize that a lot of this irreversibility is contingent: it is not *logically* impossible to go back, but for all practical purposes it is very unlikely that the appropriate variation could arise. Although animals and plants have lost sexuality in many lineages, this is never the case in gymnosperms and mammals. In the former, one gamete provides the mitochondria and the other the plastids; in mammals you have genomic imprinting. In both cases, reversion to parthenogenesis would require so many simultaneous changes that it cannot be expected on probabilistic grounds (cf. Szathmáry 2004).

MAJOR TRANSITIONS IN THE HISTORY OF THE EARTH SYSTEM

A tentative compilation of major transitions in the history of the Earth system is presented in Table 2.2, with approximate timings for each. We describe these in more detail below and discuss their contingency on one another and on major transitions in the evolution of life. We consider possible links to extrinsic or intrinsic disturbances, although evidence (e.g., for asteroid impact or mass extinction) is difficult to find in deep time.

The Emergence of Life

Life emerged over 3.8 billion years ago, if we accept the evidence from West Greenland (Mojzsis et al. 1996). For this to have occurred, the Earth must have already been in a habitable state, with liquid water present. The emergence of life was a major transition in the development of the Earth system, because it gave the system (a) a new capacity for storage and transmission of information and (b) new pathways for the transformation of energy and matter. Through the first three major transitions in the storage and transmission of information by life (Table 2.1), the impact of life at a planetary scale would have increased.

The capacity for information to be stored and transmitted can be viewed as a system characteristic, even though it is localized in the living components. Accepting that life inevitably transforms its nonliving environment, that it is constrained by the state of the environment, and hence there is feedback between life and its environment, potentially involving natural selection, then the

Table 2.2 Major transitions in the history of the Earth system.

Major Transitions	Age (Ga)
Emergence of life (abiotic to biotic)	>3.5
Emergence of Gaia (near-equilibrium to extreme disequilibrium atmosphere; sparse to abundant life; input-limited to recycling-dominated)	>2.7
Great Oxidation (reducing to oxidizing atmosphere)	2.2–2.0
The Neoproterozoic (Precambrian to Cambrian; microscopic to macroscopic life; mildly to strongly oxidizing atmosphere)	0.75–0.55

information stored in the living component contains some information about the state of the environment and the state of the system as a whole. The global gene pool also carries some "memory" of past experience: characteristics of the Last Universal Common Ancestor (LUCA) can be reconstructed to give some information about the environment in which this being lived.

With life, new chemical compounds would have begun to appear in the surface environment and unique biological signatures were left in the rock record. Compounds utilized in metabolism would have been depleted in the surface environment, while waste products of metabolism accumulated. Thus free-energy and entropy gradients between living and nonliving material emerged. Autocatalytic recycling loops must have appeared early, if life was to avoid being hopelessly limited by substrate supply. The remarkable ability of biological enzymes to catalyze chemical reactions would have accelerated many existing transformations of matter (and may thus have reduced thermodynamic disequilibrium in these cases). However, the system was limited by what free energy it could capture. Hydrothermal/geothermal energy fluxes are tiny compared with the energy flux from the Sun. The first photosynthesizers (with just photosystem I) were anaerobes and may have used H_2S as a reductant (as in the green sulfur bacteria). This substrate would have been in limited supply, thus limiting the input of free energy to the biota.

The Emergence of Gaia (Life Becomes a Planetary Force)

Life appears to have played a major role in the carbon cycle from nearly the time of its first appearance. Carbon isotopic data from sedimentary carbonates and organic carbon suggests that sedimentary carbonate isotopic composition was significantly enriched in ^{13}C relative to mantle carbon. This in turn suggests that life played a major role in the production of pools of reduced carbon from early in Earth's history (Schidlowski 1988).

Life became a planetary force capable of transforming Earth some time prior to 2.7 Ga (perhaps as early as 3.8 Ga), with the evolution and spread of oxygenic photosynthesis: an unrestricted means of free-energy capture from the Sun using water as a reductant. The use of an essentially limitless substrate (water) would have increased the supply of free energy to the biota by orders of magnitude, relative to pathways of anoxygenic photosynthesis. With a great increase in free energy at its disposal, life became a planetary force. We can expect there to have been a transition to a far-from-equilibrium atmosphere. Initially, all liberated oxygen was consumed near to its source in the oxidation of reduced materials. Carbon dioxide was drawn out of the atmosphere to form organic matter. Much of the organic matter was subject to methanogenesis, returning a mixture of methane and carbon dioxide to the atmosphere. The carbon isotope record suggests that methane was globally abundant ($pCH_4 > 0.0001$ atm) and sufficient to support widespread communities of methanotrophs (these required local $pO_2 > 0.0005$ atm to oxidize CH_4). A significant biologic hydrogen source could also have arisen. A methane-rich atmosphere with hydrogen and only traces of oxygen (chemically opposite to the situation today) may have occurred. To support high levels of global productivity, indicated by the carbon isotope record, there must have been effective recycling of essential elements. This occurred both at the scale of microbial mats (such as those that formed stromatolites) and may also have occurred at a larger scale in the open ocean (and perhaps even on the land surface). By generating local oases with significant oxygen concentrations, oxygenic photosynthesis probably had a part to play in the major transition from the metabolically diverse prokaryotes to the (predominantly) aerobic eukaryotes.

The Great Oxidation

Between about 2.2 and 2.0 billion years ago in the early Proterozoic, the partial pressure of atmospheric oxygen (pO_2) increased from < 0.0008 atm to > 0.002 atm (and possibly as high as > 0.03 atm), and the atmosphere switched from being chemically reducing to being oxidizing. This was the most significant step in the oxidation of the Earth's surface environment, and the single most striking change in the history of atmospheric oxygen (Figure 2.1) (Rye and Holland 1998). It has been described as a "Great Oxidation Event," although the difficulties of dating so far back in time mean the "event" could have covered an interval of some 200 million years and may have comprised a series of changes, including reversals. Although its occurrence is not universally accepted (Ohmoto 2003), a number of lines of evidence are consistent with a major transition having occurred. In ancient soils (paleosols) older than 2.2 Ga, iron was mobile during weathering, indicating a lack of oxygen. Disappearance of large ore deposits of uraninite by 2.0 Ga is consistent with increasing oxygen. In later paleosols, iron was immobile, indicating that O_2 had exceeded a critical concentration. Red beds (oxidized layers) deposited above paleosols aged ~2.2 Ga are indicative of more oxidizing conditions after this time. Supposed eukaryote fossils from 1.9–2.1 Ga resemble the photosynthetic algae *Grypania*, which is estimated to have required $pO_2 \sim 0.002$–0.02 atm. A large positive carbon isotope shift of

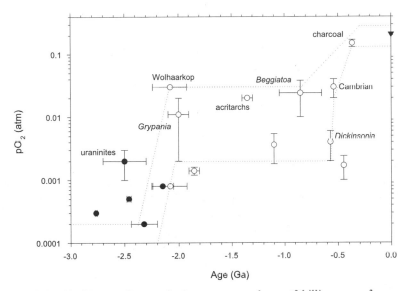

Figure 2.1 The history of atmospheric oxygen over the past 3 billion years, from geo-chemical and biological constraints. Filled circles are estimated upper limits on oxygen; empty circles indicate estimated lower limits on oxygen. The inverted triangle indicates the present partial pressure of oxygen. Age error bars indicate uncertainty in dating or the range of ages over which sediments were deposited. Partial pressure error bars indicate uncertainties in the estimates. Dashed lines are approximate upper and lower bounds on atmospheric oxygen. Unlabeled points are all paleosols, summarized by Rye and Holland (1998), who determined the pO_2 limits for each. Wolhaarkop is also a paleosol, for which the pO_2 lower limit was determined by a different technique (Holland and Beukes 1990). Uraninites upper limit is from Holland (1984). Biological lower limits on pO_2 are updated from Runnegar (1991) with an earlier record of *Grypania* (Han and Runnegar 1992) now subject to some age uncertainty, and additional estimates for *Beggiatoa* (Canfield and Teske 1996) and the Cambrian fauna (Holland 1984). These biological constraints may reflect localized dissolved O_2 levels rather than the global atmospheric pO_2 levels suggested. The appearance of charcoal (Cressler 2001) is the first convincing evidence for pO_2 nearing the present level. Note that the "eukaryotic" identification of fossils older than 0.8 Ga is disputed. Figure adapted from Lenton (2003).

carbonates 2.2–2.0 Ga suggests a massive increase in organic carbon burial, releasing an estimated 12–22 times the current atmospheric O_2 inventory, more than sufficient to generate the inferred rise in atmospheric O_2.

The consequences of the inferred rise in atmospheric O_2 would have included a large drop in the methane content of the Earth's atmosphere, once the source of oxygen became more than double the source of methane ($CH_4 + 2O_2 = 2H_2O + CO_2$). This switch from a reducing to an oxidizing atmosphere would have had profound consequences, including large reductions in the concentrations of other reduced gases (e.g., H_2), clearing of a photochemical smog of hydrocarbons that had been produced from CH_4, and the establishment of an ozone layer with effective ultraviolet protection underneath it. Reduction in the greenhouse

T. M. Lenton et al.

effect due to CH_4 would have cooled the planet; given the fainter younger Sun and estimates of relatively low atmospheric CO_2 concentrations at the time, it may have been sufficient to trigger extreme glaciations. There were indeed extensive low-latitude glaciations in the early Proterozoic but these seem to predate the Great Oxidation, being placed in the interval 2.45–2.22 Ga. They are referred to as the Makganyene and Huronian glaciations, and there were at least three of them. Perhaps they corresponded with the start of oxidation, with oxygen then taking some time to accumulate in the atmosphere. At present it is unclear what, if any, causal relations existed between these glaciations and the rise of atmospheric oxygen.

The Great Oxidation has often been portrayed as a catastrophe for anaerobic life, which would have been unable to persist in the oxic surface environments created. However, it is now thought that the bulk of the deep ocean remained anoxic long after the oxidation of the atmosphere, and that the anoxic waters may have extended into the photic zone of the ocean (Canfield 1998). Even today, regions of the ocean, such as the deep eastern equatorial Pacific, are nearly anoxic. Many habitats suitable for obligate anaerobes would thus have remained, perhaps including a region in the photic zone where anoxygenic photosynthesis could continue. The oxidation of the land surface and of surface ocean waters would have drastically reduced the availability of many biologically essential trace metals including iron and manganese. This in turn would have limited some biogeochemical processes (e.g., nitrogen fixation) that rely on these elements at the active sites of enzymes. At the same time it would have facilitated other processes (e.g., nitrification) that are fueled by oxygen. Overall, the free energy available due to the large chemical potential difference between free oxygen and reduced organic matter would have helped support a more productive biota. The assumption that CH_4 levels collapsed in the aftermath of oxidation has recently been questioned (A. Pavlov, pers. comm.). At lower-than-present oxygen levels (Figure 2.1), CH_4 could have been much more abundant than it is today if, for example, methanogens continued to thrive in an anoxic deep ocean and generated a larger than present methane source.

A number of hypotheses exist for the ultimate cause of the Great Oxidation:

1. The Earth's internal heat source decays slowly with time and this would have caused a gradual reduction in the influx of reduced matter to the Earth's surface during the Archean (Lovelock 1988). Furthermore, it has been suggested that the subduction of hydrated seafloor would lead to progressive oxidation of the mantle such that the mantle degassed increasingly more oxidized gases over time (Kasting et al. 1993). A point may have been reached when the redox scales tipped and the input of reduced matter could no longer absorb all of the oxygen being produced by the burial of reduced organic carbon, and thus oxygen began to accumulate in the atmosphere.

2. An evolutionary innovation could have caused the inferred increase in burial of reduced organic carbon (oxygen source). This may be tentatively linked to the major transition from prokaryotes to eukaryotes and, in particular the subsequent evolution of eukaryotes with chloroplasts (algae). The first algae would have occupied new niches, and they were better adapted to oxygen than cyanobacteria, thus increasing global primary productivity. Given that what was claimed to be a large alga (*Grypania*) was present during or just after the Great Oxidation (Han and Runnegar 1992), it seems plausible that the first algae could have predated it. Alternatively, the Great Oxidation could have been caused by the spread of abundant cyanobacterial populations (see below).

3. Methanogens may have inadvertently caused the Great Oxidation. By creating a methane-rich atmosphere in the late Archean they greatly increased hydrogen loss to space resulting in net oxidation of the atmosphere–ocean–sediment system (Catling et al. 2001). As oxygen rose, methanogen communities would have been disrupted because they are obligate anaerobes. Their contribution to the recycling of organic carbon may have been suppressed, and any delay before aerobic-respiring organisms filled the carbon-recycling niche would have resulted in a greater fraction of organic carbon being buried.

Was the Great Oxidation linked to a major transition in evolution? It was definitely contingent on the transition from restricted to unrestricted free-energy capture, that is, the evolution of oxygenic photosynthesis (in the cyanobacteria), although there was a considerable time delay before global oxidation occurred. If the second hypothesis is correct, the Great Oxidation was also contingent on the major transition from prokaryotes to eukaryotes and the subsequent acquisition of chloroplasts. If the third hypothesis is correct, it was contingent on the evolution of methanogenesis in the archaea, which is not considered a major transition (at least in terms of the storage and transmission of information).

Dating of the origin of eukaryotes is far from trivial. Allegedly, rhodophytic fossils notwithstanding, the oldest eukaryotic fossils that clearly indicate the presence of an endomembrane system and a cytoskeleton (shared derived traits of eukaryotes) are not older than 0.8 Ga (Porter and Knoll 2000); it is quite possible that eukaryotes, together with their sister group the Archaea (archaebacteria) are in fact not older than 0.85 Ga (Cavalier-Smith 2002a, b, c). For example, the presence of steranes in 2.7-Ga-old material has been taken as evidence for the presence of eukaryotes then, yet not only the posibacterium *Mycobacterium* but also two groups of proteobacteria make sterols; thus sterols are hardly biomarkers for eukaryotic life (Cavalier-Smith 2002c). In the case that eukaryotes are relatively young, eukaryotic algae necessarily cannot be responsible for the Great Oxidation.

In general, extreme care should be taken when dating fossils more than a billion years old. In fact, the above-quoted *Grypania* case (Han and Runnegar

1992) shows no evidence of intracellular structures, without which a eukaryotic affiliation is questionable. The case of an allegedly 3.5-Ga-old cyanobacteria (e.g. Schopf 1999), claimed on partial morphological resemblance to present-day cyanobacteria, has been incisively criticized (Brasier et al. 2002).

The Neoproterozoic

1.5 billion years after the Great Oxidation, the Earth underwent a series of extreme glaciations, a further significant rise in atmospheric oxygen, and a transition from microscopic to macroscopic life-forms during the Neoproterozoic interval ~0.8–0.54 Ga. During this time, at least two extreme low-latitude glaciations have been recognized: the Sturtian (~0.75 Ga) and Marinoan (~0.59 Ga). A third Varangerian glaciation may also have occurred. In each case, continental ice sheets appear to have reached low latitudes, and it has been suggested that the oceans were totally frozen over in so-called "snowball Earth" events, which lasted millions of years (Kirschvink 1992; Hoffman et al. 1998). The proximate cause of extreme glaciation is an inherent instability in the Earth system due to the snow/ice-albedo feedback: as land ice/snow and sea-ice cover increases, the Earth's surface albedo increases, which tends to amplify the spread. Simple one-dimensional climate models predict a critical "runaway" threshold beyond which the planet becomes completely covered in ice (North et al. 1981). Some more complex climate models predict an alternative "slushball Earth" scenario in which the equatorial ocean remains ice-free (Hyde et al. 2000). Debates continue to rage over which scenario is more consistent with observations. These observations include the association of glaciations with >10‰ negative $\delta^{13}C$ shifts (suggesting a global collapse of biological productivity), the return of large sedimentary iron formations during the glaciations (suggesting sustained anoxic conditions in the ocean), the deposition of pure carbonate layers ("cap carbonates") during postglacial sea-level rise (consistent with sudden unfreezing and weathering of alkalinity to the ocean), and the survival of eukaryotes including multicellular metazoans (indicating that there must have been oases for life). Whether or not climate change was catastrophic, it was certainly extreme.

The Neoproterozoic was a time of massive swings in the carbon isotope record. Carbon isotope studies suggest high organic carbon burial rates in the intervals of 1.1–0.8 Ga and 0.8–0.7 Ga (Walter et al. 2000). This is consistent with the evolution of nonphotosynthetic sulfide-oxidizing bacteria (*Beggiatoa*) and an increase in sulfur-isotope fractionation; both suggest that oxygen crossed a threshold of pO_2 0.01–0.04 atm in the interval 1.05–0.64 Ga (Canfield and Teske 1996). A further increase in oxygen has been suggested in a brief interval before the Cambrian boundary 0.6–0.57 Ga (Logan et al. 1995). At least three biological hypotheses exist to explain rising oxygen in the Neoproterozoic:

1. Evolution of marine animals with guts that produced fecal pellets would have led to more efficient transport of organic carbon to sediments, thus potentially increasing the oxygen source (Logan et al. 1995).

2. Biological colonization of the land surface could have increased weathering rates, especially of the essential nutrient element phosphorus, which was only available from rock weathering. There is evidence for increased weathering rates during the Neoproterozoic, and increased phosphorus weathering would have led to an increased supply of phosphorus to the ocean, increased productivity on land and in the ocean, increased organic carbon burial and a higher steady state oxygen level (Lenton and Watson 2004).

3. The origin of chloroplasts may have happened as late as 0.58 Ga when the Varangerian snowball melted. In fact, symbiotic acquisition of light harvesters could have been strongly selected for because of the increase in sunlight intensity (Cavalier-Smith 2002c). In turn, increase in photosynthesis may have triggered the Cambrian protist and animal radiations.

It is interesting to consider whether the two striking features of the Neoproterozoic transition — the rise of atmospheric oxygen and the extreme glaciations — can be mechanistically connected. A couple of recent hypotheses link the extreme low-latitude glaciations to biological evolution and changes in the carbon cycle:

1. Biological colonization of the land surface and a consequent increase in silicate weathering rates could have caused a reduction in atmospheric CO_2 and planetary cooling (Lenton and Watson 2004).

2. In the absence of planktonic calcifying organisms, the carbonate saturation state of the ocean would have been highly sensitive to sea-level change, potential creating a "runaway" positive feedback between ice sheet growth, loss of shallow water depositional environments, and atmospheric CO_2 drawdown (Ridgwell et al. 2003).

The two hypotheses are not mutually exclusive, with the second having the potential to amplify a cooling induced by the first. Parsimoniously, the biological colonization of the land surface (Heckman et al. 2001) can potentially explain both rising oxygen and climate cooling in the Neoproterozoic. We may then invoke positive feedback mechanisms to explain the extreme nature of the glaciations. Escape from extreme glaciations could come about through a reduction in weathering rates and CO_2 buildup in the atmosphere, until the ice sheets started to melt (Caldeira and Kasting 1992). This would generate positive feedback to a warmer climate state with considerable overshoot. Then the long-term negative feedback involving silicate weathering would cool the system. This combination of rapid positive feedback and slow negative feedback is a common recipe for generating self-sustaining oscillations in a system and might thus explain the recurrence of glaciations.

Can the major transition of the Earth system through the Neoproterozoic be linked to a major transition in evolution? If biological colonization of the land surface played a part, this probably involved the ancestors of plants, fungi, and lichens, and was thus contingent on cell differentiation earlier in the Protero-zoic. If the appearance of animals with guts was important, these too relied on earlier cell differentiation. Increased atmospheric oxygen was a necessary con-dition for the appearance of the first large multicellular creatures, the Ediacaran ~0.6 Ga, followed by the Cambrian fauna ~0.54 Ga. These were the first visible expression of a much earlier transition involving cell differentiation. If, in con-trast, eukaryotes themselves are not older than 0.85 Ga, then widespread glaci-ation could have selected for phagotrophy (i.e., the eukaryotic cell), which set the stage for the symbiotic engulfment of presumptive cell organelles, melting triggered the origin of algae, which in turn raised oxygen levels, and finally multicellular eukaryotes could radiate. The earliest branch off the eukaryotic lineage is the optisthokonts, including Choanazoa, animals, and fungi (Cava-lier-Smith 2002d).

PHANEROZOIC DISTURBANCES AND TRANSITIONS

The Phanerozoic aeon, approximately the last 550 million years, has been char-acterized by a series of disturbances in the form of mass extinctions (Table 2.3). Such disturbances may be internally or externally generated. There have also been transitions in the state of the Earth system that, while less extreme than those in the Precambrian, are better documented and understood.

Mass Extinctions

In the Phanerozoic, there have been at least five mass extinction events in which roughly 75% or more of extant species went extinct. Perhaps the two most stud-ied events are the Permian extinction (~250 Ma) and the end-Cretaceous extinc-tion (~65 Ma). It has been conjectured that both of these extinctions were caused, primarily, by the impact of a comet or asteroid. The evidence for such an impact is stronger for the end-Cretaceous extinction. Furthermore, both

Table 2.3 Mass extinctions during the Phanerozoic.

Interval	Age (Ma)	Loss of families (%)
Late Cambrian mass extinctions	500	?
Late Ordovician mass extinction	440	20–50
Late Devonian (Frasnian–Famennian) mass extinction	370	20–30
End Permian mass extinction	250	50
End Triassic extinction(s)	210	20–35
End-Cretaceous (K/T) mass extinction	65	15

extinction events seem to be associated with large flood basalt episodes (i.e., Siberian Traps and Deccan Traps, respectively). Relationships between impacts and the flood basalt episodes have been conjectured but remain unclear.

In the Permian extinction, over 95% of species went extinct, including over half of all animal families. Victims included trilobites, acanthodians (the earliest jawed fish), placoderms (heavily armed jawed fish), and various corals. Many other groups were affected, including sharks and bony fish. On land, pelycosaurs (finback reptiles) went extinct. With the less-severe Cretaceous extinction, about 75% of all species became extinct, including 95% of marine plankton species. Earth saw the extinction of the dinosaurs, as well as many plants (except among the ferns and seed-producers), marine reptiles, and fish. Remarkably, species of mammals, birds, turtles, crocodiles, lizards, snakes, and amphibians were largely unaffected.

It has been suggested that the evolution of marine biota can be viewed as long periods of stasis punctuated by these two mass extinction events, which produced rapid diversification leading to a new stasis (Bambach et al. 2002). Furthermore, these extinction events may have driven a secular change toward a predominance of motile organisms (Bambach et al. 2002). Only about one-third of marine species were motile prior to the Permian mass extinction. Following the Permian event, roughly half of all marine species were motile; since the end-Cretaceous mass extinction roughly two-thirds of marine species have been motile. This pattern suggests that the marine ecological/evolutionary system is characterized by multiple stable equilibria and that strong perturbations can push the Earth system to a new, relatively stable, basin of attraction.

In some cases, even a severe perturbation can be insufficient to push components of the Earth system into a very new basin of attraction and the system returns to a state similar to the pre-perturbation state. An example of this can be found in what is now the western United States. As a result of the end-Cretaceous events, the broadleaf evergreen trees that dominated some ecosystems went extinct. Within several million years after the extinction, broadleaf evergreens re-evolved. These new broadleaf evergreens were from different genera than those which dominated before the extinction; however, they played a similar ecological role.

Similarly, one could argue that mammals evolved to fill ecological niches left open by the demise of the dinosaurs: keystone predators such as *Tyrannosaurus rex* were replaced by the large cats, grazing herbivorous dinosaurs were replaced by bison, deer, and so on. Thus, the extinction event can be viewed as pushing the Earth from a quasi-stable state in which large animals were predominately dinosaurs, to one in which they were predominately mammals. Initially, there appears to have been repeated turnover of species as ecological niches were filled and challenged; however, after several million years, a new, more-stable equilibrium seems to have been reached as ecological niches became occupied by species resistant to easy turnover (Zachos et al. 1989).

Transitions

At least two transitions played a major role in the evolution of the Earth system during the Phanerozoic: the population of the land surface by vascular plants, and the population of the open ocean by calcareous plankton.

The Rise of Land Plants

Fossilized remnants of microbial mats indicate that some life existed on land as long as 2.6 or 2.7 billion years ago. However, vascular land plants did not evolve and become widespread until the Devonian, some 400 million years ago. Vascular land plants have roots that stabilize soils, while root respiration and organic matter decomposition increases the partial pressure of CO_2 in the soil atmosphere (Lovelock and Watson 1982). Many plants also secrete organic acids. These factors tend to increase the weathering rate of silicate rock. On long timescales (i.e., millions of years), carbon inputs to the atmosphere and ocean are balanced by net organic carbon burial and silicate rock weathering with subsequent carbonate mineral burial. It is thought that atmospheric CO_2 adjusts such that CO_2 consumption by silicate rock weathering balances metamorphic and magmatic CO_2 degassing less net organic carbon burial (Berner et al. 1983). Vascular land plants allowed this silicate rock weathering to occur with less CO_2 in the atmosphere. Thus, the rise of land plants led to a sustained reduction in atmospheric CO_2 content and cooler long-term temperatures.

Carbon dioxide acts as a nutrient for land plants, and higher CO_2 levels facilitate more efficient use of water. Today, the most luxurious forests can be found in warm tropical regions. The evolution of land plants resulted in lower atmospheric CO_2 concentrations and cooler temperatures. Thus, it is likely that land plants grew less well as a result of the climatic and geochemical consequences of their evolution. If CO_2 levels were to fall precipitously, the growth of land plants would be impeded, soils would become more subject to erosion, and soil atmospheres would have lower CO_2 concentration. This would slow silicate rock weathering rates and allow CO_2 to accumulate in the atmosphere, making the environment better for land plants once again. Thus, although the climate and geochemical effects of land plants on land plants may have been negative (i.e., cooling, lower CO_2), land plants play a critical role in an important negative feedback loop (i.e., higher $CO_2 \rightarrow$ more land plants $\rightarrow$ more silicate rock weathering $\rightarrow$ lower CO_2) that helps stabilize climate, allowing land plants to persist.

The rise of vascular land plants is also thought to have driven an increase in the O_2 content of the atmosphere with a peak in the Carboniferous (Berner and Canfield 1989). A number of factors probably contributed to increased organic carbon burial:

- evolution of recalcitrant organic carbon compounds such as lignin as structural materials, which enabled tree stature (Robinson 1990),
- delay in the evolution of fungi that could biodegrade these compounds (Robinson 1990),

- evolution of weathering mechanisms that increased the global weathering flux of phosphorus from rocks (Lenton 2001),
- a geographic setting that allowed the formation of vast areas of lowland coal swamps.

It has also been suggested that vascular plants are involved in negative feedback mechanisms which stabilize the oxygen content of the atmosphere, in particular buffering rises in oxygen above the present level (Lenton and Watson 2000; Lenton 2001).

The Success of Calcareous Plankton

Another major evolutionary change involved calcareous plankton, which became widespread in the Jurassic (by ~150 Ma). Calcareous plankton are photosynthetic organisms with skeletal material made out of calcium carbonate, some of which sinks to the seafloor. Sinking calcium carbonate in the open ocean is likely to be preserved if it falls on seafloor that lies in water saturated with respect to calcium carbonate. Thus after the Jurassic, there was a new pathway for carbonate accumulation on the deep seafloor.

Prior to the Jurassic, most carbonate sedimentation occurred in shallow-water coastal environments, and the chemistry of the ocean was buffered primarily by the dependence of this shallow-water carbonate accumulation on carbonate-ion concentration (Opdyke and Wilkinson 1993). After the Jurassic, the ocean chemistry was also buffered by the dependence of deep-sea carbonate accumulation on carbonate-ion concentration (Caldeira and Rampino 1993). This new pathway likely made the Earth system more stable, in that now two mechanisms buffer ocean chemistry. It has been suggested (e.g., Ridgwell et al. 2003) that when only the shallow-water mechanism functioned, ocean chemistry (and atmospheric CO_2) would have been more sensitive to changes in shallow-water area, as might occur with a glaciation, with the result that the Earth system may have been more susceptible to deep glaciation prior to the Jurassic.

The evolution of open-ocean carbonate-secreting organisms may also be increasing the flux of calcium carbonate transported to subduction zones with the seafloor (Volk 1989; Caldeira 1991a). The fate of this carbonate is unclear. The three main possible fates are: (a) accretion, (b) subduction to the mantle, and (c) subduction followed by metamorphic degassing with possible release to the atmosphere. Thus, it is possible that the evolutionary success of calcareous plankton is increasing both the transport of surficial carbon to the mantle and CO_2 degassing in subduction zone volcanism.

The Pleistocene Onset of Periodic Glaciations

For the past 2.7 million years, the Earth has been exhibiting glacial–interglacial cycles, with spectral power at, or close to, a variety of astronomically significant "Milankovitch" frequencies. Such glacial–interglacial cycles are geologically unusual, although evidence of orbital frequencies is widespread in the Phanero-zoic sedimentary record. The Quaternary Earth system was the focus of discus-sions in Group 2 (see Watson et al., this volume). Here we offer one comment: The 100-ka glacial–interglacial cycles seem to be an internal mode of climate variability paced by Earth's tilt and orbit around the Sun. The present generation of general circulation models used to predict future climate are unable to repre-sent these dynamics. These models are designed to incorporate short-term feed-backs (e.g., clouds) but are unable to simulate the long-term nonlinear behavior of the climate system. Complex and subtle interactions between the land sur-face, the oceans, ice sheets, clouds, and so on are represented crudely, if at all, in these models. Thus, predictions of CO_2-induced global warming from these models do not adequately consider how a short-term perturbation to Earth's cli-mate will propagate through the Earth system on the scale of centuries and more.

HUMAN DISTURBANCES

Humans are transforming the land surface for agricultural and other ends, unsustainably fishing the oceans, and releasing fossil carbon into the atmo-sphere. These and other such issues were the foci of Group 3's discussion (see Steffen et al., this volume). Here we try to place these changes in the context of the mass extinctions and changes in the carbon cycle discussed above.

The CO_2 that we are currently putting into the atmosphere will largely be neu-tralized by interaction with carbonate minerals on the timescale of about 10 ka, and the remaining CO_2 will be neutralized by silicate rock weathering on the timescale of 100s of ka (Archer et al. 1998). Thus, fossil-fuel releases have the potential to change significantly the climate for at least thousands of years, with unknown feedbacks on longer timescales. It has been suggested that human ac-tivities have doubled river sediment load, and thus we are undoing some of the soil stabilization provided by the evolution of land plants. If soil loss were to continue for the indefinite future, the carbonate–silicate cycle would respond with higher atmospheric CO_2 concentrations on the timescale of hundreds of thousands of years.

Since the appearance of *Homo sapiens*, the extinction rate of megafauna has been unprecedented with the exception of mass extinction events. However, there is a saying among biogeochemical oceanographers: "If you can see it, it probably doesn't matter." That is, most of the important geochemical functions are carried out by microorganisms, and most species of (at least, marine) micro-organisms have been relatively unimpaired by human activities. Similarly, whereas many species of large animals, as well as many species of both plants and animals endemic to small regions, have been driven to extinction, most of these species probably played a minor role in the basic biogeochemical func-tioning of the planet. If land-use change and the fragmentation of ecosystems continue, more extinctions will likely result. Most vulnerable would be species endemic to small regions and larger fauna that need bigger refuges to have sustainable populations.

Thus, the mass extinction we are creating will probably not have severe effects on the basic biogeochemical functioning of the planet, but it is likely to leave our planet largely depaupered of large animals and species endemic to small regions. If the past is any guide to the future, we can take solace in the thought that these ecological niches may be filled by newly evolved organisms within a few million years of our learning to live sustainably on this planet (or our extinction).

The vast impact of our species on the planet is due to our technological innovations, which arise in a society in which a novel inheritance system operates: language (Szathmáry 2003). Language with complex syntax is about 100,000 years old; agriculture and massive technology are an order of magnitude younger. It is amazing how much harm the human race has done within so little time. One reason for this is that although language allows cumulative cultural evolution, humans have evolved in relatively small populations and have not adapted to deal with large populations and timescales. Language has enabled us to enter into social contracts not available for animals: our task is now to work out a global social contract. Information technology available today (e.g., the Internet) makes this possible; whether this possibility will be realized depends on other factors. A global social contract must offer a resolution to the tension between long-term global goals and short-term, local democratic systems, based on regular elections. The Earth must *consciously* be transformed into a unit as if it had been shaped by evolution at that level, for we have only one Earth.

CONCLUSION AND OPEN QUESTIONS

Our review of the major transitions in the evolution of life and in the history of the Earth system suggests that they are usually associated, but it is not clear whether they are always associated. External disturbances (e.g., asteroid impacts, massive flood basalt eruptions) appear in some cases to have triggered significant transitions of the system between different (quasi-stable) states. However, the most major transitions in the history of the Earth system appear to have been internally generated with evolutionary innovation playing a leading role. We have raised (but not satisfactorily answered) a number of important questions, which although discussed at this Dahlem Workshop still require further consideration:

- To what extent is the Earth system self-regulating?
- What is the contribution of life to maintaining habitable conditions?
- What is the role of natural selection in planetary regulation?
- In what sense can the Earth system be said to evolve?
- Which are the more difficult major transitions in the evolution of life?

- Are the patterns of major transitions of life and of the Earth system repeatable?
- What was the relative importance of internal and external disturbance in the evolution of life and of the Earth system?
- Can astrobiology make the Gaia hypothesis falsifiable? If so, what sample size is required? What information about other systems is required?
- Which regions of phase space have been explored by the Earth system?
- Why does the Earth system possess negative, and possibly regulatory, feedbacks?
- Are there reasons to explain why regulatory feedbacks should predominate at the global scale and tend to maintain habitable conditions?
- How globally destructive can destabilizing effects of life on the environment become?

ACKNOWLEDGMENTS

We thank Julia Lupp for her patience and help in bringing this paper together.

REFERENCES

Archer, D., H. Kheshgi, E. Maier-Reimer et al. 1998. Dynamics of fossil fuel CO_2 neutralization by marine $CaCO_3$. *Global Biogeochem. Cycles* **12(2)**:259–276.

Bambach, R.K., A.H. Knoll, and J.J. Sepkoski, Jr. 2002. Anatomical and ecological constraints on Phanerozoic animal diversity in the marine realm. *Proc. Natl. Acad. Sci. USA* **99**:6854–6859.

Berner, R.A., and D.E. Canfield 1989. A new model for atmospheric oxygen over Phanerozoic time. *Am. J. Sci.* **289**:333–361.

Berner, R.A., A.C. Lasaga, and R.M. Garrels. 1983. The carbonate-silicate geochemical cycle and its effect on atmospheric carbon dioxide over the past 100 million years. *Am. J. Sci.* **283**:641–683.

Bonner, J.T. 1988. The Evolution of Complexity. Princeton: Princeton Univ. Press.

Brasier, M.D., O.R. Green, N.V. Jephcoat et al. 2002. Questioning the evidence for Earth's oldest fossils. *Nature* **416**:76–81.

Caldeira, K. 1991a. Continental-pelagic carbonate partitioning and the global carbonate–silicate cycle. *Geology* **19(3)**:204–206.

Caldeira, K. 1991b. Evolutionary pressures on planktonic DMS production. In: Scientists on Gaia, ed. S.H. Schneider and P. Boston, pp. 153–158. London: MIT Press.

Caldeira, K., and J.F. Kasting. 1992. Susceptibility of the early Earth to irreversible glaciation caused by carbon dioxide clouds. *Nature* **359**:226–228.

Caldeira, K., and M.R. Rampino. 1993. Aftermath of the end-Cretaceous mass extinction — possible biogeochemical stabilization of the carbon-cycle and climate. *Paleoceanography* **8**:515–525.

Canfield, D.E. 1998. A new model for Proterozoic ocean chemistry. *Nature* **396**:450–453.

Canfield, D.E., and A. Teske. 1996. Late Proterozoic rise in atmospheric oxygen concentration inferred from phylogenetic and sulphur-isotope studies. *Nature* **382**:127–132.

Catling, D.C., K.J. Zahnle, and C. McKay. 2001. Biogenic methane, hydrogen escape, and the irreversible oxidation of early Earth. *Science* 293:839–843.

Cavalier-Smith, T. 2002a. The neomuran origin of archaebacteria, the negibacterial root of the universal tree, and bacterial megaclassification. *Intl. J. Syst. Evol. Microbiol.* 52:7–76.

Cavalier-Smith, T. 2002b. Origins of the machinery for recombination and sex. *Heredity* 88:125–141.

Cavalier-Smith, T. 2002c. The phagotrophic origin of eukaryotes and the phylogenetic clasification of Protozoa. *Intl. J. Syst. Evol. Microbiol.* 52:297–354.

Cavalier-Smith, T. 2002d. Rooting the eukaryotic tree by using a derived gene fusion. *Science* 297:89–91.

Cressler, W.L. 2001. Evidence of earliest known wildfires. *Palaios* 16(2):171–174.

Dawkins, R. 1983. The Extended Phenotype. Oxford: Oxford Univ. Press.

Doolittle, W.F. 1981. Is Nature really motherly? *CoEvol. Qtly.* Spring.58–63.

Han, T.-M., and B. Runnegar. 1992. Megascopic eukaryotic algae from the 2.1-billion-year-old negaunee iron-formation, Michigan. *Science* 257:232–235.

Heckman, D.S., D.M. Geiser, B.R. Eidell et al. 2001. Molecular evidence for the early colonization of land by fungi and plants. *Science* 293:1129–1133.

Hoffman, P.F., A.J. Kaufman, G.P. Halverson, and D.P. Schrag. 1998. A Neoproterozoic snowball Earth. *Science* 281:1342–1346.

Holland, H.D. 1984. The Chemical Evolution of the Atmosphere and Oceans. Princeton: Princeton Univ. Press.

Hyde, W.T., T.J. Crowley, S.K. Baum, and W.R. Peltier. 2000. Neoproterozoic "snowball Earth" simulations with a couple climate/ice-sheet model. *Nature* 405:425–429.

Kasting, J.F., D.H. Eggler, and S.P. Raeburn. 1993. Mantle redox evolution and the oxidation state of the Archean atmosphere. *J. Geol.* 101:245–257.

Kirschvink, J.L. 1992. Late Proterozoic low-latitude global glaciation: The snowball Earth. In: The Proterozoic Biosphere, ed. J.W. Schopf and C. Klein, pp. 51–52. Cambridge: Cambridge Univ. Press.

Lenton, T.M. 1998. Gaia and natural selection. *Nature* 394:439–447.

Lenton, T.M. 2001. The role of land plants, phosphorus weathering and fire in the rise and regulation of atmospheric oxygen. *Global Change Biol.* 7(6):613–629.

Lenton, T.M. 2003. The coupled evolution of life and atmospheric oxygen. In: Evolution on Planet Earth: The Impact of the Physical Environment, ed. L. Rothschild and A. Lister, pp. 35–53. London: Academic.

Lenton, T.M., and A.J. Watson. 2000. Redfield revisited: 2. What regulates the oxygen content of the atmosphere? *Global Biogeochem. Cycles* 14(1):249–268.

Lenton, T.M., and A.J. Watson. 2004. Biotic enhancement of weathering, atmospheric oxygen and carbon dioxide in the Neoproterozoic. *Geophys. Res. Lett.*, in press.

Logan, G.B., J.M. Hayes, G.B. Hieshima, and R.E. Summons. 1995.Terminal Proterozoic reorganization of biogeochemical cycles. *Nature* 376:53–56.

Lovelock, J.E. 1988. The Ages of Gaia— A Biography of Our Living Earth. New York: Norton.

Lovelock, J.E., and L.M. Margulis. 1974. Atmospheric homeostasis by and for the biosphere: The Gaia hypothesis. *Tellus* 26:2–10.

Lovelock, J.E., and A.J. Watson. 1982. The regulation of carbon dioxide and climate: Gaia or geochemistry? *Planet. Space Sci.* 30(8):795–802.

Maynard Smith, J. 1986. The Problems of Biology. Oxford: Oxford Univ. Press.

Maynard Smith, J., and E. Szathmáry. 1995. The Major Transitions in Evolution. Oxford: Freeman.

Mojzsis, S.J., G. Arrhenius, K.D. McKeegan et al. 1996. Evidence for life on Earth before 3,800 million years ago. *Nature* **384**:55–59.

NASA. 1986. Earth System Science: A Closer View. Washington, D.C.: Natl. Aeronautics and Space Administration.

North, G.R., R.F. Cahalan, and J.A. Coakley. 1981. Energy balance climate models. *Rev. Geophys. & Space Phys.* **19(1)**:91–121.

Ohmoto, H. 2003. Reply to comments by H.D. Holland on "The oxygen geochemical cycle: Dynamics and stability." *Geochim. Cosmochim. Acta* **67(4)**:791–795.

Opdyke, B.N., and B.H. Wilkinson. 1993. Carbonate mineral saturation state and cratonic limestone accumulation. *Am. J. Sci.* **293(3)**:217–234.

Porter, S., and A.H. Knoll. 2000. Testate amoebae in the Neoproterozoic era: Evidence from vase-shaped microfossils in the Chuar group, Grand Canyon. *Palaeobiology* **26**: 360–385.

Ridgwell, A.J., M.J. Kennedy, and K. Caldeira. 2003. Carbonate deposition, climate stability, and Neoproterozoic ice ages. *Science* **302**:859–862.

Robinson, J.M. 1990. Lignin, land plants, and fungi: Biological evolution affecting Phanerozoic oxygen balance. *Geology* **15**:607–610.

Runnegar, B. 1991. Precambrian oxygen levels estimated from the biochemistry and physiology of early eukaryotes. *Global Planet. Change* **97(1–2)**:97–111.

Rye, R., and H.D. Holland. 1998. Paleosols and the evolution of atmospheric oxygen: A critical review. *Am. J. Sci.* **298**:621–672.

Schidlowski, M. 1988. A 3,800-million-year isotopic record of life from carbon in sedimentary rocks. *Nature* **333**:313–318.

Schopf, J.W. 1999. Cradle of Life: The Discovery of Earth's Earliest Fossils. Princeton: Princeton Univ. Press.

Szathmáry, E. 2003. Cultural Processes: The latest major transition in evolution. In: Encyclopedia of Cognitive Science. New York: Macmillan.

Szathmáry, E. 2004. Path dependence and historical contingency in biology. In: Understanding Change, ed. A. Wimmer and R. Kössler. Palgrave Macmillan, Basingstoke, in press.

Szathmáry, E., and J. Maynard Smith. 1995. The major evolutionary transitions. *Nature* **374**:227–232.

Szathmáry, E., F. Jordán, and C. Pál. 2001. Molecular biology and evolution. Can genes explain biological complexity? *Science* **292**:1315–1316.

Volk, T. 1989. Sensitivity of climate and atmospheric CO_2 to deep-ocean and shallow-ocean carbonate burial. *Nature* **337**:637–640.

Walter, M.R., J.J. Veevers, C.R. Calver, P. Gorjan, and A.C. Hill. 2000. Dating the 840–544 Ma Neoproterozoic interval by isotopes of strontium, carbon and sulfur in seawater, and some interpretative models. *Precambrian Res.* **100**:371–433.

Zachos, J.C., M.A. Arthur, and W.E. Dean. 1989. Geochemical evidence for suppression of pelagic marine productivity at Cretaceous/Tertiary Boundary. *Nature* **337**:61–64.

3

Is Life an Unavoidable Planetary Phenomenon Given the Right Conditions?

F. WESTALL[1] and F. D. DRAKE[2]

[1]Centre de Biophysique Moléculaire, CNRS, 45071 Orléans cedex 2, France
[2]SETI Institute, Mountain View, CA 94043, U.S.A.

ABSTRACT

In this chapter we argue that life, in the form of simple cells, is a common phenomenon in the Universe, given suitable conditions. By contrast, intelligent life, in the sense of organisms having technological capabilities, is probably relatively rare and, in the case of terrestrial life, appears to have been the result of a long, relatively stable geological history of the planet, punctuated by a few key events that affected the whole planet for significant amounts of time, such as the global glaciations of the Proterozoic and the giant asteroid impact at the end of the Cretaceous. Given the subject, "life as a cosmic phenomenon," we will unavoidably overlap with topics from other chapters in this volume because the conditions for the origin and evolution of terrestrial life are relevant.

WHAT IS LIFE?

There has been much discussion concerning the possibility of life forms based on elements other than carbon (e.g., silicon). However, given the enormous "flexibility" of the carbon atom, in terms of its ability to form a vast range of compounds, it is most likely that life on other planets will also be carbon-based, except in, to us, very strange environments under very different circumstances.

There are many definitions of life, and none appear to be satisfactorily complete. Cleland and Chyba (2002), for example, note that any definition is limited by the meaning of the words used and that, with respect to the word "life," it would be better to try to understand the "nature" of life rather than to define it; in fact, they believe that life cannot be defined until its nature has been deciphered (which humankind has not yet been able to do satisfactorily). We take the view that life as a whole is more than the sum of its parts, the parts being the ability to (a) store and transmit information, (b) replicate, (c) mutate (evolve), and (d) interact with the environment and coevolve with it or mutate. ("Emergent phenomena," the whole being more than the sum of its parts, is a common phenomenon throughout biology; for example, consortia of bacteria can produce effects that are far greater than the simple addition of those produced by the individual organisms on their own.)

In discussing the nature of life, what life "does" is perhaps more readily measurable and a real signature of life than what life "is" (Nealson 1997). In terms of the signatures of life (which is what we will be searching for on other planets), perhaps the most intriguing are those that are "indirect," such as the chemical disequilibria produced by life's metabolic processes and its interactions with the environment, rather than "direct" signatures, such as morphological structures, isotopic fractionations, and molecular composition. The presence of abundant free oxygen molecules in the atmosphere of the Earth over geological timescales is the most obvious example of such a chemical disequilibrium.

ORIGIN OF LIFE AND THE IMPORTANCE OF WATER

Out of necessity, we will use the history of life on Earth as a point of departure for this discussion of life as a cosmic phenomenon, in consideration of the fact that terrestrial life is, to date, the only life form for which we have evidence.

Concerning the origin of terrestrial life, there are many theories for lack of direct evidence. Those that are most in favor at present postulate the polymerization of biologically important organic molecules at the surfaces of certain minerals, possibly in a geochemically reactive environment, such as in the neighborhood of hydrothermal vents (Baross and Hoffman 1985; see also articles in Brack 1998). The latter environments are attractive sites for a number of reasons: (a) they are a source of energy, (b) they provide a source of minerals that could have acted as catalysts in the formation of biologically important molecules, and (c) the porous structure of the mineral edifices formed around the vents could have provided suitable microenvironments, in terms of element and organic molecule concentrations, as well as temperature and pressure conditions, for complex molecule formation. Furthermore, the earliest known Archaea on the tree of life are hyperthermophiles, making plausible the idea that life originated in a warm aqueous solution. However, given the frequency of lateral gene transfer among the prokaryotes, especially during the early stages of evolution, hyperthermophiles may not have been the first type of organism to develop. Indeed, Forterre et al. (1995) suggest that they evolved after thermophilic organisms.

The Ingredients of Life

It appears that the necessary ingredients for, at least, terrestrial-type life are few: organic molecules, liquid water, nitrogen and phosphorus, and a source of energy. Although these are common ingredients throughout the Universe, they are not necessarily present altogether. Both elemental carbon (in the form of graphite and diamond) and carbonaceous molecules occur in extraterrestrial materials such as interstellar dust particles (IDPs), cometary materials, and meteorites (e.g., Ehrenfreund and Menten 2002). Water is most commonly found in the form of ice, but evidence of aqueous alteration of meteorites demonstrates that there was liquid water on the parent bodies at some point in their history. Moreover, cometary passes near any solar object would lead to the melting of ice and subsequent alteration of any embedded silicate minerals. We note that water is found in abundance throughout the solar system. This is not surprising and supports the idea that any planetary system will probably contain objects on which liquid water is present. Nitrogen will be entrained in the planetesimals, which become planets and will subsequently be outgassed to become part of the planetary atmosphere. Phosphorus, similarly, is cosmically ubiquitous and will be available from the material of the planetesimals. Sulfur is another important element and will also be available from the primordial material. Finally, the energy necessary for catalyzing the chemical reactions of life could come from heat, chemical redox reactions, electrical discharges, as well as from light.

When considering the possibility of life originating on other planets, the question of which materials and molecules are essential to life becomes prominent. Which materials are necessary for life, and which are only, perhaps occasionally, selected opportunistically for use by life? This question is of great interest in its own right to biology, but it becomes of great importance in the extraplanetary context since the answers will guide us in designing instruments to detect life on other planets. Some of the answers are obvious: water, C, N, O, and H as well as molecules formed by these elements seem necessary. Less compelling is the need for P and purines and pyrimidines; could there be an alternative, perhaps crude, to be sure, genetic system (and energy delivery system) which did not require P? Another tantalizing question is whether life could exist without S. The absence of S would perhaps limit the amino acids of life to 18, omitting cysteine and methionine. Could enough functional proteins be constructed without these two amino acids? Would life compensate for the lack of these amino acids? These are puzzles. The use of molybdenum and selenium by biochemistry, however, is not a puzzle; these elements seem clearly to have been adopted by an already successful biochemistry simply to provide improved performance, and thus are a result of chemical opportunism. They would not be a good choice as biomarkers to search for life on other planets. However, they could (along with the presence of free oxygen) be measures of the level of evolution on another planet.

Water

Water is the solvent of choice, but could there be alternative solvents? Other solvents have been proposed (e.g., liquid ammonia, hydrogen sulfide, sulfur oxide, and hydrocarbons), but none has the special association of useful properties, as well as stability, with respect to heat and UV radiation, that is exhibited by liquid water (Brack 2002). Because of its polymeric H-bonds, water is stable in a wide range of temperatures whereas organic solvents are not. Brack (2002) notes that:

1. Liquid water is an ideal diffusion milieu for the exchange of organic molecules.
2. Because of the ability of water to form II-bonds with organic molecules, water is an excellent solvent for many of the biologically important organic molecules.
3. These latter molecules were probably, at least partially, templated by clay minerals that are the aqueous alteration products of silicate minerals.
4. The hydrophobic and hydrophilic properties of the hydrocarbon and CHONS portions, respectively, of prebiotic organic macromolecules produce three-dimensional structural configurations that are fundamental in the shaping of biopolymers.
5. Water molecules are good nucleophiles that provide a driving force in chemical reactions.
6. Water is an excellent heat dissipator, an attribute which is important in a deep-sea hydrothermal vent origin of life scenario, where the temperature of the exiting fluids is too high to allow organic molecular combinations but is rapidly reduced in the immediate vicinity of the vent to reach the ambient temperature of the seawater.

Thus, it appears that presence of water is one of the necessary conditions for life, and that search for life on other planets implies the search for evidence of water.

What is the origin of water in the solar system and, in particular, of the Earth? In a recent review of the various hypotheses (ranging from degassing from an Earth accreted from volatile-rich planetesimals to importation by comets and/or volatile-rich meteoritic/asteroidal bolides), Drake and Righter (2002) conclude that the water inventory on the Earth was probably largely derived from early degassing of a planet accreted from "wet" planetesimals, although a certain proportion (between 10% and 30%, depending on the particular model) was imported with bolides, such as comets, meteorites, and micrometeorites. Similar processes probably accounted for the presence of water on the other terrestrial planets, Venus and Mars, early in their histories.

The simple presence of transitory water on a planetary body, however, is not sufficient. Water needs to be stable for a minimum amount of time to permit at least the appearance of life and, as we will see below, for very long periods of time if life is to develop into some form of intelligent, technologically capable being. For instance, in our solar system, the presence of volatiles, including water, at the surface of the planets depends on (a) their distance from the Sun, (b) the size of the planet, and (c) the coevolution of the solar body and the planetary bodies (age). A factor to take into account is that there was less energy (between

25% and 30%) emanating from the young Sun than there is today (Sagan and Chyba 1997). The conditions on at least two of the terrestrial planets early in their histories, Venus and Earth, were such that liquid water was stable for geologically significant time periods (hundreds of millions of years), despite the "early faint Sun paradox" (Kasting 1988, 1997). Mercury, owing to its proximity to the Sun, probably never had water on its surface. With regard to Mars, furthest away from the fainter younger Sun, there is still debate as to whether liquid water was really stable at the surface of the planet or whether it was generally covered by ice owing to lower ambient temperatures (Clifford and Parker 2000; Baker 2001), despite the higher heat flux from the mantle in the early part of Mars' history.

The presence of aqueous alteration products in meteorites attests to the existence of liquid water on the surfaces of their parent asteroidal bodies for a minimum amount of time, even though these bodies were too far away from the Sun to have been influenced by its heat-producing radiation. Many of the asteroidal bodies, however, were apparently large enough to have undergone internal fractionation due to the heat produced by short-lived radionucleotides, such as ^{26}Al (Chapman 1999). This heat caused melting and separation within the asteroid to produce a metallic core with a silicate mineral crust, as on Earth. The same heat caused the melting of water on the asteroid (probably of mixed endogenous and exogenous origin, as on the Earth) and subsequent alteration of the silicate mineral crust.

The very existence of liquid water over geological timescales indicates a source of energy (from geothermal heat due to internal processes or to external stimulation, as in the case of Europa, which is affected by strong gravitational pull from Jupiter, or from the light of a stellar body). Heat flux on the early planets was high owing to the degradation of shorter-lived unstable radionucleotides (e.g., ^{26}Al), high enough to produce a magma ocean and separation of the metals Fe and Ni into a metallic core, leaving a silicate-rich mineral outer shell. As the planets cooled, a crust formed over the magma ocean. However, the heat flux continued to be high even after the precipitation of liquid water, as is testified by the strong volcanic activity documented by the thick piles of lava and strong hydrothermal activity on the earliest well-preserved supracrustal terrains (the 3.8-Ga-old Isua terrain in Greenland, and Barberton in South Africa and the Pilbara in Australia, both of which formed only about 3.5 Ga ago, i.e., a billion years still after the formation of the Earth). Such activity on Mars, for example, would have counteracted the effect of the fainter Sun. Geothermal heat, as well as solar radiation, would have warmed the surfaces of the planets. Stable liquid water at a planet surface also implies that the planetary body has an atmosphere "heavy" enough to prevent sublimation of the "volatile" elements and compounds. Thus, the planetary body needs to be of a minimum size so that its field of gravity can retain volatiles, such as water and gases, especially if that planet is at some distance from its local stellar body (bodies).

The simple presence of relatively stable liquid water for a minimum time period is, however, not sufficient for the origin of life; temperature of water is also critical. It is difficult to get biologically important organic compounds to associate in water above a temperature of about 80°C (60°C would be ideal) in order to form larger, more complex molecules. This puts a limit on the earliest possibility of forming a primitive cell after the formation and cooling of a planet for molecular association: temperatures on the surface of a planetary body must have cooled down sufficiently not only for the precipitation of water on its surface but even further to allow the polymerization of organic molecules. This prompts the question as to whether there was a minimum temperature for the formation of life as we know it? *A priori*, a certain amount of heat aids the haphazard contact and interaction of molecules. It would not be necessary, however, for the whole planet to have an equable temperature somewhere between 50°C and 80°C. On a planet like early Mars, for example, which may not have been particularly warm on a global scale, there would have been localized environments around volcanic and hydrothermal edifices where water temperatures would have been within the optimum limits for significant organic molecular association.

Origin of Terrestrial Life

Numerous experiments have demonstrated the facility of micelle, or cell-like membrane formation, by hydrophobic molecules (Brack 1998). This demonstrates the existence of at least one mechanism of encapsulation, and there are probably many more. It appears that neutral to slightly alkaline pH conditions favor this process whereas acidic conditions do not. The difficulty in creating a cell is to place the "machinery of life," that is, the molecules necessary for storing and transmitting information and for obtaining energy and carbon from the environment and transforming it into more complex molecules, inside the protective environment of a micelle. Not only must the micelle create and protect a suitable microenvironment for the automatic internal chemical transformations, it must also be permeable so that necessary solutes can pass through the barrier into the "cell" and unwanted waste products can be expelled. Moreover, it must not be permeable to unwanted products. Alternate drying–wetting cycles (simulating what might have happened to prebiotic molecules in a beach environment) have been successfully used in the laboratory to encapsulate RNA molecules inside a membrane (Deamer 1998). In terms of relatively "simple" molecules that could simultaneously template for reproduction and drive the metabolism of the early cell, an RNA-like molecule is thought to have been used before the machinery of life developed more sophisticated molecules, such as DNA or proteins (Schwartz 1998).

Whichever pathway was used, it has not yet been replicated in the laboratory, and we have no evidence in the geological rock record for this early stage in the history of life. In fact, the oldest intact rocks known to date that could contain evidence for life were formed only 3.5 Ga ago: the Barberton and Pilbara greenstone belts in South Africa and Australia, respectively. (The ~3.8-Ga-old Isua terrain in southwest Greenland hosts highly metamorphosed sediments whose carbon isotope signature could be intrpreted to suggest the presence of life [Rosing 1999].)These rocks from Barberton and the Pilbara were formed a long time after the origin of life on Earth, as indicated by the high degree of evolution demonstrated by the fossilized life forms contained within them (Walsh 1992; Westall et al 2001; Westall 2003, 2004). The reason for the latter is related to the phenomenon of plate tectonics, the process of formation and destruction of the outer, rigid surface of the Earth; this process has either completely destroyed the early rock record or so altered the remaining rocks through metamorphism that it is extremely difficult to identify what they originally were (thus compromising potential signatures of life). Therefore, we have no information concerning the conditions of the early Earth (e.g., the pH, salinity, and composition of the early oceans, the composition of the atmosphere, the composition, pH, and salinity of the hydrothermal fluids) and without this basic information it is difficult to try to repeat the process in the laboratory.

The timing of the origin of life on Earth is a thorny question since no unaltered rock record exists on Earth older than 3.5 Ga. Theoretically, life could have originated at any time after the condensation of water on the surface of the planet (between 4.4 and 4.3 Ga ago, according to the analyses of zircon crystals dating from that period but occurring in younger formation; Wilde et al. 2001; Mojzsis et al. 2001). However, the inner planets were subjected to an elevated flux of heavy bombardment by extraterrestrial objects during their early histories. Various scenarios for this flux exist, ranging from the catastrophic scenario that models the complete sterilization of the Earth and "frustration" of life (Maher and Stephenson 1988; Sleep et al. 1989), to the optimistic model that posits a late, relatively benign flux between about 4 and 3.85 Ga ago (Ryder 2002). According to the first hypothesis, even if life did appear (maybe more than once) within the first 500 Ma of the Earth's existence, it would have been extinguished; in fact, life is generally taken to have appeared only after 3.85 Ga ago, when the worst of the catastrophic impact storm was over. Alternatively, in the second hypothesis, life could have appeared in the first 500 Ma and simply carried on, perhaps surviving in protected habitats, such as hydrothermal vents, during particularly large impacts (that would, at maximum, have only volatilized the top 400 m of the oceans).

Our solar system hosts other contenders for a possible independent development of life: Venus, Mars, and Europa (Jakosky 1998; Jakosky et al. 2003; Chyba and Phillips 2002). Venus and Mars had somewhat similar geological histories early in their development, whereas Europa's particular relationship with Jupiter was/is such that it had/has a liquid water ocean beneath the icy crust. We will address the possibility of life appearing on these planets below.

EARLY GEOLOGICAL EVOLUTION OF THE TERRESTRIAL PLANETS AND EUROPA: APPEARANCE OF LIFE ON THESE PLANETS

Early Earth

The oldest *bona fide* evidence for life that we have to date comes from the 3.3- to 3.5-Ga-old greenstone belts of Barberton in South Africa and the Pilbara in Australia. These two, relatively small areas of well-preserved supracrustal rocks appear to have survived the ravages of plate tectonics because they represent regions that were underplated later on in their history (about 3.2 Ga ago) by a stabilizing "keel," which protected them from subduction and destruction. Other early Archaean terrains do exist, but they are either highly altered or are covered by younger geological formations. Moreover, many of the Isua rocks that were previously interpreted as sediments have recently been reinterpreted as highly metamorphosed igneous protoliths (Fedo and Whitehouse 2002; Myers 2003). (This has implications for the supposed carbon isotope evidence for life in the so-called "metasediments"[cf. Rosing 1999; van Zuilen et al. 2002; Lepland et al. 2002].)

Briefly, the early Earth on which life arose represented an "extreme" environment compared to the Earth of today. It consisted of small (proto)continental land masses in an ocean that appeared to have been more saline than at present and whose pH is unknown but estimates range from slightly acidic, neutral to alkaline. Owing to the high heat flow (cf. Franck 1998), temperatures were high (up to 85°C) despite the weaker solar radiation (Knauth and Lowe 2003). The high heat flow also resulted in much volcanism and hydrothermal activity. The atmosphere consisted mostly of CO_2 with admixtures of other gases, such as H_2O vapor, CH_4 (from volcanic exhalations), and N_2 (Pavlov et al. 2001). It is possible that the CH_4 from the volcanism could have contributed to a "greenhouse effect," thus warming the planet and compensating for the lack of energy from the faint early Sun. There is no evidence for free oxygen molecules in the atmosphere. Oxygen was definitely produced, for instance in small amounts by photolysis of H_2O vapor in the upper atmosphere and of liquid water in the upper layers of water bodies, but it immediately reacted with the reduced gas and mineral species that abounded on the planet. The lack of free O_2 raises the question of flux of UV to the Earth's surface and its deleterious effects on life (Cockell 2002). Certainly, in the absence of O_2 there could not have been an ozone layer. However, the atmosphere would probably have been quite "dirty" owing to the presence of large amounts of volcanic dust, CH_4, and water vapor that would have dissipated a part of the UV rays, thus protecting early life. In any case, life in the oceans or any water body would have been protected as the UV rays are absorbed by the top few millimeters of the water column. Finally, the flux of extraterrestrial materials (asteroids, comets, meteorites, micrometeorites) accreting to the inner planets, especially, was very high until about 3.85 Ga ago, and continued thereafter (Kyte et al. 2003). This may or may not have been dele-terious for life (Ryder 2002).

Venus and Mars

Let us turn briefly to the early geological evolution of the other terrestrial plan-ets, Venus and Mars, in an attempt to understand the possibility of the appear-ance of life on them (cf. Jakosky 1998). Like Earth, both Venus and Mars were characterized early in their development by high heat flow, owing to the decay of unstable, short-lived isotopes. Once the surfaces of the planets had cooled sufficiently, water could condense on the surface. The origin of the volatiles would have been the same as on Earth — partly from the degassing of "wet" planetesimals, partly from the accretion of volatile-rich extraterrestrial materi-als (comets, meteorites, micrometeorites). Moreover, there is plenty of geomor-phological evidence for the existence of water at the surface of Mars early in its history (Carr 1996). Having the same ingredients for life (liquid water, organic molecules, and energy sources) and similar geological conditions (hydrother-mal vents, possibly in association with water bodies), there is no reason to sup-pose that life did not appear on these planets. On Mars, for instance, even if the whole planet were not covered with an ocean, there were still many suitable en-vironments where life could have appeared and developed; a global ocean was not a prerequisite for life. It is also possible that primitive life could have been transported between the planets during the early period of heavy bombardment (Horneck et al. 2002). This gives us the possibility of cross-contamination, even if life originated on only one of the planets.

The problem is that the continued geological and environmental evolution of both planets was not conducive to the prolonged existence of life at their sur-faces. Venus became overheated because of its vicinity to the Sun and because continued volcanic activity pumped CO_2 into the atmosphere, leading to what is known as the "runaway greenhouse effect" (Kasting 1988). The vast amount of CO_2 in the atmosphere was not controlled by solution in bodies of liquid water and a resulting precipitation of carbonate salts, and the planet now has surface temperatures above 450°C that are incompatible with life. There is no water on Venus; it is a very "dry" planet. Mars, on the other hand, is further away from the Sun and its surface is basically "frozen." However, the freezing of the planet was not simply a function of distance from the Sun; it was also related to the size of the planet and its geological evolution. Mars is much smaller than either Venus or the Earth. Although it had a magnetic field very early in its history, remnants of which were recently discovered (Connery et al. 2001), the planet was too small to sustain the internal "dynamo" that produced the magnetic field (Zuber 2001). About 4 Ga ago, Mars therefore lost its magnetic field and with it, protec-tion from the ionic bombardment of the solar wind. The atmosphere on Mars was basically destroyed by attrition from the solar wind and impacts (note that the period of heavy bombardment, which lasted at least until 3.85 Ga ago, affected all of the inner planets). Thus, with the elimination of the atmosphere, there was no longer any protection for the water at the surface of the planet. Cooling led to sublimation of the water, which photodissociated in the upper atmosphere. In fact, on Mars, water bodies may have been frozen over very early in its history. (However, this is irrelevant for the existence of life because there would still have been sufficient locations where there was liquid water, e.g., in volcanically active areas or impact craters where there could have been either related volcanic or hydrothermal activity.) Furthermore, there could well have been subsurface bodies of liquid water in which life could have developed. Indeed, today it appears that a large portion of the initial water on Mars is to be found in the subsurface; there is evidence in support of the existence of a cryosphere relatively near the surface of the planet in certain latitudes (towards the poles, especially) (Malin and Edgett 2000; Boynton et al. 2002). As the planet cooled, any original life would first have retreated into the endolithic habitats of rock crevasses (Friedmann and Koriem 1989) and then possibly into the subsurface. However, it is unlikely that it could have survived at the surface continuously over the whole of the geological history of the planet (despite the fact that life is almost indestructible on Earth). This is because of the combination of extreme conditions that exist at the surface: desiccation, UV radiation, and oxidants (H_2O_2??). Although certain bacteria can resist these conditions individually, none are known to date that can resist them collectively. If, as has been hypothesized, life retreated to the subsurface and survived in a cryofrozen state, it could have been resuscitated during heating of the cryosphere during periodic volcanic and/or impact activity. Nevertheless, further extensive evolution, as occurred on Earth, would not have been possible on Mars.

Was Mars simply unlucky? Was there a history of a series of asteroid collisions, which removed so much mass from Mars that it was left unable to maintain an atmosphere suitable for surface life? Perhaps this is an example of a situation where the solar system was deprived of a possible abode of life.

Europa

The geological history of Europa is different from that of the terrestrial planets, but it has been hypothesized that life could have appeared and could still exist on that planet. The interest of Europa is that there appears to be a layer of water beneath its icy crust (Jakosky 1998; Chyba and Phillips 2002; Greenberg et al. 2001). Water accreted onto the rocky core during the formation of the planets in our solar system because of the distance of this body from the Sun. As with any of the planets and larger planetesimals in the early formation of the solar system, decay of short-lived radionucleides would have warmed at least the icy layers next to the rocky core. This warming would have been complemented by the warming due to tidal frictional heating caused by variations in the gravitational pull of Jupiter (the volcanism on Io is a good example of the effects of the frictional heating of Jupiter on its moons). The ice mass on Europa would have been "dirty" in the sense that it would have also contained organic molecules, i.e., the same organic ingredients as occurred on early Earth. It is possible that there could have been submarine volcanic activity at the bottom of Europa's ocean, which could have produced a source of energy. Thus, Europa could potentially have given rise to an independent form of life. Could it still exist? If liquid water has always existed, it is possible that life could have continued (psychrophilic organisms developed on Earth from thermophilic/mesophilic ones). Moreover, life forms could have been cryofrozen, as potentially happened on Mars.

IMPORTANCE OF THE GEOLOGICAL EVOLUTION OF THE EARTH IN THE EVOLUTION OF TERRESTRIAL LIFE

Above we discussed the relative simplicity of the conditions necessary for the appearance of primitive life, in the sense of single cells: conditions that are so simple that life probably arose on a number of bodies in our solar system, despite their distance from the Sun. Such primordial conditions must have been and still may be common on extrasolar planets, given the number of solar systems in the Universe, even if distance from the center of a galaxy and age of the galaxy/stellar system are also factors to take into account. What are probably not so common are the factors that lead to further evolution with the possible appearance of technologically intelligent life. On Earth, the evolution of life went hand in hand with the evolution of geological processes, as we list in Table 3.1.

The salient information from Table 3.1 is that Earth has had a long and relatively peaceful development over the last 4.5 Ga, apart from several global glaciations, e.g., in the Neoproterozoic (Hoffmann et al. 1998), and at least one, and probably more, major impacts, all at critical moments in the evolution of life. This is in no small part the result of the "protection" afforded by the presence of a giant planet, such as Jupiter, at the outer edges of the solar system. Jupiter basically acts as the "sweeper up" of marauding extraterrestrial objects, thus partially protecting the inner planets from deleterious impacts. However, it was not always successful, as demonstrated by the Late Cretaceous impact.

The appearance of eukaryote life about 1.9 Ga ago (Han and Runnegar 1992) seems to be linked to the rise of O_2 in the atmosphere, which occurred at the same time as the first global glacial episode (Evans et al. 1997; Ward and Brownlee 2000; Lindsay and Brasier 2002). The rise in O_2 appears to have been the result of two processes: (a) removal of CO_2 in the atmosphere by the burial of C in the mantle by the subduction of carbonate and organic remains in sediments deposited on the oceanic floors, and (b) gradual increase in O_2 in the atmosphere after oxidation of all the reduced mineral species, the O_2 probably originating mainly from oxygenic photosynthesis. In fact, the first global glaciation may have been triggered by the input of O_2 into the atmosphere and the resultant reduction in

Table 3.1   The evolution of life and of geological processes.

Timing (Ga)	Geological Context	Life
4.4–4.0?	Small protocontinents High heat flow Much volcanism/hydrothermal activity $> CO_2$, some H_2O, CH_4,N_2 impacts	Origin

3.5–3.3	Decreasing heat flow but
	   $T \leq 85°C$
		Well-developed prokaryotes
		Thermophiles
		Halophiles
		Anoxygenic photosynthesis
		   widely distributed?
		Aqueous/littoral/evaporitic
		   environments

3.2–2.5	Developing continents Developing stable carbonate Platforms Decreasing heat flow Plate tectonic burial of C, CO_2	
> 2.7		Oxygenic photosynthesis
2.5	Global glaciation?	
~2.2	Start of buildup of O_2 in atmosphere	
? 1.9		First eukayotes?
? 1		First multicellular life
1–0.546	Increasing O_2	Wide development of life without hard parts
0.8–0.6?	Global glaciation	
0.546	Increasing O_2	Explosion of life with hard parts
0.546+		Development of life Diversification of life on land
0.065	Asteroid impact	Extinction of large reptiles
0.065+		Rapid evolution and the development of mammals
0.002+	Glaciations	Development of technological intelligence
Future		Self-destruction of technologically intelligent organisms?
? > 1	Gradual heating of Sun	Disappearance of species on Earth
> 5.5	Disappearance of the Earth	Colonization of the solar system?

greenhouse warming. The hypothesized kilometer-thick ice crust would essentially have blocked the availability of nutrients, as well as exchange between the atmosphere and ocean (Kirschvink et al. 2000). However, continued pumping of CO_2 gas into the atmosphere by volcanic output would have gradually restored greenhouse warming and increased temperatures such that the ice melted. The vast amount of nutrients suddenly released into the biosphere would have led to a bloom of photosynthetic organisms at the surface of the ocean and the production of vast amounts of O_2, a poison for most organisms. However, much of the O_2 produced was taken up by the oxidation of reduced metal species in solution (most of the banded iron formation deposits date from this epoch, as well as of organic matter). Organisms needed to adapt to the presence of the poisonous O_2 that was occurring in their habitats in ever-increasing amounts (e.g., the production of enzymes to react with the O_2, the compartmentalization of the cell to do certain specific functions.) Thus, the appearance of eukaryotes seems to have been related both to the aftereffects of the "snowball" Earth and the rise in free oxygen.

The Neoproterozoic saw a further evolution of animal life from the initial single-celled eukaryotes to the first multicellular individuals by about 0.6 Ga ago (Xiao et al. 1998), followed by a plethora of strange yet wonderful forms of life (the Ediacaran fauna), which still did not exhibit an exoskeleton. At the beginning of the Cambrian (546 Ma ago), animals with exoskeletons first appeared. However, as with the first fossil evidence for life from the 3.5-Ga-old Early Archaean sediments, these first hard-shelled organisms (including trilobites and brachiopods) were already highly evolved. Their ancestors can be traced back into the soft-bodied fauna of the Neoproterozoic. The appearance of a skeletal component to life was fundamental in enabling larger organisms (eventually with larger brains) to evolve. Why did organisms suddenly develop exoskeletons 546 Ma ago? There is perhaps a relationship with changes in the geological environment. The Neoproterozoic saw another global glaciation between about 800 and 600 Ma ago. As with the glaciation that occurred in the Paleoproterozoic, there would have been destruction of those photosynthetic organisms that lived at the surface of the oceans, followed by massive blooms created by the huge increase in nutrient levels at the end of the glaciation. Perhaps this event stimulated evolution among shell-less animals (Hoffman et al. 1998). It is widely believed that organisms with exoskeletons became predominant because they were better protected against predatory attacks.

Finally, if an asteroid had not hit the Earth at the end of the Late Cretaceous, resulting in the annihilation of large reptiles, then mammals could not have diversified and developed species of a size that could support a brain large enough to be intelligent. The human species is a result of an extraterrestrial accident. Could a technologically intelligent organism have developed from the reptile or some other lineage? This is one of the more interesting unanswered questions in our field. It is clear that organisms with a certain amount of manual dexterity, apart from a suitable brain, are needed for technological development.

FURTHER EVOLUTION OF LIFE ON OTHER PLANETS IN THE SOLAR SYSTEM?

We have seen above the importance of the continued geological evolution on Earth in the evolution of life. This point is fundamental because it conditions the possibility of evolution of single-celled organisms into multicellular and, eventually, technologically capable organisms. Plate tectonics, free oxygen in the atmosphere, the development of a subaerial biosphere, chance impacts that selectively eliminated certain species while giving others the chance to develop — these are all ingredients in the formula (as far as we understand it) for evolving multicellular organisms capable of evolving into beings with large brains. Thus, even though life could have appeared on other planets in our solar system (Mars, Venus, and Europa), it is unlikely that it ever developed beyond the single-celled stage. Venus rapidly overheated and lost any water, one of the fundamental ingredients of life. Mars, because of its small size, probably never got to the stage of developing active plate tectonics before the loss of most of its surface volatiles. Although life may still be preserved in the cryofrozen state in the subsurface, it will be as single-celled organisms. Likewise, Europa, covered with water and ice, never had plate tectonics, the evolution of an oxygen-rich atmosphere, and the development of subaerial life that gave rise to mammals. Even though life may still be present in the cold ocean, it would be represented by single-celled anaerobic organisms.

CONCLUSION AND TRENDS IN THE FIELD

Earth seems to have been privileged since it is large enough to retain an atmosphere and has a magnetic field that protects it from solar wind; it is just the right distance from the Sun to be able to have liquid water for geologically significant periods of time, is generally protected from massive asteroid impact by Jupiter (except for at least one, beneficial in some respects, impact), and has had a long period of geological (tectonic) evolution in combination with a general accumulation of the waste products of life's metabolic pathways (e.g., O_2). A few key geological events appear to have aided evolution through the elimination of certain species, thus favoring the further development of others. These events include two global periods of glaciation and the Late Cretaceous asteroid impact.

The search is now on for extraterrestrial planets in other solar systems with ever-increasing numbers being detected, more than a hundred at last count, since the first observations (Mayor and Queloz 1995). Owing to the difficulties involved in the search, those found to date tend to be giant planets, similar to Jupiter. However, a number of specialized space observation missions are being planned to continue the search under better conditions (see, e.g., Foing 2001). Terrestrial-based research is presently oriented toward understanding the limits of life under extreme conditions that could exist on other planets (e.g., Horneck and Baumstark-Khan 2001).

With the myriad of solar systems in the Universe, it seems inconceivable that other systems do not have planets with similar geological histories. They could have developed, or could develop in the future, more evolved species than simple single cells, resulting possibly in technologically intelligent life forms. The latter, however, are very unlikely to be humanoid in form.

The only way to determine the existence and, optimistically, the abundance of technological life forms is to search for evidence of technological activity (Drake 1965). No theoretical analysis can predict the likelihood of technological species appearing; the path to technology is too complicated and dependent on a number of events of unknown and unpredictable probability. The most promising method of search is, as it has been for more than forty years, a search for radio signals from engineered radio transmitters. However, the paradigm guiding the radio searches has changed in a way that is somewhat discouraging. Over the years, in designing our searches, we have used our own technology as a guide; there has been no alternative. Until recently, this produced a paradigm in which we expected a continuous growth in number of transmitters and transmitter power, making detection increasingly possible. However, a combination of technical inventions and, in particular, sociology, has changed the predictions of the paradigm. We now have systems that can communicate effectively with much lower transmitter powers. For example, the satellite transmission of television to homes requires transmitters whose power level is 100,000 times less than the power of traditional television stations. Perhaps of more importance has been the fast-growing social demand for personal communications of all kinds: telephones, internet, etc. These high information rate systems require the use of much frequency bandwidth, but there is only a limited total amount of bandwidth available in the electromagnetic spectrum. To meet the demands of the public, bands and bandwidths must be shared. This only works without interference between users if transmitter powers are low. The prime example of this approach is the "mobile" or "cell" phone systems. The impact of this paradigm shift on the detectability of civilizations may be major: the power levels being transmitted in any given frequency band may well decrease markedly.

It is an interesting twist to this trend that this evolution has resulted not so much from technical developments but from social developments. One can only wonder if this technical and social evolution is also a common in other civilizations. In any case, it suggests that extraterrestrial civilizations may be more difficult to find than previously thought, a surprising result of social pressures. No matter what the actual situation, the only way to succeed is to search. Success may come at a higher price than previously assumed.

KEY QUESTIONS AND CONTROVERSIAL ISSUES

1. What is life?
2. Could life in the form of viable cells or spores travel interstellar distances to seed life on planets with suitable environments in distant solar systems?

3.   Which aspects of biology are "essential" or "necessary," and which are "contingent" or "opportunistic"?
4.   What are the influences of geological and cosmic processes on the evolution of terrestrial life/life in general?
5.   Is our species really an accidental product of rare chance phenomena occurring on Earth (e.g., a giant impact eliminating the large reptiles and permitting the mammals to develop), or are there so many alternative pathways to intelligent life that it is essentially inevitable, given sufficient time? Is evolution "convergent" given sufficient time?
6.   What is intelligent life? (Is it possible to have intelligent life that is based on good sense rather than simple "survival of the fittest" — the fittest not necessarily being the most intelligent or the individual with the best sense?)
7.   Would "technologically intelligent" life survive and remain detectable for long enough periods to be discovered remotely by another "technologically intelligent" race?

REFERENCES

Baker, V.R. 2001. Water and the Martian landscape. *Nature* **412**:228–236.
Baross, J.A., and S.E. Hoffman. 1985. Submarine hydrothermal vents and associated gradient environment as sites for the origin and evolution of life. *Orig. Evol. Life* **15**:327–345.
Boynton, W.V. ,W.C. Feldman, S.W. Squyres et al. 2002. Distribution of hydrogen in the near surface of Mars: Evidence for subsurface ice deposits. *Science* **297**:81–84.
Brack, A. 1998. The Molecular Origins of Life: Assembling Pieces of the Puzzle. Cambridge: Cambridge Univ. Press.
Brack, A. 2002. Water, the spring of life. In: Astrobiology, the Quest for the Conditions of Life, ed. G. Horneck and C. Baumstark-Khan, pp. 279–288. Berlin: Springer.
Carr, M.H. 1996. Water on Mars. Cambridge: Cambridge Univ. Press.
Chapman, C.R. 1999. Asteroids. In: The New Solar System, ed. J.K. Beatty, C.C. Petersen, and A. Chaikin, pp. 337–350. Cambridge: Cambridge Univ. Press.
Chyba, C.F., and C.B. Phillips. 2002. Europa as an abode of life. *Orig. Life & Evol. Biosphere* **32**:47–68.
Cleland, C., and C. Chyba. 2002. Definition of life. *Orig. Life & Evol. Biosphere* **32**: 387–393.
Clifford, S.M., and T.J. Parker. 2001. The evolution of the Martian hydrosphere: Implications for the fate of a primordial ocean and the current state of the Northern Plains. *Icarus* **154**:40–79.
Cockell, C.S. 2002. The ultraviolet radiation environment of Earth and Mars: Past and present. In: Astrobiology, the Quest for the Conditions of Life, ed. G. Horneck and C. Baumstark-Khan, pp. 219–232. Berlin: Springer.
Connerney, J.E.P., M.H. Acuña, P.J. Wasilewski et al. 2001.The global magnetic field of Mars and implications for crustal evolution. *Geophys. Res. Lett.* **28**:4015–4018.
Deamer, D.W. 1998. Membrane compartments in prebiotic evolution. In: The Molecular Origins of Life: Assembling Pieces of the Puzzle, ed. A. Brack, pp. 189–205. Cambridge: Cambridge Univ. Press.

Drake, F. 1965. The radio search for intelligent extraterrestrial life. In: Current Aspects of Exobiology, ed. G. Mamikunian and M.H. Briggs, pp. 323–345. New York: Pergamon.

Drake, M.J., and K. Righter. 2002. Determining the composition of the Earth. *Nature* **416**:39–44.

Ehrenfreund, P., and K.M. Menten. 2002. From molecular clouds to the origin of life. In: Astrobiology, the Quest for the Conditions of Life, ed. G. Horneck and C. Baumstark-Khan, pp. 7–24. Berlin: Springer.

Evans, D.A., N.J. Beukes, and J.L. Kirschvink. 1997. Low-latitude glaciation in the Paleoproterozoic era. *Nature* **386**:262–266.

Fedo, C.M., and M.J. Whitehouse. 2002. Metasomatic origin of quartz-pyroxene rock, Akilia, Greenland, and implications for Earth's earliest life. *Science* **296**:1448–1452

Foing, B.H. 2001. Space activities in exo-astrobiology. In: Astrobiology, the Quest for the Conditions of Life, ed. G. Horneck and C. Baumstark-Khan, pp. 389–398. Berlin: Springer.

Forterre, P., F. Confalonieri, F. Charbonnier, and M. Duguet. 1995. Speculations on the origins of life and thermophily: Review of available information on reverse gyrase suggests that hyperthermophilic procaryotes are not so primitive. *Orig. Life & Evol. Biosphere* **25**:235–249.

Franck, S. 1998. Evolution of the global mean heat flow over 4.6 Ga. *Tectonophysics* **291**:9–18.

Friedmann, E.I., and A.M. Koriem. 1989. Life on Mars: How it disappeared (if it was ever there). *Adv. Space Res.* **9(6)**:167–172.

Greenberg, R., R. Tufts, P. Geissler, and G. Hoppa. 2001. Europa's crust and ocean: How tides create a potentially habitable physical setting. In: Astrobiology, the Quest for the Conditions of Life, ed. G. Horneck and C. Baumstark-Khan, pp. 111–124. Berlin: Springer.

Han, T.-M., and B. Runnegar. 1992. Megascopic eukaryotic algae from the 2.1-billion-year-old Negaunee Iron-Formation, Michigan. *Science* **257**:232–235.

Hoffman, P., A. Kaufman, G. Halverson, and D. Schrag. 1998. A Neoproterozoic snowball Earth. *Science* **281**:1342–1346.

Horneck, G., and C. Baumstark-Khan. 2001. Astrobiology, the Quest for the Conditions of Life. Berlin: Springer.

Horneck, G., C. Mileikowsky, H.J. Melosh et al. 2002. Viable transfer of micro- organisms in the solar system and beyond. In: Astrobiology, the Quest for the Conditions of Life, ed. G. Horneck and C. Baumstark-Khan, pp. 57–78. Berlin: Springer.

Jakosky, B.M. 1998. The Search for Life on Other Planets. Cambridge: Cambridge Univ. Press.

Jakosky, B.M., F. Westall, and A. Brack. 2003. Mars. In: Astrobiology, ed. W. Sullivan and J.A. Baross. Cambridge: Cambridge Univ. Press.

Kasting, J.F. 1988. Runaway and moist greenhouse atmospheres and the evolution of Earth and Venus. *Icarus* **74**:472–494.

Kasting, J.F. 1997. Warming early Earth and Mars. *Science* **276**:1213–1215.

Kirschvink, J.L., E.J. Gaidos, L.E. Bertani et al. 2000. Paleoproterozoic snowball Earth: Extreme climatic and geochemical global change and its biological consequences. *Proc. Natl. Acad. Sci. USA* **97**:1400–1495.

Knauth, L.P., and D.R. Lowe. 2003. High Archaean climatic temperature inferred from oxygen isotope geochemistry of cherts in the 3.5 Ga Swaziland Supergroup, South Africa. *Geol. Soc. Am. Bull.* **115**:566–580.

Kyte, F.T., A. Shukolyukov, G.W. Lugmaorm D.R. Lowe, and G.R. Byerly. 2003. Early Archaean spherule beds: Chromium isotopes confirm origin through multiple impacts of projectiles of carbonaceous chondrite type. *Geology* **31**:283–286.

Lepland, A., G. Arrhenius, and D. Cornell. 2002. Apatite in early Archean Isua supracrustal rocks, southern West Greenland: Its origin, association with graphite and potential as a biomarker. *Precambrian Res.* **118**:221–241.

Lindsay, J.F., and M.D. Brasier. 2002. Did global tectonics drive early biosphere evolution? Carbon isotope record from 2.6 to 1.9 Ga carbonates of Western Australian basins. *Precambrian Res.* **114**:1–34.

Maher, K.A., and D.J. Stevenson. 1988. Impact frustration of the origin of life. *Nature* **331**:612–614.

Malin, M., and K.S. Edgett. 2000. Evidence for recent groundwater seepage and surface runoff on Mars. *Science* **288**:2330–2335.

Mayor, M., and D. Queloz. 1995. A Jupiter mass companion to a solar-type star. *Nature* **378**:355–358.

Mojzsis, S.J., T.M. Harrison, and R.T. Pidgeon. 2001. Oxygen-isotope evidence from ancient zircons for liquid water at the Earth's surface 4,300 Ma ago. *Nature* **409**: 178–181.

Myers, J. 2003. Isua enigmas: Illusive tectonic, sedimentary, volcanic, and organic features of the >3.8–>3.7 Ga Isua greenstone belt, south-west Greenland. *Geophys. Res. Abstr.* **5**:13,823.

Nealson, K.H. 1997. The limits of life on Earth and searching for life on Mars. *J. Geophys. Res.* **102**:23,675–23,686.

Pavlov, A.A., J.F. Kasting, L.L. Brown, K.A. Rages, and R. Freedman. 2001. Greenhouse warming by CH_4 in the atmosphere of early Earth. *J. Geophys. Res.* **105**: 11,981–11,990.

Rosing, M.T. 1999. ^{13}C-depleted carbon microparticles in >3700-Ma seafloor sedimentary rocks from West Greenland. *Science* **283**:674–676.

Ryder, G. 2002. Mass influx in the ancient Earth–Moon system and benign implications for the origin of life on Earth. *J. Geophys. Res.* **107**:10.1029/2001JE001583.

Sagan, C., and C. Chyba. 1997. The early faint Sun paradox: Organic shielding of ultraviolet-labile greenhouse gases. *Science* **276**:1217–1221.

Schwartz, A.W. 1998. Origins of the RNA world. In: The Molecular Origins of Life: Assembling Pieces of the Puzzle, ed. A. Brack, pp. 237–254. Cambridge: Cambridge Univ. Press.

Sleep, N.H., K.J. Zahnle, J.F. Kasting, and H.J. Morowitz. 1989. Annihilation of ecosystems by large asteroid impacts on the early Earth. *Nature* **342**:139–142.

Van Zuilen, M., A. Lepland, and G. Arrhenius. 2002. Reassessing the evidence for the earliest traces of life. *Nature* **418**:627–630.

Walsh, M.M. 1992. Microfossils and possible microfossils from the Early Archean Onverwacht Group, Barberton Mountain Land, South Africa. *Precambrian Res.* **54**:271–293.

Ward, P.D., and D. Brownlee. 2000. Rare Earth. New York: Copernicus/Springer.

Westall, F. 2003. Stephen Jay Gould, les procaryotes et leur évolution dans le contexte géologique. *CR Acad. Sci. Paris Palevol.* **2**:485–501.

Westall, F. 2004. The geological context for the origin of life and the mineral signatures of fossil life. In: The Traces of Life and the Origin of Life, ed. B. Barbier et al. New York: Springer, in press.

Westall, F., M.J. De Wit, I. Dann et al. 2001. Early Archaean fossil bacteria and biofilms in hydrothermally influenced, shallow water sediments, Barberton greenstone belt, South Africa. *Precambrian Res.* **106**:93–116.

Wilde, S.A., J.W. Valley, W.H. Peck, and C.M. Graham. 2001. Evidence from detrital zircons for the existence of continental crust and oceans on Earth 4.4 Gyr ago. *Nature* **409**:175–178.

Xiao, S., Y. Zhang, and A.H. Knoll. 1998. Algae and embryos in a Neoproterozoic phosphorite. *Nature* **391**:553–558.

Zuber, M.T. 2001. The crust and mantle of Mars. *Nature* **412**:220–227.

4

What Are the Necessary Conditions for Origin of Life and Subsequent Planetary Life-support Systems?

S. A. FRANCK[1] and G. A. ZAVARZIN[2]

[1]Potsdam Institute for Climate Impact Research (PIK), 14412 Potsdam, Germany
[2]Russian Academy of Sciences, Institute of Microbiology, 117811 Moscow, Russia

ABSTRACT

In this chapter, some questions regarding the origin of life and the coevolution of the biosphere and geosphere are discussed. According to the Gaia theory, the environment is influenced by and for the benefit of the biosphere, but whether this occurs on all timescales is uncertain. Feedback from the biota to the geosphere is particularly strong for the composition of the atmosphere.

Coevolution of the geosphere and biosphere may be investigated using a simple Earth system model. The scenario of an Archaean hothouse with a thermophilic early biosphere is presented. The life span of a photosynthesis-based biosphere is limited by the atmospheric content of carbon dioxide. The simplest living being is a prokaryotic cell organism. The interactions between biosphere and geosphere are realized via the cycling of elements such as carbon, phosphorus, nitrogen, and sulfur, where the carbon cycle is the most important. To evaluate the present Earth system, which is in a state of optimum self-regulation, an extended model is used that includes three different types of biosphere: prokaryotes, eukaryotes, and complex life, where prokaryotes are always the base for higher types of biosphere.

INTRODUCTION

At the conclusion of Chapter 7 in his book, *What Is Life? The Physical Aspect of the Living Cell*, Schrödinger (1944) posed the question: "Is life based on the laws of physics?" If we view the hereditability principle that he formulated, whereby *order derives from order*, we see that it is applicable not only to aperiodic crystals but to living organisms as well, as has been confirmed via the discovery of the structure of nucleic acids. Schrödinger's second principle, *order from disorder*, considers the thermodynamics of open systems, whereby living beings stay alive by absorbing energy from the environment and producing a highly organized state through the increase of entropy in their environment.

By applying this principle to the Earth system, we are in a better position to address Schrödinger's question. The basic yield/energy relationship, introduced in the nineteenth century through the discovery of chemosynthesis, experimentally established the approximate proportionality of biomass yield to the heat of reaction. Thus, primary photosynthetic producers introduce highly organized biomass into the biosphere at the expense of an extraterrestrial energy source and according to the biomass/energy relation. This has important consequences for the geosciences, since living beings impose a thermodynamic gradient on their surroundings, which in turn acts as the driving force for biogeochemical reactions.

Life can be defined as a self-sustained system of organic molecules immersed in liquid water contained within a source of free energy. This definition is useful in calculating the habitable zone, that is, the region around a given central star within which an Earth-like planet might possess surface temperatures that are moderate enough to support life. It is well known that organic molecules are rather common in the solar system, even in interstellar clouds, and it is usually not difficult to locate a source of free energy. Thus, the simplest definition of the habitable zone relates to the possibility that liquid water exists. On planetary timescales, the variation of CO_2 in the atmosphere plays an important role in investigating the relationship between the evolution of a sun as a main-sequence star and the stabilization of the surface temperature of an Earth-like planet, since the burning rate of such stars increases over time. For example, over the course of Earth's history, the Sun's luminosity has increased 30% to its present level. If during this time the atmospheric composition and planetary albedo had remained the same as today, the surface temperature of the Earth would have been below 0°C until ca. 2 billion years (Ga) ago. This is the so-called "faint young Sun paradox" first described by Sagan and Mullen (1972). However, there is no geological or biological evidence to suggest that Earth was in a permanent snowball state at this time.

Cross-disciplinary studies that address the origin, distribution, and future of life on our planet examine life as a part of cosmic evolution. Some fundamental questions in current Earth-related astrobiological research include: How do habitable worlds form, and how do they evolve? How did living systems emerge? How did the physical Earth and its biosphere influence each other over time? How do changes in the environment affect emergent ecosystems and their evolution?

As we begin the new millennium, answers to some of these questions seem within reach. We expect that new, important insights will result as technological innovations are implemented in astronomical, planetological, and biological research. The issues we discuss in this chapter closely follow the early ideas of

Alexander von Humboldt in terms of the interdependence of the biosphere and climate. In addition, we ascribe to the concept of *virtual biospheres*, as formulated by Svirezhev (see, e.g., Svirezhev and von Bloh 1997), which states that the current biosphere of the Earth is only one of many possible virtual biospheres that correspond to different equilibria of the nonlinear geospherc–biosphere system. Such a concept is nondeterministic: If life were to evolve again, the results could be radically different.

HABITABILITY AND HABITATION

According to the Gaia theory, the biosphere influences the environment for its own benefit. The question that arises, however, is whether this influence acts as a driving force to induce habitability. We do not believe that the process of habitation can transform an inhabitable planet to a habitable one. Such a statement would be contradictory. From a biological perspective, it has been established (Zavarzin 2000) that the development of biota within an ecosystem might result in a dynamic series of adaptive changes on different scales, beginning with habitat-specific changes on a landscape scale (known as succession) to global changes in the biogeochemical succession. The switch from a nonoxic to oxygenic atmosphere did not benefit anaerobes. Less obvious is the development of sedimentary cover by biotically mediated weathering, which reduced the availability of mineral biogenic elements and indicates that the inner forces in a dynamic system prevail.

On geological timescales, this suggests that the evolutionary patterns of biogeochemical cycles were controlled and generated by global tectonic evolution rather than by a Gaian mechanism. In other words, the Earth system acts like a hierarchical system, whereby the biosphere operates (except on short timescales) within the constraints established by the higher levels of the hierarchy, that is, the long-term geosphere (Veizer 1988). Accordingly, the Hadean environment exerted the primary influence on the origin and first evolutionary steps of life. (In this chapter, we focus on the conditions for the origin of life on Earth. For a discussion on panspermia, see Horneck and Schellnhuber, this volume.)

To determine the habitable zone, a quantitative model was used to calculate the evolution of the terrestrial atmosphere over geologic time at various distances from the Sun. It was found that a habitable zone between "runaway greenhouse" and "runaway glaciation" is surprisingly narrow for G2 stars like our Sun:

$$R_{inner} = 0.958 \text{ AU}, R_{outer} = 1.004 \text{ AU}, \qquad (4.1)$$

where AU denotes the astronomical unit. These calculations, however, failed to account for the negative feedback between atmospheric CO_2 content and mean global surface temperature. Full consideration of this feedback provided the interesting result of an almost constant inner boundary, yet a remarkable extension of the outer boundary. Later, calculations of the habitable zone were improved and extended to other main sequence stars (for a comprehensive overview, see Doyle 1996).

Recent studies (Franck, von Bloh et al. 2000; Franck, Block et al. 2000a, b) have generated a general characterization of habitability, based on the possibility of photosynthetic biomass production under large-scale geodynamic conditions, whereby the availability of liquid water on a planetary surface as well as the suitability of CO_2 partial pressure are taken into account. The habitable zone for Earth-like planets in our solar system was calculated by Franck, Block et al. (2000a). They found it to have been much broader in the past, and to have shifted and narrowed as time passed (Figure 4.1). According to their calculations, the inner boundary of the habitable zone will reach the Earth's orbit in about 500 million years from now, at which time the life span of the biosphere (Caldeira and Kasting 1992) will end. Concurrently, the outer boundary will decrease in a strong nonlinear way as a result of geodynamic effects. The existence of biological productivity is the only criterion. It is greater than zero within the so-called temperature tolerance window and above the minimum value of 10 ppm for the atmospheric CO_2 partial pressure. The lower temperature limit is 0°C; the maximum temperature is usually 50°C but might be extended to 100°C. The results show that the inner boundary of the habitable zone is primarily determined by the 10 ppm limit and the outer boundary by the 0°C limit.

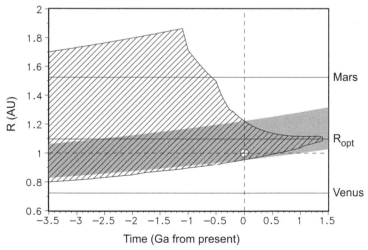

Figure 4.1 Evolution of the habitable zone for a geostatic model (gray area) and a geodynamic model (hatched area). In the geostatic model there is no continental growth, and the spreading rate does not change with time. Both effects are included in the geodynamical model (Franck, Block et al. 2000a, b). The optimal position for an Earth-like planet is at $R_{opt} = 1.08$ AU. In this case, the life span of the biosphere is at its maximum. Venus was never positioned within the habitable zone, whereas in the past, Mars was within the habitable zone for the geodynamic model. (Figure taken from Franck, Block et al. 2000a.)

Although some forms of chemolithoautotrophic hyperthermophiles might survive in a future environment of higher temperatures, independent of atmospheric CO_2 pressures, all higher forms of life would certainly be eliminated under such conditions. The described model is only relevant to photosynthesis-based life. Therefore, the upper temperature threshold is below 70°C. The upper temperature does not significantly affect the results for the habitable zone. As shown in Figure 4.1, the optimal position for an Earth-like planet would be at $R_{opt} = 1.08$ AU. In this case, the biosphere would realize a maximum life span of about 1.4 Ga.

The success of highly sophisticated detection techniques for extra-solar planets has led astronomers to forecast the imminent dawning of a golden age of astronomy. Thus far, more than 100 extra-solar planets have been discovered. Questions once considered beyond reach may soon find at least partial answers. In particular, is there life on planets outside of our solar system, or are they at least habitable? Fundamental work was performed by Lovelock (1965), who discussed general interactions between life and its planetary environment. Later, when considering the possibility of self-regulation by and for the biosphere, Lovelock and his colleagues put forward the Gaia hypothesis. Beyond the discussion of extraterrestrial life, there is an ongoing debate about other civilizations outside our solar system (see Horneck and Schellnhuber, this volume). The Drake Equation, which identifies the relevant factors for a statistical estimation, can provide further information about the abundance of possible extraterrestrial civilizations. Although several factors are highly speculative, a subset (e.g., determining the selection of contemporary biospheres that interact with their environment on a global scale or "Gaias") can be investigated rather rigorously. Based on investigations of the habitable zone for extra-solar planets, we can calculate the probability for the existence of an Earth-like planet in the habitable zone as one factor of this subset. Accordingly, there should exist about 500,000 Gaias in the Milky Way (Franck et al. 2004).

THE HADEAN AND ARCHAEAN ENVIRONMENT

The Hadean is defined as the aeon from the formation of the planet Earth (ca. 4.56 Ga ago) to the origin of life. The origin of life is now dated at 4.0 ± 0.2 Ga (Nisbet and Fowler 1996). When investigating the Hadean environment, one must take into account the energy influx caused by repeated bombardment of planetesimals, including the massive Moon-forming impact ca. 4.5 Ga ago. Even smaller impacts could have provided enough energy to heat up the oceans to more than 100°C, which would have sterilized all water.

Recent investigations provide a detailed scenario of the geospheric processes during the Hadean: a steam atmosphere was probably formed by impact degassing during accretion of the Earth; the Earth's surface was covered with a magma ocean, which promoted the degassing of N_2 and inert gases from basalts.

Investigations of extremely old zircon grains provide information on the state of our planet until 4.4 Ga ago (Wilde et al. 2001). There is evidence that continental crust and liquid water were surface features of the earliest Earth.

As the impact energy flux decreased, the steam atmosphere became unstable and water condensed to form the proto-ocean. After the formation of the ocean, the Earth's atmosphere consisted primarily of water vapor, N_2, and CO_2; from the composition of the chondritic meteorites, we know that CO_2 was the second most abundant volatile in the accreting material. As a greenhouse gas, CO_2 variation in the atmosphere influences surface temperature. Hence, it is important to study the carbon cycle among various reservoirs at the surface and in the interior of the Earth.

Two distinct scenarios exist for the Hadean and early Archaean environment: a hot scenario and a cold scenario with sporadic heating caused by impacts. The hot scenario is based on the fact that the early atmosphere contained up to 5 bar CO_2, which would have resulted in a greenhouse effect strong enough to have compensated for the faint young Sun. Consequently, surface temperatures of up to 80°C could have existed, which would suggest the existence of a thermophilic biosphere in early Earth's history.

Decrease of global mean surface temperatures can result from a strong reduction of atmospheric CO_2 content (e.g., by weathering or enhanced CO_2 uptake in the ocean). A major CO_2 sink results from strong weathering of oceanic impact ejecta, as their fresh surface is vulnerable to chemical attack. In addition to subaerial weathering, a strong CO_2 sink might result, at least in part, from oceanic impact ejecta and the subsequent binding of dissolved carbonate species into alkali-earth carbonates. Furthermore, a high pH of the ocean (*soda ocean*) as well as the so-called fast hydrothermal reaction kinetics can cause an enhanced solution of CO_2 (Franck et al. 2002).

The cold scenario for the Hadean Earth entails an iced-over planet where liquid water, which is heated episodically during large impacts, is available only near active volcanoes. Even under these extreme conditions, the Earth could have served as an environment for the origin of life (Bada et al. 1994).

THE ORIGIN OF LIFE AND THE
INITIAL CONDITIONS

The origin of life is defined as an emergent feature in a system of interacting components. The level of organization characterizes the boundary between living and nonliving. The level of organization in prokaryotic cell organisms is a minimal feature for life. The components, including plasmids and viruses, may be self-assembled. There is no "living" matter in the sense of an individual substance. Over the last century, attempts have been made to construct a "staircase of substances," from simple to complex, as a prerequisite for the origin of life. Ideas such as *life is protein*, *life is DNA*, and *life is RNA* are not valid as they ignore the supporting system for self-replicating molecules. Such a supporting system is the cell organism, delineated from the energy-supporting environment by a membrane boundary. "The idea that biological organization is fully determined by molecular structures is popular, seductive, potent and true up to a point — yet fundamentally wrong" (Harolds, 2001, p. 56). Life is represented by discrete objects: the organisms. Biochemists use a well-known logic machinery known as the substitution of the subject. The key step in understanding the origin of life is not the origin of compounds but rather the origin of living beings. The crucial question involves the transition from the components to the self-supporting cell. Until the simplest replicating cell (not a virus crystal or particle) is assembled from a mixture of components, speculations about the origin of the compounds are misleading. The simplest known living being is the prokaryotic cell organism, known as a bacterium. It includes the interaction of four major components: the informational DNA-replicating unit, the protein-synthesizing RNA unit (ribosome), the energy- and transport-producing membrane, and the cytoplasm enzymatic bag that produces precursors and metabolites. None of these components can be reproduced by itself despite the ability to self-assemble (Zavarzin and Kolotilova 2001).

The possible existence of life has been demonstrated from the beginning of the traceable geological record. Bacterial paleontology records remnants of living beings within certain ecosystems. An actualistic reconstruction of the past might be accomplished by studying microbial communities in ecosystems within extreme environments devoid of higher forms of life. These communities might be regarded as relics analogous to a past microbial biosphere. Survival in extreme environments indicates a shrinking basis of life during the increase of complexity. The persistence of microbial communities leads us to doubt whether environmental fitness could increase during phylogenetic lines. Bacterial communities have maintained a planetary life-support system from the beginning to the present.

All other forms of life are superimposed on the system formed by the interaction of fundamentally diverse bacteria with their environment and between themselves. Diversification is typical for a planetary life-support system. Therefore, a universal common ancestor cannot exist. A bacterial community depends on the energy and material balance for the prime producers. Energy may come from an external source, such as light for photosynthesis, or from endogenous geological sources, such as chemosynthesis.

A common model for the origin of life involves conditions near a hydrothermal system and is strengthened by the argument that hyperthermophiles belong to the most deeply rooted organisms. Furthermore, the geochemistry of hydrothermal systems provides opportunities to extract energy from the environment by processing various redox states. Sulfates and sulfites in the ambient water could be evolved in energy-yielding reactions of more reduced chemical species emerging from the vent fluids.

THE CYCLING OF CARBON

It is well known that life greatly affects the global environment and that the environment constrains life and can drive natural selection. In the Gaia theory, cycling ratio is a quantitative measure that describes the ratio of the flux of any specific element within the system relative to the flux of that element across the boundaries (Volk 1998). Thus, cycling ratio measures life's amplification process. One of the most important cycling processes is the carbon cycle. The global cycling ratio of carbon can be calculated in the following way: about 100 billion tons of carbon are fixed annually through global photosynthesis but only half a billion tons per year are supplied by rocks and volcanoes into the Gaian system. Therefore, the cycling ratio of carbon is about 200.

Cycling of matter is the main interactive process within the Vernadskian concept of the biosphere. Direct pathways lead to the accumulation of the products. Such a pure accumulation cannot operate for a long time on a large scale. Cycling of matter occurs under the rule *wheels within wheels*. The most important biotic cycle is the organic carbon cycle. It includes a productive branch operated by autotrophic primary producers as well as a destructive branch operated by organotrophic destructors. Inorganic carbon enters the cycle in the form of assimilated CO_2 and exits as respired CO_2. The balance of organic carbon depends on its decomposition by organotrophs, which is less than production. The difference between composition and decomposition of organic carbon is compensated for by excess oxygen that remains in the atmosphere; excess organic carbon is buried in sedimentary rocks (e.g., kerogen and its derivates). Oxygenation of the atmosphere depends on the inability of the organotrophs to decompose all organic matter. Chemosynthetic life depends on the endogenic production of substances, whereby chemosynthetic organisms develop when there is thermodynamic stability of the product (Zavarzin 1972). However, hydrogenotrophic life, using various oxidants, is most probable; it can develop in porous space of rocks in the subterranean hydrosphere. Transformation of inorganic carbon includes a pathway from atmospheric CO_2 to carbonates. This pathway is initiated by subaerial weathering of igneous rocks by the atmospheric hydrological cycle with water equilibrated with CO_2. The result is a leaching of Ca, Mg, and Fe into bicarbonate solutions. Leaching leads to transformation of source rocks into clays with a decreased amount of metals. Consumed atmospheric CO_2, carbonates, and clays are involved in the material balance. Weathering is regarded as the major process of CO_2 depletion from the atmosphere and mitigates the greenhouse effect. Subaquatic leaching of rocks is not relevant to the problem since dissolved CO_2 is involved only indirectly in the greenhouse effect, that is, after exchange with the atmosphere through a temperature-dependent carbonate–bicarbonate equilibrium in water. Subaerial weathering leads to alkalinization as a result of the release of cations into a bicarbonate solution, while anions remain as aluminum silicates. Alkalinity is controlled by the washout of sodium from the watershed. The inorganic carbon cycle is linked to the calcium cycle. On a planetary scale, CO_2 is found in sedimentary Ca and Mg carbonates. Its formation depends on the alkaline barrier layer. It is hypothesized that this barrier evolves at the border between alkaline terrestrial water and Ca/Mg-bearing marine water. The biogeochemical calcium cycle is responsible for the evolution of the neutral environment on the Earth's surface. During the early Precambrian, the primary carbonate accumulation operated in the absence of skeleton-forming organisms (Zavarzin 2002).

Prokaryotic organisms can transform compounds of nonbiogenic elements in biologically mediated processes. Biota participate in the inorganic carbon cycle by (a) autotrophic concentration of CO_2 into organic carbon with a concomitant pH increase, and (b) organotrophic liberation of CO_2 from the organic carbon locally concentrated in the dead matter. This mechanism is reflected by the isotopic ratio inherited by the pedogenic carbonates from the biomass. The biotic enhancement of weathering depends on the production of acidic substances. The paleontological record of stromatolites, representing organic sedimentary structures formed by the benthic cyanobacterial communities, demonstrates the participation of prokaryotes in carbonate formation. Eukaryotic organisms participate directly in the inorganic cycle and cause the concentration of calcium. However, this type of carbonate skeleton formation occurred late in geological history and is important for the recycling of calcium carbonate. Calcium is remobilized from carbonates through its biotically mediated dissolution during the organotrophic decomposition of dead organic matter.

Such a descriptive approach to the carbon cycle emphasizes the biogeochemical feedbacks. To date, however, no complete model exists to connect all of these biogeochemical processes to the evolution of other components of the Earth system. Thus, we propose a minimal model for the global carbon cycle between the mantle and five surface reservoirs (ocean floor, continental crust, continental biosphere, ocean + atmosphere, and the kerogen) as follows.

Kerogen results from a very small portion of organic carbon that is not transformed back to CO_2 by respiration and decay, but is instead accumulated in a remarkable reservoir. The size of the carbon reservoir in the current biosphere is less than the actual amount of carbon in the present atmosphere. Nevertheless, the biosphere plays a main role in controlling the terrestrial climate through its influence on the chemical composition of the atmosphere. The biosphere is also necessary for kerogen accumulation.

From the point of view of carbon cycling, kerogen could be important, even if it is relatively inert and can be only brought back into the carbon cycle through oxidation, with the help of atmospheric O_2 or bacteria. Kerogen occupies presently about 10–20% of the total amount of carbon in the surface reservoirs and therefore must be included in a minimal model. The model describes the evolution of the mass of carbon in the mantle, C_m, in the oceanic crust and seafloor, C_f, in the continents, C_c, in the continental biosphere, C_{bio}, in kerogen, C_{ker}, and in the combined reservoir of ocean and atmosphere, C_{o+a}. The equations for the efficiency of carbon transport between the reservoirs take into account degassing and regassing of the mantle, carbonate precipitation, carbonate accretion, evolution of the continental biomass, the storage of dead organic matter, and weathering processes. Kerogen is probably the least important reservoir from the point of view of carbon cycling because it is relatively inert. However, kerogen weathering and kerogen formation do occur, and the present size of the kerogen reservoir is obviously the net result of these processes. The main constraint for the reservoir size results from isotopic geochemistry. Since kerogen is isotopically light, due to its biological origin, it preferentially sequesters ^{12}C, whereas the continental carbon reservoir must be enriched by the heavier isotope [13]C. The isotopic composition of the two carbon reservoirs, kerogen and continental crust, might have been constant over the last 3.5 Ga. In this case, the ratio of kerogen carbon to carbon of carbonates would also have been constant at a value of 1:4, taking into account the isotopic signature of the mantle carbon. Figure 4.2 illustrates the main reservoirs and fluxes, and depicts the cycling of the two main

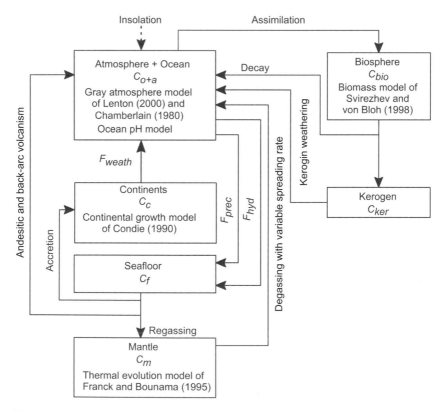

Figure 4.2   Basic mechanisms and interactions of the global carbon cycle (from Franck et al. 2002). Fluxes to and from the different pools are indicated by arrows.

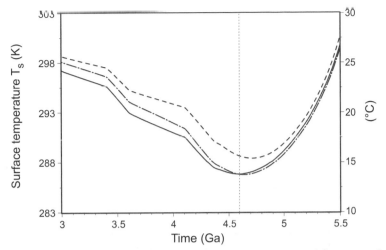

Figure 4.3 Evolution of the global mean surface temperature around the present showing the cooling effect of the biosphere and kerogen. The dashed line represents the model results for constant hydrothermal flux without biosphere; the dashed-dotted line with biosphere but without kerogen; the solid line the full model. (From Franck et al. 2002.)

volatiles in the Earth system: water and CO_2. Although the mantle water content exerts the primary influence on mantle rheology and the volatile exchange between mantle and surface reservoirs, atmospheric CO_2 is the main determinant of climate. The total amount of both volatiles is a topic of ongoing controversial discussion.

The influence of the biosphere on the mean global surface temperature (with an Archaean hothouse start) is shown in Figure 4.3, where the results for three different models are plotted: the full model (with both a biosphere and kerogen pool), a model with a biosphere but without a kerogen pool, and a model without a biosphere (and thus also without kerogen pool). Cooling of the biosphere is clearly shown: in the presence of a biosphere and kerogen pool, surface temperature is always lower than in the abiotic case. At present, cooling is about 1.8 K and is mainly caused by the biotic enhancement of weathering. The present geological epoch is characterized by the lowest global mean surface temperatures and a remarkable cooling of the biosphere. This suggests that the present state of the ecosphere is at the optimum in the sense that the present state has resulted in the emergence of higher life forms, including intelligence (Schwarzman and Middendorf 2000). At the same time, however, the Mesoproterozoic optimum of sustainability has been replaced by increasing complexity and instability.

CONNECTION TO OTHER CYCLES

The organic carbon cycle depends on the autotrophic synthesis of biomass by primary producers, with a ratio of organic carbon to organic nitrogen greater than 6. Excess organic carbon depends on the formation of the supporting organic structures, which leads, for example, to ratios greater than 100 for trees. A ratio of 6 is characteristic for proteins and cytoplasm. It is generally assumed that biomass is composed of $(CH_2O)_{106}(NH_3)_{16}(H_3PO_4)_1$. Cyanobacteria could initiate an organic carbon accumulation since they are able to assimilate both CO_2 and N_2. Their evolution would be limited by O_2 accumulation as well as by buried organic carbon. This, however, is not demonstrated in the Precambrian geological record. Iron oxides were deposited as banded iron formations and represent an inorganic sink for O_2. Phototrophic CO_2 assimilation depends on the area of wet light-exposed surfaces. Annual organic carbon production (NPP, net primary production) is on the order of several gigatons of C, with a possible two- to threefold fluctuation. These fluctuations depend on the area of habitable surfaces, which is limited by the chlorophyll concentration of wet light-exposed surfaces, and remained approximately constant during Earth history with a NPP of about 145 kg C_{org} per 1 kg of the chlorophyll cover.

Organic carbon cycling excludes the accumulation of CO_2 assimilation products due to the destructive pathway performed by organotrophs. Decomposition includes two major pathways: proteolytic and saccharolytic. Both pathways have aerobic or anaerobic modes. In the nineteenth century, Winogradsky (see, e.g., Zavarzin 2000) formulated a general rule for microbial decomposition: each natural compound has its specific microbial decomposer. This means that in the organotrophic microbial community there is a set of trophic pathways. Insoluble components in the prokaryotic world are subject to extracellular hydrolysis by hydrolytic microorganisms, whereas dissipating soluble compounds are utilized by dissipotrophs. The specificity of the decomposing substrates predicts a functional diversity of organotrophs. Under aerobic conditions, complete destruction might occur, which is usually performed by the concert of microorganisms. Under anaerobic conditions, a cascade of reactions dealing with metabolites takes place.

The proteolytic pathway is responsible for the regenerative cycle that produces combined nitrogen and phosphorus. When assimilation is incomplete, ammonia is produced during secondary biomass production. Energy is obtained from the metabolism of deaminated carbon skeletons. Ammonia production is a net measure of the proteolytic pathway. In an anaerobic environment, NH_3 is a final product. It is utilized in an aerobic environment by nitrifiers, where nitrate is a thermodynamically stable compound in an oxygenated atmosphere. Nitrate is either reutilized in the assimilatory pathway or enters anaerobic sites and acts as an oxidizer by denitrifiers.

The saccharolytic pathway is linked to the nitrogen cycle as combined nitrogen is required for secondary biomass production. Combined nitrogen is obtained either from the products of the proteolytic pathways or through nitrogen fixation. Nitrogen fixation is very prevalent between prokaryotes, especially among anaerobes, but it is an energy-consuming process. The quantitative nitrogen fixation in ecosystems usually offsets the losses of combined nitrogen that result from denitrification.

The major groups involved in decomposing dead organic matter are hydrolytics and dissipotrophs, which utilize soluble molecules dissipating from the sites of hydrolysis. The kinetic characteristics of hydrolytics are surface dependent, whereas for dissipotrophs, they depend on concentration. The cascade of reactions in anaerobic communities includes steps of sequential degradation of organic matter. The first step is performed by fermentative primary anaerobes. The second step involves oxidizing the metabolites of fermentative organisms with the key intermediates H_2 and acetate by secondary anaerobes. Methanogenic archaea use CO_2 that is internally produced during decomposition as an oxidant. However, methane is inert in an anaerobic environment.

For complete decomposition in an anaerobic environment, the sulfur cycle is essential. It starts with sulfidogenesis by sulfate reducers, which are capable of oxidizing H_2 and acetate or a number of simple organic compounds that are either directly or syntrophically in cooperation with other microbes due to the interspecies hydrogen transfer. Hydrogen sulfide produced by sulfidogens may be oxidized anaerobically by anoxygenic sulfur phototrophs into sulfate. In this way, the catalytic sulfur cycle functions independently from the oxygen supply to complete the destruction cycle. It is unclear to what extent this mechanism could contribute to the reservoir of sulfates in the ocean (*purple ocean*). Anoxygenic nonsulfur phototrophs oxidize most nonfermentable compounds and represent an O_2-saving mechanism in the destructive pathway.

The products of anaerobic decomposition (methane, sulfide, reduced organic compounds) are transported into the aerobic zone and oxidized by a group of aerobes representing an oxidative microbial filter. They belong to the so-called *gradient organisms*. Microaerobes (or microaerophilic organisms) are the most important representatives of this group.

It is evident that the microbial community operates in a cooperative manner and results in a trophic network. Such a community is sustainable for a geologically significant time. By contrast, the prokaryotic community is sufficient to fulfill all system needs for sustainability.

ENVIRONMENTAL FORCING AND THE EVOLUTION OF EUKARYOTES AND COMPLEX LIFE

On geological timescales, biogeochemical succession is based on the accumulation of products from the microbial community. This succession occurs because of the imbalance in biogeochemical cycles caused by the inability of microbes to decompose completely under special conditions. The major obstacle to decomposition is the physical burial of dead organic matter in finely dispersed mineral sediments. Organic matter in such a state is unattainable to bacteria and subject to geological transformations during sedimentogenesis under high pressure and temperature in the so-called catagenesis step. Furthermore, it is involved in the process of coal formation with a loss of hydrogen, followed by crystallization to graphite, and is an important driving force for biogeochemical succession on a planetary scale. Biologically mediated reactions are primarily caused by collateral reactions in the organic carbon cycle and result in a transformed environment with high O_2, H_2S, and CO_2 contents, whereby new thermodynamic domains of mineral stability are created. The loop from subaerial weathering to clay formation and to sedimentary rocks leads to a net geospheric production of reduced carbon due to organic carbon burial in finely dispersed sediments.

The interactions within the biotic system and the resulting geological changes drive the large-scale geosphere–biosphere system. There are two aspects to the change of the biosphere system: (a) the intrinsic force caused by the evolution of biota with the leading role of primary producers (cyanobacteria, algae, plants) and (b) the exogenic geosphere forces mainly caused by tectonics. External forcing caused by the evolution of our Sun is relevant to long-term climate changes.

Superposition of other life forms on the prokaryotic system is an additional event (Zavarzin 2000). Functional biodiversity of prokaryotes allows them to build up and support a complete geosphere–biosphere system. Newcomers can only establish themselves if they are functionally compatible with the already existing cooperative system, that is, conservation of the old system is a prerequisite for the evolution through the inclusion of new members. Environmental fitness means compatibility to the larger system. Evolution thus occurs through addition rather than substitution, which is particularly evident on the scale of a biosphere–geosphere system. Evolution into free or new ecological niches is a secondary process. On a large scale, cooperation is more important than competition, which is indicative of a fine regulatory mechanism in the adaptive dynamics of functional groups. This explains why substitution of certain components of the system does not change its overall functional structure. As a result, the prokaryotic system that catalyzes biogeochemical cycles remains as the base of the habitable planetary system during the time of its existence. In the Earth system, biogeochemical succession induced the formation of an oxygenated atmosphere. A 24-hr aerobic metabolism proved more effective than the transition from oxygenic photosynthesis during the day to fermentation in the night, as in cyanobacterial mats. The aerobic mode of life was thus a prerequisite for the development of larger eukaryotic organisms and required such organisms to digest particulate organic matter intracellularly. The O_2 reserve in the atmosphere and the changes of the mean global surface temperature forced the transition from prokaryotes to eukaryotes. Protista spread out during the Neoproterozoic and decomposition within the organic cycle was enhanced. Subsequent major steps include the intracellular skeleton formation in the Cambrian and the development of a terrestrial plant cover in the Late Silurian–Early Devonian, with a concomitant spread of fungi capable of decomposing lignin in the oxygenic atmosphere. Vascular plants produced a new environment for the terrestrial hydrological cycle, which resulted in the terrestrial organic carbon cycle being controlled primarily by plants and fungi. The terrestrial plant–soil cover provides the background for habitability of higher life forms (Zavarzin 2001).

Just what precipitated the major evolutionary transitions has been the subject of intense scientific debate (Lenton, Caldeira, and Szathmáry, this volume; Maynard Smith and Szathmáry 1995). The increase in biosphere complexity may have been achieved through a series of major evolutionary transitions, which brought changes to the way information is stored and transmitted. Franck et al. (2002) applied a simple Earth system model to test the coevolution of the geosphere and biosphere. They introduced three different types of biosphere: prokaryotes, eukaryotes, and complex multicellular life. The biosphere types have various temperature tolerance windows within which their biological productivity is greater than zero: prokaryotes, from 2°–100°C; eukaryotes, from 5°–40°C; and complex life, from 0°–30°C. Furthermore, biotic enhancement of weathering by complex life adds an additional feedback to the system. The upper graph in Figure 4.4 shows the results for the evolution of the mean global surface temperature (solid line); the lower graph illustrates the corresponding cumulative biosphere pools. A prokaryotic biosphere has always existed, from the Archaean to the present. Eukaryotic life first appeared 1.8 Ga ago when global surface temperature reached the tolerance window for eukaryotes. According to Cavalier-Smith (2002), however, eukaryotes first began to disperse about 1 Ga ago. In contrast to eukaryotes, the first appearance of complex life began with an explosive increase in biomass in the Vendian, connected with a strong decrease in the Cambrian global surface temperature 540 million years ago. The appearance of metaphyta in the Devonian further increased biogenic enhancement of weathering. In this way, the Cambrian explosion, that is, the instantaneous emergence of complex life, could have been triggered by nonlinear geosphere–biosphere interactions. The strength of these interactions can be described by the biotic enhancement factor. In principle, two stable solutions exist for a certain interval of this biotic enhancement factor: one with and one without complex life. A transition from one solution to the other can be triggered by environmental perturbations. Such a phenomenon is typical for the concept of virtual biospheres mentioned earlier. Neoproterozoic snowball Earth events, for example, could have initiated an earlier appearance of complex life. It can be shown (von Bloh et al. 2003) that the Cambrian explosion was mainly driven by extrinsic environmental causes. After the Cambrian explosion, there was a continuous change in the amount of biomass in all pools. However, the environment itself was actively changed by the biosphere to maintain the conditions necessary for its existence. At present, biomass is almost equally distributed between the three types. Over the long term, we will experience a continuous decrease of cumulative biomass. Thus, the prokaryotic biosphere must be present during the whole life span to act as the base for more complex biosphere types.

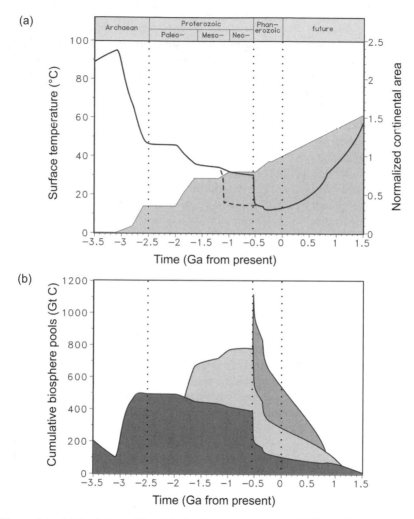

Figure 4.4   (a) Evolution of the global surface temperature (solid line). The dashed line denotes a second possible evolutionary path triggered by a temperature perturbation in the Neoproterozoic era. The shaded area indicates the evolution of the normalized continental area according to Condie (1990). (b) Evolution of the cumulative biosphere pools for prokaryotes (dark shading), eukaryotes (light shading), and higher metazoa (medium shading) (von Bloh et al. 2003).

OPEN QUESTIONS REGARDING THE CONDITIONS FOR THE ORIGIN OF LIFE AND SUBSEQUENT PLANETARY LIFE-SUPPORT SYSTEMS

- Did life first exist thermophilically in an Archaean hothouse or mesophilically as the result of the temperature limitations in photosynthesis?

- If the "tape" of life's history could be played again, would biological evolution result in the same biosphere or in a radically different one?
- What are the minimal requirements for a living system?
- What are the minimal requirements for an environment to be habitable?
- What is the longest timescale upon which the biosphere may dominate the evolution of the Earth system?
- Does the present state represent an optimum of self-regulation in the Earth system, and will the long-term future be a renaissance of the past?
- What causes major evolutionary transitions within the biosphere?
- Do prokaryotes provide the necessary foundation for the succession of other life forms?
- How does fitness apply to the environment, and what does it mean?
- What is evolution?
- How does the evolution of a geosphere–biosphere system correlate with phylogeny?

ACKNOWLEDGMENTS

We would like to thank C. Bounama, W. von Bloh, and I. Dorofeeva for many stimulating discussions and for their help in the preparation of the manuscript.

REFERENCES

Bada, J.L., C. Bigham, and S.L. Miller. 1994. Impact melting of frozen oceans on the early Earth: Implications for the origin of life. *Proc. Natl. Acad. Sci.* **91**: 1248–1250.

Caldeira, K., and J.F. Kasting 1992. The life span of the biosphere revisited. *Nature* **360**: 721–723.

Cavalier-Smith, T. 2002. The phagotrophic origin of eukaryotes and phylogenetic classification of Protozoa. *Intl. J. System. & Evol. Microbiol.* **52**:297–354.

Condie, K.C. 1990. Growth and accretion of continental crust: Inferences based on Laurentia. *Chem. Geol.* **83**:183–194.

Doyle, L.R. 1996. Circumstellar Habitable Zones. Proc. 1st Intl. Conf. Menlo Park: Travis House Publ.

Franck, S., A. Block, W. von Bloh et al. 2000a. Habitable zone for Earth-like planets in the solar system. *Planet. Space Sci.* **48**:1099–1105.

Franck, S., A. Block, W. von Bloh et al. 2000b. Reduction of biosphere life span as a consequence of geodynamics. *Tellus* **52B**:94–107.

Franck, S., W. von Bloh, C. Bounama et al. 2000. Determination of habitable zones in extrasolar planetary systems: Where are Gaia's sisters? *J. Geophys. Res.* **105(E1)**: 1651–1658.

Franck, S., K.J. Kossacki, W. von Bloh, and C. Bounama. 2002. Long-term evolution of the carbon cycle: Historic minimum of global surface temperature at present. *Tellus* **54B**:325–343.

Franck, S., W. von Bloh, C. Bounama, and H.-J. Schellnhuber. 2004. Extraterrestrial Gaias. In: Scientists Debate Gaia: The Next Century, ed. S.H. Schneider, J.R. Miller, E. Crist, and P.J. Boston. Cambridge, MA: MIT Press, in press.

Harold, F.M. 2001. The Way of the Cell, Molecules, Organisms and the Order of Life. Oxford: Oxford Univ. Press.

Lovelock, J.E. 1965. A physical basis for life detection experiments. *Nature* **207**: 568–570.

Maynard Smith, J., and E. Szathmary. 1995. The Major Transitions in Evolution. Oxford: Freeman.

Nisbet, E.G., and C.M.R. Fowler. 1996. Some liked it hot. *Nature* **382**:404–405.

Sagan, C., and G. Mullen. 1972. Earth and Mars: Evolution of atmospheres and surface temperatures. *Science* **177**:52–56.

Schrödinger, E. 1944. What Is Life? The Physical Aspect of the Living Cell. Cambridge: Cambridge Univ. Press.

Schwarzman, D., and G. Middendorf. 2000. Biospheric cooling and the emergence of intelligence. In: A New Era of Bioastronomy, ed. G. A. Lemarchand and K.J. Meech, ASP Conf. Series, vol. 213, pp. 425–429. San Francisco: Astron. Soc. of the Pacific.

Svirezhev, Y. M., and W. von Bloh. 1997. Climate, vegetation, and global carbon cycle: The simplest zero-dimensional model. *Ecol. Model.* **101**:79–95.

Veizer, J. 1988. The Earth and its life: System perspective. *Orig. Life* **18**:13–39.

Volk, T. 1998. Gaia's Body: Toward a Physiology of Earth. New York: Copernicus/ Springer.

von Bloh, W., C. Bounama, and S. Franck. 2003. Cambrian explosion triggered by geosphere–biosphere feedbacks. *Geophys. Res. Lett.* **30(18)**:1963.

Wilde, S.A., J.W. Vally, W.H. Peck, and C.M. Graham. 2001. Evidence from detrital zircons for the existence of continental crust and oceans on the Earth 4.4 Gyr ago. *Nature* **409**:175–178.

Zavarzin, G.A. 1972. Lithotrophic microorganisms. Moscow: Nauka (in Russian).

Zavarzin, G.A. 2000. The non-Darwinian domain of evolution. *Herald Russ. Acad. Sci.* **70(3)**:252–259.

Zavarzin, G.A. 2001. The rise of the biosphere. *Herald Russ. Acad. Sci.* **71**:988–1001.

Zavarzin, G.A. 2002. Microbial geochemical calcium cycle. *Microbiology* **71**:1–17.

Zavarzin, G.A., and N.N. Kolotilova 2001. Introduction to Environmental Biology: For Students. Moscow: Univ. Publ. House (in Russian).

5

Destiny of Humankind from an Astrobiology Point of View

Does Astrobiology Provide an Exit Option for Terrestrial Mismanagement?

G. HORNECK[1] and H. J. SCHELLNHUBER[2]

[1]German Aerospace Center (DLR), Institute of Aerospace Medicine,
51147 Cologne, Germany
[2]Tyndall Center for Climate Change Research, School of Environmental Sciences,
University of East Anglia, Norwich NR4 7TJ, U.K. and Potsdam Institute for
Climate Impact Research (PIK), 14412 Potsdam, Germany

ABSTRACT

Astrobiology is a multidisciplinary approach used to study the origin, evolution, distribution, and future of life on Earth and in the cosmos and is based on the assumption that life is a universal planetary phenomenon if the right environmental conditions prevail. Such an approach helps to transform our *Weltbild* from a geocentric to a more Universe-oriented one. This chapter focuses on migration as a central attribute of life within the astrobiological context. Migration has been observed for terrestrial life throughout its more than 4 billion years (Ga) of history, across all domains of life. It appears, however, that migration is not confined to the planetary realm.

It is argued that microbial life-forms can be transported by meteorites between the planets of our solar system. In addition, humans have now developed the technology to leave their home planet and to explore other bodies in our solar system. This evolutionary step is still in its infancy, with the Moon being the only celestial body visited so far, yet it is demonstrated that technology exists for more ambitious projects, such as a visit to the planet Mars. The issue of whether human spaceflight has the potential to promote peaceful cooperation on a global scale is discussed briefly. Finally, forced migration due to environmental degradation is considered, with a special emphasis on hypothetical scenarios of a future exodus from the Earth as a consequence of anthropogenic collapse of the planet's life-support systems. The conclusion is drawn that sustainable development on *Terra* should be achievable using only a tiny fraction of the efforts required for interplanetary mass transportation.

ASTROBIOLOGY: A NEW APPROACH TO EARLIEST QUESTIONS ABOUT HUMANKIND

The emerging science of astrobiology is a multidisciplinary approach to study the origin, evolution, distribution, and future of life on Earth and in the Universe (Blumberg 2002). Its central focus is directed toward questions that have intrigued humans for a long time: Where do we come from? What is life? Are we alone in the Universe?

These questions have been tackled jointly by scientists converging from widely different fields, ranging from astrophysics to molecular biology and from planetology to ecology, among others. By spilling over the traditional boundaries of classical science, completely new opportunities for research have been opened, a state described by some contemporaries as the "astrobiology revolution of the sciences." In this chapter, we address the last two research topics of astrobiology, namely the study of the distribution and future of life on Earth and elsewhere. We conclude with a discussion of the pros and cons of an extension of the human *Lebensraum* beyond the limits of our own planet Earth to neighbors in the solar system, especially to the Moon and to Mars.

MIGRATION: A CENTRAL ATTRIBUTE OF LIFE

Wherever and whenever life has established itself, it has exhibited a natural impulse to expand the area from which its resources are drawn. The driving forces for migration to new settlements can be endogenous (e.g., hormones) or exogenous (e.g., deprivation of food or space, overcrowding by other species, unfavorable environmental conditions, light stimulus). Migration has certainly played a role in the evolution of our biosphere. In the following, we outline briefly how the main life-forms known to us have perpetually managed to push beyond their habitat boundaries.

MIGRATION OF LIFE ON ITS HOME PLANET

Migration of Microbial Life on Earth

The history of life on Earth reaches back about 4 Ga (Figure 5.1). The fossil record gives evidence that life started very early as a "microbial world" on the juvenile Earth, that is, quite certainly earlier than 3.5 Ga ago and possibly earlier than 3.8 Ga ago (see chapters by Lenton, Caldeira, and Szathmáry as well as Drake and Westall, this volume). Yet because the oldest known microfossils are already structurally complex and stem possibly from autotrophic organisms, as inferred from the isotopic ratio of carbon, the first appearance of life on Earth should be dated even earlier than this. Hence, microbial life has persisted on Earth almost since the cooling of the planet's crust and, evidently, until today.

Figure 5.1   In this artistic rendition of the history of life on Earth, ingredients for life (e.g., organics and water) arrived via planetesimals and comets, and life began as a "microbial world."

Prokaryotes (i.e., unicellular organisms belonging to the two domains Bacteria and Archaea of the phylogenetic tree) dominated the Earth's biosphere during the first two billion years of life's history, before unicellular mitotic eukaryotes appeared. Prokaryotes developed sophisticated strategies to conquer the whole surface of the Earth as well as subsurface regions. Many environments within this empire could not sustain eukaryotic life (especially higher-developed multicellular forms, including humans), and thus we refer to these niches as "extreme" domains (e.g., hot vents, permafrost areas, "eternal" ice, subsurface regions down to several kilometers, the high atmosphere, rocks, and salt crystals). Microbial life stretched its limits to invade nearly all regions of Earth, spreading down below the surface several kilometers and including other life-forms, such as symbionts or parasites. Traces of microbial activities have been left everywhere on our planet, especially in soils, rocks, and atmosphere.

Migration of Plants and Animals on Earth

Long ago in Earth's history, during the first 1.5–2 Ga of life's existence, the atmosphere was virtually devoid of oxygen and hence ozone, which allowed the highly mutagenic and lethal solar UV-C (200–280 nm) and UV-B (280–315 nm) radiation to reach the surface of the planet virtually unattenuatedly (Kasting et al. 1992). To cope with this harmful UV radiation climate, life was forced either to withdraw into UV refuges or to develop strategies to tolerate the intense UV radiation influx. The situation changed drastically with the advent of photosynthesis, whereby oxygen was enriched in the atmosphere about two billion years ago and ozone was photochemically formed in the stratosphere as an effective UV shield. Only then did the conditions on Earth become favorable enough for life to invade the upper layers of the ocean and the surface regions of the continents. However, it still took a long time, until about four hundred million years ago, for animals and plants to expand globally over the continents (see Lenton, Caldeira, and Szathmáry, this volume).

Although most are of sessile nature, plants developed several mechanisms to increase their habitat. Most of their spores or seeds can be transported easily

over long distances through air or water; other seeds use animals for passive
movement, either through attachment to furs or feathers or by passage through
the digestive system. Phytoplankton, one of the primary providers to the food
chain and a proven CO_2 sink on a global scale, uses the circadian light cycle for
its vertical migration in the ocean to optimize photosynthesis.

By contrast, most animals are capable of active movement. Migrating birds
change their habitat following the seasonal rhythms. One of the champions is the
Arctic tern *Sterna parasidea*, which covers distances of 17,000 km during its annual quest for better conditions. Food depletion, limited space, or unfavorable
climate can cause mass migrations; combined, these factors may prohibit a species' return to its original point of departure. Examples can be found in the migration patterns of insects, birds, lemmings, and horses. The evolution of the
latter group, Equidae, described by Ernst Haeckel as the *Paradepferd* (parade
horse, showpiece) of paleontology, clearly demonstrates the role of migration in
phylogeny. Whereas the evolution of the Equidae proceeded predominantly in
America, early forms, such as the small *Urpferde*, migrated several times to Eurasia via the North Atlantic bridge between the two continents, only to become
extinct in the new habitat. However, about three million years ago, the genus
Equus migrated to Eurasia over the Bering bridge. This time it survived at its
new destination, although it eventually became extinct on the American continent. Later it was domesticated in Europe and brought back to the New World
after its "discovery" by the Iberian conquistadores. Subsequently, the horse developed into an integral part of the civil and martial culture of the American Indian tribes of the Great Plains and elsewhere (e.g., the Comanches). Thus, in a
sense, coevolution had come full circle.

Human Migration on Earth

No-nonsense debates about the origin of humans and their subsequent distribution over the globe require fossil evidence from different habitats and samples
from many countries. In fact, fossil records have grown enormously over the
past fifty years, in particular, through rich finds in the African Rift valley (Gamble and Stringer 1997).

The course of human evolution can briefly be divided into four periods:

- 5 million years to 1 million years ago,
- 1 million to 300,000 years ago,
- 300,000 to 30,000 years ago, and
- 150,000 to 10,000 years ago.

The salient discoveries for each period are as follows. During the first period,
hominids split from the last common ancestor. There is both genetic and ana-
tomic evidence that this was a chimpanzee-like primate. Most estimates date the
evolution of a distinct hominid line at five million years ago. Early evolution is
exclusively African, and most of it happened south of the Sahara. Some time

between two and one million years ago, hominids moved north and colonized large parts of Asia and Europe.

In the second period (1 million to 300,000 years ago), there seem to exist only representatives of the genus *Homo*. Regional diversity, however, was considerable, as demonstrated by the shape of fossil skulls and teeth. These hominids were still confined to parts of the Old World. They were strongly built, and recent finds have shown them to be considerably taller than previously thought.

For much of the following period (300,000 to 30,000 years ago), we see the regional evolution of *Homo*. The best fossil evidence comes from Europe and Western Asia, where the Neanderthals formed a powerful population.

As early as 150,000 years ago in Eastern Africa, we find fossil skulls of essentially modern type. Africa has now been identified both genetically and anatomically as the evolutionary center for the appearance of humans who looked, and eventually acted, like us. Obviously, successive migrations out of Africa replaced the regional populations, such as the Neanderthals, in all parts of the inhabited Old World (Figure 5.2). One of the novelties associated with the modern humans is their drive for global colonization. In the last 50,000 years these people invaded Australia, the Pacific islands, the northern parts of Asia, the American continents, and the Arctic. This great prehistoric migration was achieved by people with a hunting and gathering lifestyle; it drew the demographic and geographical map for the later development of ethnic and regional populations of *Homo sapiens*.

It is well known that massive tribal migrations occurred in Europe since the third millennium B.C. (Figure 5.3). Best investigated are the wanderings of Germanic peoples, probably pioneered by the Goths going south and west from Scandinavia, and becoming more profound after the invasion of the Huns in A.D.

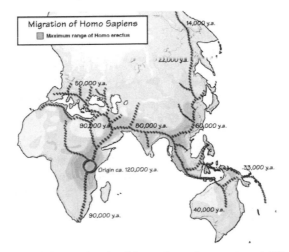

Figure 5.2 State-of-the-art sketch of the major pathways toward "*Homo sapienization*" of our planet.

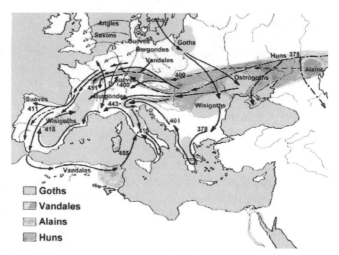

Figure 5.3 Tribal migration patterns in Europe during the decline phase of the Roman Empire.

375. Whereas the original settlement area was sometimes simply abandoned (as exemplified by East Germanic tribes like the Vandals), in many cases migration resulted in a territorial expansion whereby the connections to the traditional settlement areas were maintained (as exemplified by West Germanic tribes like the Alemanni, Franks, etc.). Human migration has continued until today, particularly in response to expulsions from their homelands caused by armed conflicts.

The motivational forces behind such migration are manifold. Escaping natural and anthropogenic deprivations or disasters are certainly dominant reasons in many instances (see below). However, human curiosity, the spirit of research, and the zest for discovery and exploration constitute other powerful forces driving migration. Recently, the rapidly growing tourism industry has led to a burst of periodic short-term "migrations," some of them over quite long distances, spurred by the desire for relaxation, a change of scenery, and for visiting climatically pleasant regions. Technical and technological progress has provided fast means of transportation and communication, and thus has facilitated this kind of wanderings. Globalization in business, science, and lifestyle has led to the daily mass transportation of millions of people between the continents. In short, jet-setting has become a way of life for many.

The urge to explore seems to be ingrained in human nature. In the endeavor to unveil the secrets hidden beyond the perpetually receding horizon, humans have crossed the seas, climbed the highest mountains, visited the poles, and studied the abysses of the ocean. Hence, humans have identified and reached the outer limits of habitability on our planet. Exploration has also shaped the scientific culture of our civilization. A prominent example is Charles Darwin's voyage on board the Beagle from 1832 to 1837 and his discoveries during this trip, which laid the foundations for the theory of evolution.

The important role of exploration as a driver in science and culture has been well described by M. E. DeBakey (2000, p. XI): "We know from earliest recorded history, some 5000 years ago, that human beings have always sought to learn more about their world. The century just ended has witnessed stunning advancements in science and medicine, including the launching of space exploration. There is a danger, however, that the new century may usher in an age of timidity, in which fear of risks and the obsession with cost-benefit analysis will dull the spirit of creativity and the sense of adventure from which new knowledge springs." Has exploration, however, reached a dead end at the outer limits of our planet now? Or will new frontiers be opened beyond the limits of our Earth to enable an understanding of the unexplored in our solar system and the Universe? After having reached the most distant and hostile places on Earth, is it a logical step to start exploring the neighborhood of the Earth, that is, the Moon and the terrestrial planets? We approach the answer to these questions by first addressing the potentially most fundamental mechanisms for the transplanetary expansion of *Lebensraum*.

MIGRATION OF LIFE BEYOND ITS PLANET OF ORIGIN

Panspermia

The hypothesis of panspermia, as formulated by Arrhenius (1903), postulates that microscopic forms of life (e.g., spores) can be propagated in space, driven by the radiation pressure from the Sun, thereby seeding life from one planet to another or even between solar systems. At the time when Arrhenius published this bold idea, panspermia was suggested as a way out of a major dilemma confronting the scientific community — one that resulted after Pasteur experimentally disproved the spontaneous generation of life. However, when it was realized that abiotic synthesis of organic compounds — the precursors of life — was a rather common process, either on the primitive Earth (e.g., in some protected niches of reducing gas mixtures) or elsewhere in the Universe, panspermia was no longer considered a likely route toward the emergence of life on our planet. The hypothesis was criticized, in particular, for not being experimentally testable, and it was argued that spores would not survive long-term exposure to the hostile environment of space, especially characterized by vacuum and radiation. Recent experiments in space have demonstrated clearly that isolated bacterial spores, as postulated by panspermia, will be killed in space within seconds as a consequence of the high lethality of extraterrestrial solar UV radiation (Horneck, Rettberg et al. 2001). Finally, it was argued that panspermia does not solve the problem of the origin of life but merely shunts the question aside to another location in the Universe — a compelling counterargument indeed. To cope with some of these objections, alternate ideas of panspermia developed toward the end of the twentieth century (e.g., Hoyle and Wickramasinghe 1978; Melosh 1985).

Directed Panspermia

F. H. C. Crick proposed a form of directed panspermia, suggesting that other intelligent beings from some other planet — or some other solar system — purposely sent protected packages of special microbes on interstellar flights (Crick and Orgel 1973). Although he agreed that the arguments he employed in favor of directed panspermia — such as that the history of all life on Earth goes through a bottleneck, the "last common ancestor," and that the genetic code is universal for all living beings of our biosphere — were somewhat sketchy, the possibility of life having reached the Earth via deliberate "infection" should not be ruled out as one of the many possibilities under consideration, as long as the question of the origin of life on Earth is still far from being answered: "the historical facts are important in their own rights" (Crick and Orgel 1973, p. 341). Although Crick's ideas of directed panspermia never found widespread support, schemes for "terraforming" Mars (i.e., intentionally modifying the Martian environment such that it becomes more habitable and eventually more Earth-like), as are presently under development, will attempt to prove the technical feasibility of directed panspermia, at least within our own solar system (McKay et al. 1991).

To prevent the introduction — intentional or otherwise — of microbes from the Earth to another celestial body or vice versa, an international treaty was established (UN Doc. A/6621, Dec. 17, 1966 and the follow-up agreement UN Gen. Ass. Resol. A/34/68, Dec. 5, 1979). Based on this treaty, a concept of contamination control has been elaborated by the Committee of Space Research (COSPAR) under consideration of specific classes of mission/target combinations (Table 5.1) and recommended to be followed by each space-faring organization. The intention of planetary protection is twofold: (a) to protect the planet being explored and to prevent jeopardizing search for life studies, including those for precursors and remnants, and (b) to protect the Earth from the potential hazards posed by extraterrestrial matter carried on a spacecraft returning from another celestial body (Rummel 2002). These restrictions should definitely be observed until intense research has completely ruled out the possibility of indigenous life on that celestial body.

Litho-Panspermia

Encouragement for revisiting panspermia came recently from the detection of a class of meteorites, which were clearly identified as space travelers sent out from planet Mars. Of the 28 Martian meteorites scrutinized thus far, some are quite massive, the heaviest one weighing about 18 kg. Furthermore, there are indications that several of them stem from the same, even larger, parental body that fragmented after colliding with other rocks. How could these heavy rocks ever leave their home planet? The most plausible process capable of ejecting surface material from a planet or a moon into space is the hypervelocity impact of a large object, such as a kilometer-sized asteroid or comet (Melosh 1985).

Table 5.1 Proposed categories (I–V) for planetary protection with regard to missions to solar system bodies and types of missions (after Rummel 2002).

	Degree of Concern	Representative Range of Requirements
I[1]	None	None
II[2]	Record of planned impact probability and contamination control measures	Documentation only (all brief): • PP plan • Pre-launch report • Post-launch report • Post-encounter report • End-of-mission report
III[3]	Limit on impact probability Passive bioload control	Documentation (Category II plus): • Contamination control • Organics inventory (as necessary) Implementing procedures such as: • Trajectory biasing • Cleanroom • Bioload reduction (as necessary)
IV[4]	Limit on probability of non-nominal impact Limit on bioload (active control)	Documentation (Category II plus): • P_c analysis plan • Microbial reduction plan • Microbial assay plan • Organics inventory Implementing procedures such as: • Trajectory biasing • Cleanroom • Bioload reduction • Partial sterilization of contacting hardware (as necessary) • Bioshield Monitoring of bioload via bioassay
V[5]	If restricted Earth return: • No impact on Earth or Moon • Returned hardware sterile • Containment of any sample If unrestricted Earth return: None	*Outbound*: Same category as target body/outbound mission *Inbound:* If restricted Earth return: • Documentation (Category II plus) • P_c analysis plan • Microbial reduction plan • Microbial assay plan • Trajectory biasing • Sterile or contained returned hardware • Continual monitoring of project activities • Project advanced studies/research

[1]Category I   *Mission type*: Any except Earth return (flyby, orbiter, lander)
   Target: Venus, undifferentiated metamorphosed asteroids, others to be determined
[2]Category II   *Mission type:* Any except Earth return (flyby, orbiter, lander)
   Target: comets, carbonaceous chondrite asteroids, Jupiter, Saturn, Uranus, Neptune, Pluto and Charon, Kuiper belt objects, others to be determined
[3]Category III   *Mission type:* No direct contact (flyby, some orbiters)
   Target: Mars, Europa, others to be determined
[4]Category IV   *Mission type:* Direct contact (lander, probe, some orbiters)
   Target: Mars, Europa, others to be determined
[5]Category V   *Mission type:* any Earth return mission, "restricted Earth return"
   Target: Mars, Europa, others to be determined
PP = planetary protection; P_c = probability of contamination

Material located close to the surface and at the rim of the impact crater could even be ejected at quite moderate temperatures, below 100°C, as a result of spallation, which occurs when the reflected shock wave is superimposed on the direct one (Melosh 1985). This impact mechanism would work not only on Mars (escape velocity 5 km s^{-1}), but also on the much larger planet of Earth (escape velocity 11.2 km s^{-1}). Estimates suggest that within the last four billion years, several millions of fragments ejected from Mars, with a diameter of 2 m or more and temperatures below 100°C, have arrived on Earth. Hence, the 28 Martian meteorites, mentioned above, represent probably only an infinitesimal fraction of matter imported from Mars within Earth's history.

The next question to address is whether this impact scenario allows for a transfer of life-forms from their home planet to a neighboring planet, in other words, whether "hitchhikers" would survive such a "litho-panspermia" (Figure 5.4). For this to happen, a number of barriers must be overcome:

1. The initial impact site must be on an inhabited planet at a location where living organisms, preferably microorganisms, are present.
2. The material ejected from the spall zone of the impact crater at velocities sufficiently high to overcome planetary gravitation must contain living (micro)organisms.
3. The (micro)organisms enclosed in the ejecta must survive the ejection process as well as the extended periods of traveling across interplanetary space.
4. When the meteorite arrives at the planet of destiny, the (micro)organisms must cope with the landing process, which might include entry through an atmosphere as well as thermal and pressure attacks during impact landing on the surface.
5. The new planet must provide a habitable environment accessible to the invading (micro)organisms.

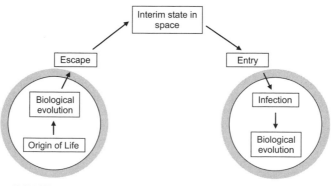

Figure 5.4 Scenario of an interplanetary transfer of life in the solar system by litho-panspermia.

The following formula (Clark 2001) summarizes the uncertainties associated with a viable transfer of life within our solar system:

$$P_{AB} = P_{biz} \times P_{ee} \times P_{sl} \times P_{ss} \times P_{se} \times P_{si} \times P_{rel} \times P_{st} \times P_{sp} \times P_{efg} \times P_{sc} \quad (5.1)$$

where P_{AB}, the probability for a successful litho-panspermia event, is calculated as the product of the following factors:

P_{biz} = probability that the impactor hits a biologically inhabited zone;

P_{ee} = probability that a rock from the biologically inhabited zone is ejected onto an escape orbit;

P_{sl} = probability that an organism ejected with the rock material survives the launch;

P_{ss} = probability that an organism in the ejecta survives in space;

P_{se} = probability that an organism in the ejecta survives entry through the atmosphere of the target planet;

P_{si} = probability that an organism in the ejecta survives the impact onto the surface of the target planet;

P_{rel} = probability that the organism is released from the meteorite;

P_{st} = probability that the environment of the target planet is not toxic to the organism;

P_{sp} = probability that the organism survives potential predator attacks from the target biosphere (if the latter exists at all);

P_{efg} = probability that the environment is favorable for growth and development of the organism;

P_{sc} = probability that the organism and its descendants compete successfully with the indigenous biosphere.

Several of these factors have been studied in space experiments or in the laboratory simulating conditions in space. Shock recovery and ballistic experiments with microorganisms have shown that bacterial spores can survive a simulated meteorite impact, at least to a certain extent (Horneck, Stöffler et al. 2001). The probabilities of microorganisms surviving the complex interplay of the parameters of space (e.g., vacuum, UV and ionizing radiation, temperature extremes) when traveling in space over extended periods of time have been assessed from experiments in space, at simulation facilities on ground, as well as from model calculations. Mileikowsky et al. (2000) concluded that radiation-resistant microbes could survive a journey from one planet to another in our solar system if they were located inside a meteorite, and thus shielded against cosmic radiation. If travel times of one to several million years are assumed — as deduced from the analyses of the Martian meteorites — a shielding thickness of 1 m or more is required (Mileikowsky et al. 2000). However, viable transport between solar systems by litho-panspermia, assuming impact ejection as the first step, does not seem to be possible (Horneck et al. 2002). Recent simulation studies indicate, on the other hand, that the Earth was created in the perfect time window for galactic infection with life (von Bloh et al. 2003; Cuntz et al. 2003); thus it may be worthwhile to pursue Arrhenius's intriguing idea further.

Human Spaceflight

The history of human spaceflight reaches back some forty years to April 12, 1961, when the spacecraft Vostok 1, carrying Yuri Gagarin, was launched to orbit the Earth. Since the environment in the Earth's orbit is not habitable per se, spaceflight entails many life-threatening risks. To survive in space, humans must carry their own life-support systems, either within the spacecraft itself or individually (e.g., during extravehicular activity). Even so, humans have not been able to escape all adverse conditions of spaceflight. Above all, cosmic radiation and weightlessness appear to impair human health and efficiency in space. What then has motivated humankind to leave the habitable regimes of its home planet and to expose some members to the hostile conditions of outer space?

In the beginning, competition between two large political blocks provided strong motivation for human space enterprises. In the 1960s, a race ensued between the United States and the Soviet Union, as each strived to achieve "firsts" in human spaceflight. This race culminated on July 20, 1969, when Neil Armstrong took the first step on the surface of the Moon during the Apollo 11 mission. Altogether the Apollo program placed 12 men on the lunar surface, and each landing was hailed as a major national event (Johnston et al. 1975). Were political and ideological confrontations, however, the sole driving force behind human spaceflight, or were other factors involved in our attempt to explore space? Clearly, other aspects of science, technology, culture, and economy contributed — and continue to do so — to inspire humankind to reach behind the boundaries of Earth in pursuit of, above all, the Moon and the terrestrial planet Mars (Horneck et al. 2003).

Scientific Aspects of Human Exploratory Missions

The Moon represents a crucial research outpost for humankind to advance the following fields:

1. Science *of* the Moon, including its geophysical, geochemical, and geological aspects, to lead to a better understanding of the origins and evolution of the Earth–Moon system;
2. Science *from* the Moon, which takes advantage of the stable lunar surface, the atmosphere-free sky, and the radio-quiet environment, to allow us to perform superb astronomical observations, particularly on the far side of the Moon; and
3. Science *on* the Moon, with a special emphasis on studies of the stability of biological and organic systems under hostile conditions (especially strong radiation), on the regulation of autonomous ecosystems, and on the cognitive and biomedical preparation of human missions to other planets of the solar system (Balsiger et al. 1992).

Mars is also a major target in the search for life since it is the only other planet besides Earth that is located within the habitable zone of our solar system (see

Franck and Zavarzin, this volume). Dry river beds indicate that huge amounts of water and a denser atmosphere were present there about three billion years ago. During this warmer and wetter period, life may have originated on Mars and may even subsist today in special "oases" or "ecological refuges" (e.g., geological formations below the surface with favorable conditions for life). The published, but controversial, discoveries of possible fossil life-forms in Martian meteorites may warrant optimism. Therefore, the search for morphological or chemical signatures of life or its relics is one of the primary and most exciting goals of Mars exploration (Brack et al. 1999; Nealson 1999; Westall et al. 2000). In addition to exobiology, disciplines such as geology, mineralogy, and atmospheric research play a central role in the scientific exploration of Mars. The general goal is to understand planetary formation and evolution processes including, if possible, the evolution of life itself. Mars is the planet most similar to Earth, and the question of climatic changes, especially the loss of water and atmospheric gases, is fascinating. There is good reason to think that the study of the evolution of the Martian climate will also contribute to the understanding of the history and future of the terrestrial climate. Furthermore, the study of the evolution of Mars may even contribute to the understanding of the evolution of the whole solar system. Human spaceflight may be important in this research context, since a number of items have been identified where the action of astronauts *in situ* could be beneficial to reach the scientific goals: site identification by local analysis, sample acquisition at these sites, and, if a laboratory is available on Mars, supervision of sample analysis (Brack et al. 1999).

Technological Aspects of Human Exploratory Missions

It has frequently been observed that new and demanding situations require new technologies and that this need spurs on development. Moving humans away from their home planet and establishing a new habitable environment on the Moon or on Mars may pose such a new and demanding situation.

The Moon is a suitable test bed for technologies to be applied in future space exploration initiatives, such as a mission to Mars. Numerous aspects of the environments on the Moon and Mars are similar — above all, the radiation field, thin atmosphere, reduced gravity, and dust. Even assuming a significant improvement in space technology in the coming years, intermediate steps, such as a manned mission to the Moon, would be extremely useful in preparing a journey to Mars. In addition, there are speculations about the Moon providing resources for on-site exploitation, such as oxygen for life support and propellant production, or even ^{3}He for fusion reactors.

Meeting the scientific objectives of a Mars mission will require autonomous and smart tools, such as intelligent sample selection and collection systems on a very high level of automation and robotics. As soon as human travelers are involved, the need for integrated advanced sensing systems will become obvious, such as for biodiagnostics, medical treatment, and environmental monitoring and control. Furthermore, the development and test of technologies for *in-situ* resource utilization (for producing propellant from atmospheric CO_2 or from water ice, but also for life-support purposes as mentioned above) may turn out to be a powerful technological stimulus.

Cultural Aspects of Human Exploratory Missions

The Moon, as the natural companion of the Earth, has substantially shaped conditions on our planet to make it habitable and to sustain a biosphere. It can also be considered as a natural space station orbiting the Earth. Human missions to the Moon have already moved from the realm of science fiction writers, like Jules Verne's *From the Earth to the Moon*, to reality. They seem to meet the natural human need to explore and expand to new regions. Hence, establishing a first habitable outpost beyond the Earth on a natural body of our solar system might be an important cultural event. By redirecting our view from an Earth-oriented toward a more Universe-oriented one, this may even contribute to solving terrestrial conflicts. A lunar base initially established for scientific and technological purposes might develop later into an attractive object for tourism. Once relatively cheap access to space becomes available, space tourism might become a major business.

Of the terrestrial planets, Mars is by far the most attractive and fascinating one. When the first telescopes were directed toward Mars, channels and shaded areas were interpreted as huge agricultural plantations or lichens covering the surface. This latter conception was not ruled out until the two Viking spacecrafts landed on Mars in 1976 and encountered a hostile and chemically highly reactive surface. However, Viking and the follow-on missions also revealed that the early Mars had probably experienced a climate similar to that of the early Earth, when life started here. Hence, Mars has been considered as a suitable and attractive target for terraforming, for instance, by using modern techniques of planetary and genetic engineering (McKay et al. 1991).

After the realization of the International Space Station (ISS), human exploratory missions to the Moon or Mars, that is, beyond the Earth orbit, may be the next step in the human desire to conquer the outer limits of habitability. Does human spaceflight have the potential of promoting peaceful cooperation on a global scale? The ISS is the first example of an international cooperative venture for the joint development, operation, and utilization of a permanent space habitat in low Earth orbit, involving nearly all space-faring nations. Hence, with the ISS, a new era of peaceful international cooperation in space has started. Major partners are the U.S.A., Russia, Japan, Europe, and Canada, with the U.S.A. taking the leading role. Lessons learned from this experience may help all nations to be engaged in future large international space projects, creating harmony through common scientific endeavor.

Economic Aspects of Human Exploratory Missions

Human exploratory missions targeting the Moon or Mars are very expensive. The cumulative costs of a human Moon program, lasting for about twenty years, have been estimated at 50 billion US$, with a mean annual budget of about 2.5 billion US$ (Reichert et al. 1999). For an exploratory Mars program, comprising three human missions over twenty years, nearly the same cumulative costs have been estimated: 52 billion US$ with an average budget per year of 2.6 billion US$. On the other hand, a comparable annual budget is presently spent by the U.S.A. for the operation of the Space Shuttle fleet alone. Hence, a lunar base or a Mars program might become financially affordable, if it is carried out following the construction phase of the ISS, especially if it is done through an international cost- and task-sharing scheme. Synergies with terrestrial applications are expected in various fields, such as nanotechnology, autonomous health control systems, and/or telemedicine systems, to mention just the biomedical aspects. However, even considering the possibility of *in-situ* resource exploitation and the faint prospects of a future boom in lunar tourism, it will require a major investment which probably cannot be provided by just one nation.

FORCED MIGRATION DUE TO ENVIRONMENTAL DEGRADATION

Thus far, we have only briefly touched upon the environment, which is actually one of the overriding factors involved in pushing or pulling all sorts of living entities across space. Note that paleontologists, archaeologists, historians, ethnologists, biogeographers, and other researchers have piled up overwhelming evidence to suggest that species, ecosystems, and civilizations have been forced by regional environmental change to move for survival throughout the Earth's distant and recent past. Some of these moves — like the migration of European plants and animals to benign refuges (such as Sicily or the Iberian peninsula) during the last glaciation ten thousands of years ago (Frenzel 2002) — were caused by natural variability, whereas others — like the exodus of U.S. farmers from the Midwest during the "Dust Bowl" era of the last century (WBGU 1994) — were caused by socioeconomic mismanagement. Almost certainly, however, both physiogenic and anthropogenic factors are involved in most human migration processes, and it is rather difficult to disentangle these factors by careful analysis, even in contemporary cases such as the Sahel catastrophe, which emerged in the early 1970s. Erudite speculations about the driving forces behind the Germanic *Völkerwanderung* mentioned earlier, the disappearance of the Maya and Anasazi cultures, or the Mongolian invasion of Eurasia during the thirteenth century, for instance, still lack a fully convincing empirical foundation relying on high-quality data for an entire list of key indicators (e.g., soil erosion rates, frequency of extreme events, average nutrition status, or institutional coping capacity). Yet there are a number of convincing case analyses available now (see, e.g., Pringle 1997) and the overall research field is moving forward quickly (for a recent review, see Redman et al. 2004).

In the context of *Earth System Analysis for Sustainability* (see, e.g., Schellnhuber and Wenzel 1998), that is, the grand topic of this Dahlem Workshop, current and future environmental migration at the planetary scale — and, in fact, beyond that scale — needs to be considered. Anthropogenic global change, as epitomized by the ongoing transformation of the Earth's atmosphere (Crutzen 2002), creates two qualitatively different tangles of driving forces that cause people, animals, and plants to leave their home grounds and to invade remote domains:

1. The industrial metabolism that supports humanity's modern lifestyle is altering many important environmental conditions in a continuous, if not insidious, way. Life is, in principle, highly adaptive to such "regular" changes (e.g., in temperature, humidity, salinity, acidity), but there are limits, depending on the physiological characteristics of the individual organism affected as well as on the ecosystem structure or the social fabric into which this organism is embedded. The notorious question about the "carrying capacity" of a specific environmental entity (e.g., a forest, a biome, or the entire biosphere) addresses this very problem. A striking observation in this context is that the scientific assessment of the number of human passengers that the total Earth system might be able to support, does not seem to converge to a robust value thus far (e.g., Cohen 1995).

2. Humanity-induced global change may trigger several "singular" events, changing regional or even planetary environmental conditions abruptly and significantly, thus causing, in a domino-like fashion, singular impacts in the anthroposphere (e.g., pandemics, market crashes, or political destabilization of entire countries). In particular, the current human interference with the climate system may well trigger a number of disruptions within the planetary machinery, such as the shutdown of thermohaline ocean circulations and disturbance of other geophysical teleconnection patterns (Schellnhuber and Held 2002; Alley et al. 2003). In addition, however, major phase transitions in the biosphere (e.g., the emergence of novel, highly invasive virus families) may be already underway in response to subtle nonlinear processes induced by our civilization. Nobody has really looked yet into this frightening matter with the scientific dedication it deserves. Thus, there may or may not be huge icebergs below the tips that we now perceive, or there may be icebergs that we never even thought of encountering.

Both types of global change stresses, that is, regular and singular ones, will heavily affect the distribution of life on this planet in the decades and centuries to come. Whether there are already now millions of environmental refugees on the run, as UN organizations keep on claiming, remains to be corroborated by research; however, there is little doubt that millions of people will ultimately follow the shifting regional climates in the wake of global warming, or will yield to large-scale "natural" threats, such as sea-level rise. Note that recent climate simulations (Gregory et al. 2004) indicate that the Greenland ice sheet will almost certainly melt down under the radiative forcing of future greenhouse gas emissions caused by our civilization. This process alone will raise the oceans by about 7 m and inundate large parts of the densely populated coastal areas of the planet. Thus, a planetary reallocation of habitats is going to occur with all of its (mostly unpleasant) repercussions and reverberations.

Will it ever be necessary to export human beings to habitats beyond the Earth, for instance, to terraformed moons and planets inside (or even outside?) our solar system? In contradistinction to the considerations made above, where spaceflight for individuals was perceived almost as a sophisticated caprice of a mature civilization, what we are now talking about is compulsory stellar mass transportation! This is, in fact, the hypothetical "exit option for terrestrial mismanagement," highlighted by the title of this chapter, and is often put forward by techno-optimists to cope with the apocalyptic environmental scenarios painted by culture pessimists. Although a thorough feasibility study for future transmigration into space is beyond the scope of this chapter and therefore requires a separate platform, we can anticipate here the main commonsense conclusion:

Whatever the future holds for scientific, technological, and logistical progress in human spaceflight, it seems crystal clear that it is much wiser to try to preserve the terrestrial environment rather than to first ruin the planet (e.g., by a runaway greenhouse process) and then to send its inhabitants into outer space. In actuality, a tiny fraction of the money needed for interplanetary mass transportation could be used instead to purchase global sustainability through emission and imission control, waste management, resource protection, participatory lifestyle coaching, etc.; and a tiny fraction of the governmental and organizational power needed for interplanetary mass transportation could instead establish international environmental agreements that really make a difference (thus transcending the rather weak Kyoto Protocol).

To illustrate our point, we provide a back-of-the-envelope estimate of the orders of magnitude of some crucial factors involved in comparing the "exit option" with the "sustainability option": Imagine that an environmental apocalypse would force humanity to leave *Terra* during the 22nd century. This would incur costs of roughly 10^{16} US\$, assuming the evacuation of about 10^{10} people and transport costs of 10^{6} \$ per stellar refugee (an extremely optimistic assessment). By way of contrast, our civilization could decide to stabilize the global climate in a tolerable state during this century. Although recent macroeconomic calculations allowing for policy-induced innovative dynamics (Edenhofer et al. 2004) support more optimistic estimates, let us assume that the necessary mitigation activities (as described, for instance, in WBGU 2004) would cost 3% of global GDP over the next 100 years. This would amount to an overall investment of roughly 10^{14} \$. (Note: All costs are adjusted to present economic values in this hyper-simplistic analysis.) The bottom line is that the sustainability option is cheaper by a factor of 100 than the exit option! Accounting for the avoided damages as a consequence of environmental policy (such as the preservation of human lives and ecosystems) renders the balance even more unfavorably for the exit option.

Therefore, the best advice appears for us to remain on Earth, to consolidate our efforts to secure a sustainable existence on this planet, and to use space travel to expand our quest in search of who we are and from where we originated.

REFERENCES

Alley, R.B., J. Marotzke, W.D. Nordhaus et al. 2003. Abrupt climate change. *Science* **299**:2005–2010.

Arrhenius, S. 1903. Die Verbreitung des Lebens im Weltenraum. *Die Umschau* **7**:481–485.

Balsiger, H., J. Beckers, A. Carusi et al. 1992. Mission to the Moon. Report of the Lunar Study Steering Group, ESA SP-1150. Noordwijk: ESA-ESTEC.

Blumberg, B.S. 2002. Introduction. In: Astrobiology, the Quest for the Conditions of Life, ed. G. Horneck and C. Baumstark-Khan, pp. 1–4. Berlin: Springer.

Brack, A., B. Fitton, and F. Raulin. 1999. Exobiology in the Solar System and the Search for Life on Mars. ESA SP-1231. Noordwijk: ESA-ESTEC.

Clark, B.C. 2001. Planetary interchange of bioactive material: Probability factors and implications. *Orig. Life & Evol. Biosphere* **31**:185–197.

Cohen, J.E. 1995. How Many People Can the Earth Support? New York: Norton.

COSPAR (Committee of Space Research). 1984. Internal Decision No. 7/4, Promulgated by COSPAR letter 84/692-5.12-G. 18 July 1984.

COSPAR. 1994. Decision No. 1/94. *COSPAR Info. Bull.* **131**:30.

Crick, F.H.C., and L.E. Orgel. 1973. Directed Panspermia. *Icarus* **19**:341–346.

Crutzen, P.J. 2002. Geology of mankind — The Anthropocene. *Nature* **415**:23.

Cuntz, M., W. von Bloh, C. Bounama, and S. Franck. 2003. On the possibility of Earth-type habitable planets around 47UMa. *Icarus* **162**:215–222.

DeBakey, M.E. 2000. Foreword. In: Challenges of Human Space Exploration, ed. M. Freeman, pp. XI–XII. Chichester: Springer and Praxis.

DeVincenzi, D.L., P.D. Stabekis, and J.B. Barengoltz. 1983. A proposed new policy for planetary protection. *Adv. Space Res.* **3**:13.

DeVincenzi, D.L., P.D. Stabekis, and J.B. Barengoltz. 1994. Refinement of planetary protection policy for Mars missions. *Adv. Space Res.* **18**:314.

Edenhofer, O., N. Bauer, and E. Kriegler. 2004. The impact of technological change on climate protection and welfare. Insights from the MIND model. *Ecol. Econ.*, in press.

Frenzel, B. 2002. History of the flora and vegetation during the Quaternary. In: Springer Series Progress in Botany, vol. 63, ed. K. Esser, U. Lüttge, W. Beyschlag, and F. Hellwig, pp. 368–385. Berlin: Springer.

Gamble, C., and C. Stringer. 1997. Potential fossil hominid sites for inscription on the World Heritage List — A comparative study. ICOMOS.

Gregory, J.M., P. Huybrechts, and S.C.B. Raper. 2004. Threatened loss of Greenland ice-sheet. *Nature* **428**:616.

Horneck, G., R. Facius, M. Reichert et al. 2003. HUMEX, a Study on the Survivability and Adaptation of Humans to Long-duration Exploratory Missions. ESA SP 1264. Noordwijk: ESA-ESTEC.

Horneck, G., C. Mileikowsky, H.J. Melosh et al. 2002. Viable transfer of microorganisms in the solar system and beyond. In: Astrobiology, the Quest for the Conditions of Life, ed. G. Horneck and C. Baumstark-Khan, pp. 57–76. Berlin: Springer.

Horneck, G., P. Rettberg, G. Reitz et al. 2001. Protection of bacterial spores in space, a contribution to the discussion on Panspermia. *Orig. Life & Evol. Biosphere* **31**:527–547.

Horneck, G., D. Stöffler, U. Eschweiler, and U. Hornemann. 2001. Bacterial spores survive simulated meteorite impact. *Icarus* **149**:285–193.

Hoyle, F., and C. Wickramasinghe. 1978. Lifecloud: The Origin of Life in the Universe. London: J.M. Dent and Sons.

Johnston, R.S., L.F. Dietlein, and C.A. Berry. 1975. Biomedical Results of Apollo. NASA SP-368. Washington, D.C.: NASA.

Kasting, J.F., H.D. Holland, and L.R. Kump. 1992. Atmospheric evolution: The rise of oxygen. In: The Proterozoic Biosphere, a Multidisciplinary Study, ed. J.W. Schopf and C. Klein, pp. 159–163. Cambridge: Cambridge Univ. Press.

McKay, C.P., O.B. Toon, and J.F. Kasting. 1991. Making Mars habitable. *Nature* **352**:489–496.

Melosh, H.J. 1985. Ejection of rock fragments from planetary bodies. *Geology* **13**:144–148.

Mileikowsky, C., F.A. Cucinotta, J.W. Wilson et al. 2000. Natural transfer of viable microbes in space. Part 1. From Mars to Earth and Earth to Mars. *Icarus* **145**: 391–427.

Nealson, K.H. 1999. Post-Viking microbiology: New approaches, new data, new insights. *Orig. Life & Evol. Biosphere* **29**:73–93.

Pringle, H. 1997. Death in North Greenland. *Science* **275**:924.

Redman, C.L., S.R. James, P.R. Fish, and J.D. Rogers, eds. 2004. The Archaeology of Global Change: The Impact of Humans on Their Environment. Washington, D.C.: Smithsonian Institution Press.

Reichert, M., W. Seboldt, M. Leipold et al. 1999. Promising Concepts for a First Manned Mars Mission. ESA Workshop on Space Exploration and Resources Exploitation (ExploSpace), Cagliari, Sardinia, Italy, Oct. 1998. ESA Report WPP-151.

Rummel, J.S. 2002. Report on COSPAR/IAU Workshop on Planetary Protection. Paris: COSPAR.

Schellnhuber, H.J., and H. Held. 2002. How fragile is the Earth system? In: Managing the Earth: The Eleventh Linacre Lectures, ed. J. Briden, and T. Downings. Oxford: Oxford Univ. Press.

Schellnhuber, H.J., and V. Wenzel, eds. 1998. Earth System Analysis. Berlin: Springer.

von Bloh, W., S. Franck, C. Bounama, and H.J. Schellnhuber. 2003. Maximum number of habitable planets at the time of Earths's origin: New hints for Panspermia? *Orig. Life & Evol. Biosphere* **33**:219–231.

WBGU 1994. World in Transition: The Threat to Soils. German Advisory Council on Global Change (WBGU). Bonn: Economica.

WBGU 2004. Renewable Energies for Sustainable Development: Impulses for Renewables 2004. Berlin: WBGU.

Westall, F., A. Brack, B. Hofmann et al. 2000. An ESA study for the search for life on Mars. *Planet. Space Sci.* **48**:181–202.

Back: Siegfried Franck, Ken Caldeira, Heike Zimmermann-Timm, Tim Lenton, and Gyorgy Zavarzin

Front: Eörs Szathmáry, Frances Westall, Elke Rabbow, John Schellnhuber, and Gerda Horneck

6

Group Report: Long-term Geosphere–Biosphere Coevolution and Astrobiology

T. M. LENTON, Rapporteur

K. G. CALDEIRA, S. A. FRANCK, G. HORNECK, A. JOLLY,
E. RABBOW, H. J. SCHELLNHUBER, E. SZATHMÁRY, F. WESTALL,
G. A. ZAVARZIN, and H. ZIMMERMANN-TIMM

ABSTRACT

This discussion group attempted a qualitatively new synthesis of long-term geosphere–biosphere coevolution, with the aim of understanding and presenting to the other groups the broadest possible context in which to consider Earth system analysis for sustainability. This included the prospects for detecting life and intelligence elsewhere in the Universe, as debated by astrobiology. The chemoton model of life comprising three autocatalytic subsystems (boundary, metabolic, genetic) was adopted. The topology of evolution was characterized as a network in the prokaryote realm and as a tree (or bush) in the eukaryote realm. It was agreed that prokaryotic life is common in the Universe but that eukaryotic life is rare and intelligent life is extremely rare. The appearance of intelligent life on a planet might theoretically involve four or five difficult evolutionary transitions along the way. These probably include the origins of the genetic code, of oxygenic photosynthesis, of eukaryotes, and of language. Optimistic and pessimistic scenarios for the long-term coevolution of the geosphere–biosphere were contrasted. A key finding was that dating of the major transitions in evolution and, to a lesser extent, dating of the major transitions in the state of the environment are subject to large error bars that need to be reduced in order to address the causal relationships of coevolution. A major output was a visualization of a time line of coevolution that includes these uncertainties. New suggestions of coevolutionary connections were also made. The feasibility of unequivocal life detection on extrasolar planets was questioned, but it was recognized that astrobiology is already encouraging a useful broadening of Earth system analysis. The failure of the search for extraterrestrial intelligence (SETI) and the apparent difficulty of the transition to natural language support the view that intelligence (or at least natural language) is extremely rare in the Universe. Habitation was defined as a first-order influence of life on the geochemical cycling of a planet, and it may be important for the maintenance of habitability. Theoretical considerations as well as Earth history suggests that there are limits (albeit rather broad ones) on how globally destructive life can become. A proposal was made to extend an existing model of global coevolution to address this and other Gaia questions. An "autocatalytic Gaia" hypothesis was put forward to suggest that autocatalytic recycling is an almost inevitable planetary phenomenon, once there is life. This is a natural extension of the autocatalytic theory of life (the chemoton model). Some broad lessons of sustainability can be learned from Gaia and the unfolding coevolution of life and its environment on Earth, in particular, the importance of avoiding long time lags for maintaining system stability.

INTRODUCTION

> Co-Causality
> Of Gaia
> You're the only theory
> I adore
> But save me
> Saints Darwin and Dawkins
> From asking "What is coevolution for?"
> — Alison Jolly

The word "evolution" comes from the Latin *evolutio* and means to roll out or unfold. Our group explored the unfolding coevolution of life and the Earth and how things might unfold elsewhere. The discussion was richly nonlinear and followed a quite different order to that imposed in this report. Here we attempt a qualitatively new synthesis rather than a review of the state-of-the-art, which can be found in other chapters of this volume and the recent pertinent literature.

ORIGINS AND EVOLUTION OF LIFE

Our discussions of the origins and evolution of life focused on three fundamental questions: What is life? What is the topology of life's evolution? Is life an unavoidable cosmic phenomenon?

What Is Life?

There is no consensus definition of life; however, there is agreement on the basic characteristics of life. As we know it, all life is based on coupled oxidation–reduction reactions that are far from thermodynamic equilibrium and occur in metastable domains (Schrödinger 1944). As such, on long timescales, biogeochemistry of a planet requires an open, thermodynamic system. Life is self-replicating and adaptive, and contains synthetic machinery that produces polymers. Life-forms are bounded. On Earth, all life-forms catalyze phase-state transitions in which gases in the environment are converted to solutes and vice versa.

A possible strategy to define life is as follows:

1. Assume a tacit (instinctive) knowledge of life.
2. Viruses must be excluded because they do not metabolize or propagate on their own (and hence are not organisms).

3. Abandon purely verbal criteria for life.
4. Look for minimum models of living *systems*.
5. Abstract essential properties from the simplest living systems of the present day: bacteria.
6. Make the simplest possible abstract model.
7. Ask what kinds of real chemistry can manifest the abstract system.

This strategy hinges on Gánti's seminal contribution to understanding the principles of the living state (Gánti 1979, 2003). He observed that for the living state, reproduction is neither necessary nor sufficient. Many cells and organisms are commonly regarded as being alive even if they cannot reproduce (any longer). So-called potential life criteria must be met only if the population of units is to be maintained and evolved. The correct relation, then, between units of evolution and units of life is that of two, partially overlapping, sets. Gánti's model of a *minimal* living system, satisfying also the potential criteria, is illustrated in Figure 6.1. The chemoton is a chemical supersystem composed of three autocatalytic subsystems: a metabolic network, a replicating template, and a boundary membrane. Stoichiometric coupling among them ensures regulated reproduction of the system as a whole. Spatial reproduction happens because the growth of membrane surface outstrips the increase in mass of internal material, since a doubling of the surface area of a sphere requires more than a doubling in its volume. It is important to emphasize that the membrane is also autocatalytic: building block T, produced by the metabolic network, is spontaneously inserted by virtue of the fact that there is a preexisting membrane surface. It is, however, the presence of the template molecule pV_n that qualifies this system as a unit of evolution (with unlimited hereditary potential). If one supposes that these templates are the abstract versions of ribozymes (RNA molecules acting as enzymes), then they catalyze steps of the metabolic cycle and membrane growth using their inherited information.

The chemoton model is useful because it combines two traditions: the genetical and the systems theoretical approaches to the problem of defining life. The concept of discrete life rules out "soup" ideas. Our model of what is life affects how we go about detecting life, either on Earth (in the rock record) or elsewhere (astrobiology). Traditionally, the geologist's approach has been to look for morphology (which is potentially flawed), for signs of redox chemistry, and more recently for biomarkers. This is in line with the "order-from-disorder principle" (Schrödinger 1944) for describing the entropy production in the environment of life. Our model of what is life also has implications for what are viewed as the major transitions of life. These issues recur later in the report.

Evolutionary Line: Bush, Tree, or Network?

A productive example of "physics babble" being applied to biology was the question: What is the topology of life's evolution? Although outmoded in

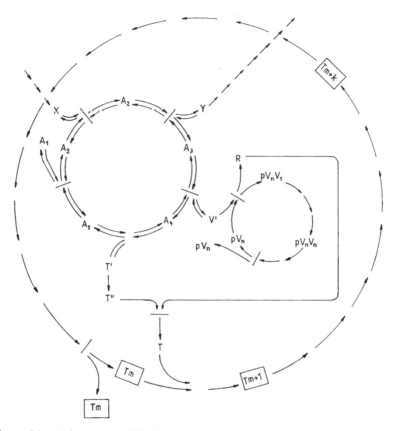

Figure 6.1   A chemoton model of a minimal living system. The chemoton (Gánti 1979) is a chemical supersystem composed of three autocatalytic subsystems: a metabolic network, a replicating template, and a boundary membrane. The metabolic subsystem, with intermediates A_i, is an autocatalytic chemical cycle, consuming X as nutrient and producing Y as waste material. Template molecule pV_n is a polymer of n molecules of V, which undergoes template replication; R is a condensation byproduct of this replication, needed to turn T' into T, the membranogenic molecule. T_m represents a bilayer membrane composed of m units made of T molecules. Building block T is spontaneously inserted by virtue of the fact that there is a preexisting membrane surface. The chemoton system can grow and divide spontaneously.

science, some popular conceptions still see evolution as a line, often progressing toward us. This is untenable. At the opposite end of the spectrum is the notion of an evolutionary network or tangle to describe the prokaryote realm. "Down there" frequent transfer of genetic information means that the various lines of descent may often have connections between them. In contrast, in the eukaryote realm, there is little horizontal gene transfer and an evolutionary "tree" or "bush" may reasonably describe the divergence of lineages. A "tree" can appear as a result of bottleneck event(s).

In our discussions, the concept of a Last Universal Common Ancestor (LUCA) was criticized and even described as a "phantom." A glimpse at a typical tree of life drawn from LUCA at the center can create the erroneous impression that early life was not diverse. However, LUCA is really just an operational idea — the source of all the genetic information today — and it may more realistically represent a community that freely exchanged genetic information. There was much additional genetic diversity in that early population that has just not made it to the present.

Is Life an Unavoidable Cosmic Phenomenon?

The likelihood of the origin of life is difficult to assess, not least because our existence is predicated upon it. This is a problem for Bayesian inference. If the interval to the first prokaryote life on Earth is short, then this suggests that prokaryote life is common in the Universe (Carter 1983). Although suggested morphological evidence for life at 3.8 Ga is no longer accepted, our group agreed that prokaryotic life was present by 3.5 Ga (Westall et al. 2001). This is a relatively short time after the Earth became habitable; hence, prokaryotic life is probably common.

If prokaryotic life is relatively easy to evolve, this raises the question: Did life begin once or many times? There is no clear evidence that life started more than once on Earth (if it did it would of course alter the evolutionary topology); however, absence of evidence is not evidence of absence. There are many extant cases of multiple origins within the network/tangle/bush of life, for example of multicellularity (~20 times) and social organization (in insects ~12 times). However, the simplest life we know of on the planet is very complex, hinting that the origin of such life involved at least one difficult transition.

The sequence of major transitions (Szathmáry and Maynard Smith 1995) is summarized in Table 2.1 (Chapter 2, this volume). Of the three transitions that precede prokaryote life, it is the third — the origin of the genetic code — that is thought to have been particularly difficult. Subsequently, the origin of the eukaryotic cell (cells with a well-formed cell nucleus, such as our cells) and the emergence of language are good candidates for further difficult transitions. A transition can be difficult for either or both of the following reasons: limitation by genetic variation or by natural selection. A variation-limited transition is difficult if the set of the requisite genetic variations is very unlikely to arise. Limitation by selection means that the right conditions for the spread of the appropriate genetic variation is very special and unlikely. A subcategory within selection-limited transitions is the case of "preemption": though the first transition by itself is not so difficult, it modifies conditions to such an extent that a second, independent trial becomes virtually impossible (Szathmáry 2004).

With the premise that there are certain bottlenecks in evolution that represent "difficult" transitions, one can calculate how many such transitions there might be in the history of our planet until observers arise. Carter (1983) argued that if observers find themselves arising about half way through the habitable life span of a planet, then there were probably two really difficult transitions to get to observers. Updating this, Andrew Watson noted that humans arise about four-fifths of the way through the habitable region of the planet, suggesting when one follows Carter's reasoning that there are four or five difficult transitions. In addition to the three mentioned above (origins of the genetic code, eukaryotes, and language), the origin of oxygenic photosynthesis may have been a difficult transition.

LONG-TERM COEVOLUTION OF THE GEOSPHERE–BIOSPHERE

We adopted the basic premise of a coupling between the evolution of the biota and of the global environment. A number of hypotheses for connections between the evolution of life and of the environment are discussed by Lenton, Caldeira, and Szathmáry, Westall and Drake, and Franck and Zavarzin (this volume). A key outcome of this workshop was the realization that dating of the major transitions in evolution and, to a lesser extent, dating of the major transitions in the state of the environment are subject to large error bars. These need to be reduced (see final section) to address the potential causal relationships of coevolution.

A summary diagram of the time line of coevolution, with age error bars and arrows from one side to the other indicating necessary conditions or potential causal connections, is presented in Figure 6.2. Ages are given in billions of years (giga annum: Ga). This forms the basis for the following sections, where we set the geological scene for the emergence of life on Earth. Thereafter we offer two alternative narratives for the development of life and of the environment that highlight the current degree of uncertainty: (a) an optimistic narrative, which takes the earliest dates for various major transitions in the evolution of life; (b) a pessimistic storyline, taking the latest dates for major transitions in evolution.

The Earth system cannot be considered in isolation. Instead it is embedded within the dynamics of our solar system and our galaxy. Examples are impacts by large meteorites, which may result in severe insults to the biosphere with mass extinctions (e.g., the K/T event), or supernova explosions, which may cause a depletion of the stratospheric ozone layer and thus increase the flux of solar UV radiation at the surface of the Earth.

The Planetary Setting

Planet Earth formed 4.56 Ga ago and the Moon was formed in an impact event 4.52 Ga ago. There are few rocks dating from 4.5–4.0 Ga and those in the range 4.0–3.5 Ga are highly metamorphosed. This meager record, however, tells us

that there was water on the surface of the Earth 4.4–4.2 Ga and that some fractionation of hydrated crust took place to form protogranitic material. The early
subaerial portions of the crust probably resembled Iceland and were rapidly recycled. They were not continents with broad continental shelves. Evidence suggests that up until the Moon's Late Impact Cataclysm at ~3.85 Ga, planet-steril-
izing impacts repeatedly rendered the surface of the Earth inhospitable for life.
However, there is one model that suggests that, even during the largest impacts,
only about the uppermost 400 m of water from the oceans were vaporized
(Ryder 2003). There were continued impacts after 3.85 Ga, notably at 3.47 Ga
and 3.26–3.24 Ga, and these were 10–100 times more massive than the K/T
event (Byerly et al. 2002; Kyte et al. 2003). Despite these, the Earth was in a relatively protected position within the Solar System, with Jupiter acting as a
"sweeper up" of rogue asteroids. The prebiotic atmosphere was probably composed principally of CO_2 with some N_2, CH_4, and H_2O. Ancient rocks indicate
that temperatures, at least close to the seafloor, were very warm, the latest estimate being $70 \pm 15°C$ (Knauth and Lowe 2003). There is still debate about the
pH of the early oceans.

The oldest, best-preserved rocks come from the Early Archaean formations
of Barberton in South Africa and the Pilbara in NW Australia, aged 3.5–3.2 Ga.
These two relatively small areas of supracrustal rocks appear to have survived
the ravages of plate tectonics because they represent regions which were
underplated by a stabilizing "keel" that protected them from subduction and destruction. Making global extrapolations from just two locations is problematic,
but there is no other material available at present. In terms of continental growth,
important for providing suitable, stable habitats for the evolution of life, the period to about 3.0 Ga saw the production of a more rigid crust through continued
fractionation to produce buoyant granitic material leading to the lateral plate tectonic style that we know. Tectonics caused cratonization whereby supracrustal
masses were thrust together, eventually producing true continents with broad
continental platforms and shallow shelves.

Carbonate formation in the Early to Late Archaean was related to the alteration of freshly formed lavas (and breccias formed by volcanic explosions and
impacts) by seawater, a process called carbonatization, for which there is ample
evidence. Carbonates were also deposited as hydrothermal exhalations. However, it was only from the Late Archaean onwards that large-scale carbonate formations started to appear, because this relied on the development of continental
platforms.

THE OPTIMISTIC VIEW

The following coevolutionary scenario assumes an early date for each of the major transitions in evolution. The scenario is one of a stable development of life
(after the early asteroid bombardment) with occasional environmental crises (of
terrestrial or extraterrestrial origin) giving a spurt to evolution.

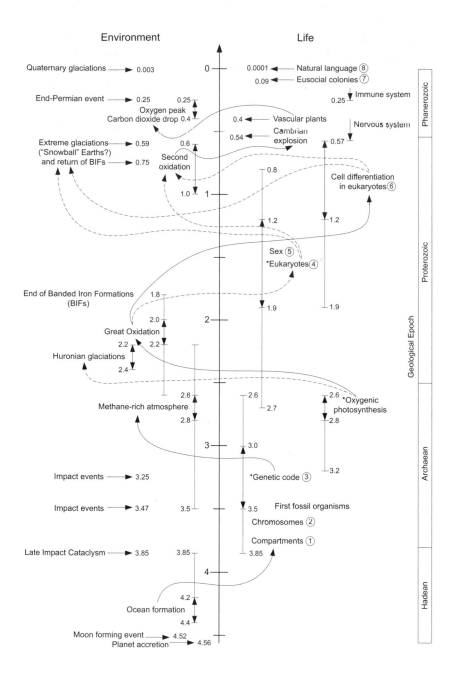

Planetary Stability → Evolution of Life

The evolution of life on Earth was contingent on the acquisition of oceans of liquid water and a relaxation of asteroid bombardment, at least to the point that the oceans were no longer being vaporized. This means that the earliest possible date for the emergence of life was 3.85 Ga. The first *bona fide* evidence for life comes from the rocks from Barberton and the Pilbara, 3.5–3.2 Ga (Westall et al. 2001) and shows that life was already highly evolved, having the same characteristics as modern prokaryotes in terms of cell morphologies, colony formation, biofilm formation on sediment surfaces, and carbon and nitrogen isotopic signature (Beaumont and Robert 1999; Westall et al. 2001). Stromatolites were biologically formed at this time (Byerly et al. 1986), but abiogenic stromatolites also existed (Lowe 1994). It is possible that anoxygenic photosynthesis may already have arisen (filamentous microbial mats in shallow water/littoral environments). There is, however, no mineralogical evidence for oxygenic photosynthesis. Pigments, which evolved originally as UV shields, could have been influential in the development of photosynthesis (Nisbet and Sleep 2001).

Figure 6.2   Time line of geosphere–biosphere coevolution on Earth. The time line runs from bottom to top, starting with the accretion of planet Earth and ending at the present with numbers indicating ages in billions of years (giga annum: Ga). The major geological epochs are indicated in the scale on the right. Left of the time line are major features of and changes in the state of the environment (geosphere), including some perturbations from outside the system. Right of the time line are major transitions in the evolution of life (biosphere), plus some other significant appearances. The major transitions in evolution are given abbreviated descriptions and numbers in circles following the scheme of Szathmáry and Maynard Smith (1995; see Table 2.1, Lenton, Caldeira, and Szathmáry, this volume), oxygenic photosynthesis is included as an additional transition, and asterisks are used to indicate difficult transitions. Two types of age error bar are used where major changes or transitions took sufficiently long or their timing is sufficiently uncertain to be resolvable at this scale. On the "Environment" side, the double arrowed lines indicate the time range over which we are confident that a feature or change was occurring, whereas the dotted lines indicate wider time ranges over which it may have occurred. On the "Life" side, the doubled arrowed lines indicate the time range over which we are confident a particular transition or group of transitions occurred, whereas the dotted lines indicate the absolute earliest and latest times that a particular transition or group of transitions could have occurred. The first three major transitions in the evolution of life are grouped together with the error bars being for the appearance of prokaryotic life with a genetic code. The fourth and fifth transitions are grouped together because sex is an ancestral trait of eukaryotes (i.e., it appeared at the same time). Arrows from one side of the time line to the other indicate historical contingencies (i.e., necessary conditions) or potential causal connections between the evolution of life and changes in the environment or *vice versa*. Solid arrows indicate connections that are better established and/or less controversial. Dashed arrows indicate less well established and/or more controversial connections. Some further connections are speculated upon in the text. For a full explanation of the various changes, transitions, and interconnections, see the text.

Evolution of Life → Methane-rich Atmosphere

A methane-rich atmosphere has been proposed in the late Archaean (~2.7–2.8 Ga) on the basis of observed extreme carbon isotope shifts indicative of methanotrophy (biogenic methane consumption) (Hayes 1994), and to counteract the faint young Sun at the time (Pavlov et al. 2000). The greenhouse effect of methane (CH_4) could have been especially important in the Archaean if one accepts a low limit on CO_2 from paleosols (Rye et al. 1995). In the anoxic Archaean atmosphere, abiotic, volcanic sources could have supported CH_4 < 0.0001 atm, whereas today's biological CH_4 flux would have generated CH_4 > 0.001 atm (Pavlov et al. 2000). The methanogens responsible for this flux today are close to the universal common ancestor. Hence they were probably present soon after 3.5 Ga. Methane could have been a major component of the atmosphere from then on, but methanogens alone would not have left a detectable geochemical signal. That had to wait for the evolution of methanotrophs that oxidize methane with oxygen or sulfate. Their appearance at ~2.7–2.8 Ga, while indicating an abundance of methane, would actually have lowered the methane content of the atmosphere by consuming it.

Oxygen was extremely scarce in the Archaean atmosphere (pO_2 ~10^{-11} atm) prior to the origin of oxygenic photosynthesis. This suggests high UV radiation levels reaching the surface of the Earth that would have had a strong influence on mutation rates in early life. However, the flux of UV radiation would have been mitigated by the vast amount of aerosols and dusts in the atmosphere produced by the volcanic activity, as well as a hydrocarbon smog from the photochemical polymerization of CH_4 in the atmosphere (Pavlov et al. 2001). Moreover, early marine life would have been protected by water and the exposed mats were thick, robust polymer sheets where the outer layers would have protected the living organisms beneath, as happens today in exposed environments.

Oxygenic Photosynthesis → Glaciations and Great Oxidation

The appearance of continental shelves and shallow water platforms over the period 3.2–2.5 Ga provided a suitable environment for the development of oxygenic photosynthesizers. The occurrence of abundant methanotrophy at ~2.7–2.8 Ga indicates that oxygen reached a modest concentration (pO_2 ~0.0005atm) at least locally, which was likely due to the evolution of oxygenic photosynthesis. The first direct evidence for oxygenic photosynthesis comes from chemical biomarkers in the 2.7-Ga Hammersley Basin rocks, Australia (Brocks et al. 1999;Summons et al. 1999). The oldest fossils that strongly resemble colonial cyanobacteria are found in 2.6-Ga formations from the Transvaal in South Africa (Altermann and Schopf 1995; Kazmierczak and Altermann 2002). Large stromatolites, generally thought to be constructed by cyanobacteria, occur in shallow sea, continental platform settings and are common in the Proterozoic.

Oxygen may have risen sporadically from ~2.7–2.8 Ga onwards, culminating in the Great Oxidation at 2.2–2.0 Ga. At some point the oxygen flux to the atmosphere exceeded twice the methane flux and there would have been a catastrophic decrease in the partial pressure of methane causing global cooling. That may explain the onset of the extreme global (Huronian) glaciations 2.4–2.2 billion years ago. During the glaciation, productivity was suppressed and oxygen should have dropped. If a "snowball Earth" occurred, all nutrient input to the ocean and atmosphere–ocean exchange would have been blocked by the presence of a thick ice crust. Continued volcanic input of CO_2 gas into the atmosphere would have gradually increased temperatures until the ice melted. The vast amount of nutrients suddenly released into the biosphere would have led to a bloom of photosynthetic organisms at the surface of the ocean and a sudden increase in the source of oxygen. Perhaps the Earth went through a series of cycles of rising oxygen triggering glaciation, falling oxygen during glaciation, oxygen rising again afterwards, until the Great Oxidation put the system into a more stable state. Much of the O_2 produced (e.g., after glaciation) was taken up by the oxidation of reduced metal species in solution thus potentially explaining the abundance of banded iron formation (BIFs) at the time.

The evolution of oxygenic photosynthesis was a necessary but not a sufficient condition for the Great Oxidation. In addition, the right environment was required for the burial of organic carbon (which provides a long-term net oxygen source). Kerogen could be buried in sedimentary deposits or subducted to the mantle (Lindsay and Brasier 2002). Furthermore, the net input flux of oxygen had to exceed the input flux of reduced chemical species to the ocean and atmosphere (e.g., from volcanoes and methanogenesis) before oxygen could rise.

Oxygenic Photosynthesis, Glaciations, and Great Oxidation → Evolution of Eukaryotes

The origin of oxygenic photosynthesis and the subsequent rise of oxygen in the atmosphere increased the energy available to the biota, aerobic metabolism being about five times more efficient than anaerobic metabolism. This may have increased the overall rate of mutation, such that there was an increased capacity for evolution. The Great Oxidation has often been portrayed as a disaster for the anaerobic biota. However, in the end it may have been advantageous for them, as it greatly increased their supply of food (organic matter).

The appearance of eukaryote life may be linked to local accumulation of oxygen beginning ~2.7 Ga, to the Huronian glaciations 2.4–2.2 Ga, and/or to the Great Oxidation 2.2–2.0 Ga (Evans et al. 1997; Ward and Brownlee 2000; Lindsay and Brasier 2002). Sterols from 2.7 Ga could indicate the presence of eukaryotes (Brocks et al. 1999) but are no longer accepted as unequivocal eukaryotic biomarkers. Eukaryotes are mostly aerobic and have a higher oxygen requirement than aerobic prokaryotes, because the mitochondria are encased within a larger cell that limits oxygen diffusion. One hypothesis is that the compartmentalized eukaryote cell is a response to the accumulation of O_2 in the environment, following the evolution of oxygenic photosynthesis. Eukaryotic algae with photosynthesizing chloroplasts appeared later, but in a most optimistic scenario they could have predated and contributed to the Great Oxidation (Lenton, Caldeira, and Szathmáry, this volume). In particular, planktonic algae provide a much greater potential organic carbon sink than cyanobacteria in mats or open water (Paul Falkowski, pers. comm.).

Whether eukaryotes were a cause or a consequence of the Great Oxidation, the rise of oxygen provided a necessary condition for the later transition of cell differentiation in eukaryotes. The timing of this transition is very uncertain. *Grypania* was originally described as a multicellular eukaryotic algae aged 2.1 Ga (Han and Runnegar 1992). However, the age has now been revised to 1.9 Ga and it is unclear whether *Grypania* is a multicellular eukaryote or prokaryote. The oldest eukaryote that can be assigned to an extant phylum is a red alga aged 1.2 Ga (Butterfield et al. 1990; Butterfield 2000). Trace fossils and possible metazoan fecal pellets are widespread from ~1 Ga onwards.

Multicellular Eukaryotes → Neoproterozoic Environmental Changes

A secondary Neoproterozoic rise in oxygen has been linked to either the origin of animals with guts and a consequent change in sedimentology (Logan et al. 1995), or an early evolution of phototrophic communities on the land surface (Lenton, Caldeira, and Szathmáry, this volume). The latter could also be implicated in cooling the Earth, thus priming it for extreme glaciations ~0.75 Ga and ~0.59 Ga (see Lenton, Caldeira, and Szathmáry, this volume).

Neoproterozoic Environmental Changes → Cambrian Fauna

The secondary rise in oxygen was a necessary condition for the evolution of the Cambrian fauna. The contribution of an increase in calcium content of the oceans to the origin of biogenic calcification in the Neoproterozoic was addressed (*Cloudina* being the first calcifying organism). However, calcium seems unlikely to have been a limiting factor, the evolution of hard shells being attributed instead to protection from predation.

Vascular Plants → Permo-Carboniferous Environmental Changes

It is reasonably well established that the rise of vascular land plants triggered a rise in atmospheric oxygen and an order of magnitude drop in carbon dioxide in the Permo-Carboniferous. The evolution of lignin resistant to biodegradation as well as an increased efficiency of phosphorus weathering from rocks may both have driven a rise in atmospheric oxygen. An amplification of bulk silicate weathering due, for example, to increased depth of the soil profile, rock-splitting by plant roots, and amplification of the hydrological cycle drove a reduction in carbon dioxide.

A PESSIMISTIC SCENARIO FOR THE MAJOR TRANSITIONS IN EVOLUTION

This scenario always assumes the latest possible dating for the appearance of different levels of organization. Such pessimism may be justified by (a) the general underestimation of the difficulty of certain transitions; (b) the contesting of the claim (Schopf 1993) of fossil cyanobacteria 3.5 Ga (a shift of ~1 Ga); and (c) the invalidity of sterols (Brocks et al. 1999) as eukaryotic biomarkers 2.7 Ga (at least 1 Ga shift).

Replicators → Compartments

Fossils from 3.5 Ga indicate that some form of cellular life was present then. A long phase of chemical/replicator evolution is likely to have occurred between late bombardment and this date.

Compartments → Linked Genes (Chromosomes)

Linkage and active segregation of genes is highly advantageous once the mutation rate is low enough. This allows a longer genome with a higher potential for complexity. If 3.5 Ga fossils are genuine, then they are likely to have had chromosomes, even made of DNA.

RNA Organisms → Encoded Proteins (Translation)

It was speculated that many of the earliest fossils could be those of cells without a genetic code, but with actively catalytic genes (Szathmáry, pers. comm.). Whether there was something before RNA (requiring genetic takeover) is unknown, but its existence is possible. If one accepts the earliest unequivocal evidence for cyanobacteria (2.6 Ga), then the divergence of bacteria must have happened before that. Subjectively, the origin of translation is estimated to have happened in cells about 3 Ga (Szathmáry, pers. comm.).

Prokaryotes → Eukaryotes

The major innovation giving a high selective advantage to early eukaryotes was phagotrophy, which also required formation of a cytoskeleton and an endomembrane system, including the nuclear membrane. Thousands of mutations must have been positively selected. Many eukaryotic genes have absolutely no prokaryotic homologue. Archaebacteria and eukaryotes are sister groups. Mitochondria came in very early, just after the origin of primitive phagocytosis. The eukaryotic nucleocytoplasm descended from some Gram-positive bacterium, similar to present-day actinobacteria (with sterols, proteasomes, and histon H1). The loss of the eubacterial cell wall triggered the revolution leading to eukaryotes (with a cytoskeleton) and archaebacteria (with a rigid ether membrane) (Cavalier-Smith 2002).

Eukaryotic fossils must have a clearly identifiable trace of a nucleus, an endomembrane system, and a cytoskeleton (possibly all three together): mere size and shape resemblance to the cell/filament boundaries of present-day eukaryotes do not guarantee anything. The earliest unequivocal eukaryotic evidence is then 0.8 Ga (Porter and Knoll 2000). The eukaryotic revolution may be related to the major glaciations that began ~0.75 Ga. Even if the ancestral eukaryotic host cell was aerophilic by itself, the acquisition of mitochondria offered an advantage at a time when oxygen levels would have dropped because of the inhibition of photosynthesis during extreme glaciation.

A corollary to this is that the first major oxidation (2.2–2.0 Ga) was due to cyanobacterial activity; the second (1.0–0.6 Ga) was due to the late arrival of eukaryotic algae.

Clones → Eukaryotic Sex

Sex increases exchange of genetic information within the species level and is a primary character state for all known eukaryotes. It must have been selected for shortly after the origin of cytosis, in the form of an ancient ploidy cycle with incidental recombination. Genes for meiosis are related in all eukaryotes. Thus sex must have emerged at the latest 0.8 Ga.

Protists → Plants, Animals, Fungi

The origin of complex forms of eukaryotic multicellularity was presumably triggered by oxidation (due to the rise of eukaryotic algae) after the extreme Neoproterozoic glaciations. Once again, only fossils clearly of multicellular eukaryote origin can be accepted: anything that can be interpreted as mere colonial protists must be rejected. This leaves us with 0.57 Ga for metazoan origin. These three kingdoms of plants, animals, and fungi have independently evolved a complex epigenetic inheritance system allowing for a large number of cell types and complicated development.

Solitary Individuals → Eusocial Colonies (Societies)

Whereas reproductive skew is common, eusociality is much rarer. It means reproductive division of labor, cooperative brood attendance, and several generations living together. The first social insects date back to ~92 million years ago.

Primitive Societies (Protolanguage) → Complex Society (Natural Language)

Natural language is unique to humans. It produces a cultural inheritance system with indefinitely large potential. It could have emerged sometime between 100–200 thousand years ago.

NEW CONNECTIONS

A couple of tentative new coevolutionary connections were suggested in our discussions:

1. End-Permian Extinction → Immune System: It was speculated whether the evolution of the immune system occurred in the aftermath of the End Permian extinction.
2. Quaternary Environmental Changes → Intelligence and Syntactic Language: The possibility of links between rapid changes in the environment during the Quaternary and the evolution of intelligence in our species was discussed. This includes the most recent difficult transition (the origin of natural language). Intelligence is a costly innovation in terms of the energy demand of the brain. The question is whether the environment may have changed rapidly enough to make it beneficial to have intelligence? Previous links have been made between Quaternary environmental change and human development, including the influence of 23 ka precession cycles on the monsoon region (Africa, Asia). However, any link to language development is unclear at this stage. If there is a link between rapid environmental change and human evolution, it was noted that our current inducing of global change could stimulate our own evolution. We return to discuss the evolution of intelligence in more depth below.

ASTROBIOLOGY

Astrobiology is a multidisciplinary approach to study the origin, evolution, distribution, and future of life on Earth and in the Universe, based on the assumption that life is a cosmic phenomenon. This includes the study of the formation of the elements, molecules, and processes that are involved in the formation of habitable planets and life. This holistic approach is based on Earth and terrestrial life as analogues for habitable planets and extraterrestrial life. Research includes:

1. The search for extrasolar planets, including potential signatures of life, such as water and oxygen (ozone).
2. Exploration of our solar system with special emphasis on Mars and Europa (or other moons of the giant planets, e.g., Calypso) that are potential candidates for having hosted or hosting still life.
3. Study of ancient life on Earth and potentially in extraterrestrial materials.

4. Studies on Earth of the strategies of microbial communities to adapt to extreme environments as terrestrial analogues for potential extraterrestrial life.
5. Laboratory studies simulating space conditions or conditions of other planets on chemical processes and the survivability of microorganisms under these conditions.
6. Modeling the habitable zones around stars or galactic centers.

Of the more than 100 extrasolar planetary systems discovered thus far, there are two (47 Uma and 55 Cancri) that may in some respect (mass of central star, giant planets at larger distances) be good candidates for systems with Earth-like planets in the habitable zone.

Is Astrobiology Feasible?

It is clear that one can do some "astro"-biology on Earth (points 3–6 above), but what is our capability to detect life on other planets? Two missions are currently being designed to try and detect life on extrasolar planets: Darwin (ESA) and Terrestrial Planet Finder (NASA). Both plan to use a technique of interferometry to cancel out the light from the parent star and thus see any planets in orbit around them. Of particular interest are terrestrial planets (those with hard surfaces, of modest size, in inner orbits). The missions plan to use Lovelock's technique of looking at the absorption spectrum of infrared light coming from such planets to try and deduce their atmospheric composition (Lovelock 1975). If gases coexisting in extreme thermodynamic disequilibrium (e.g., oxygen and methane) could be detected, this would be evidence for the presence of life. Although the detection of ozone (and thus oxygen) appears feasible, simultaneous detection of methane (which has narrower absorption bands in the infrared) does not, at present. Unfortunately, oxygen/ozone alone is not necessarily indicative of life, whereas extreme atmospheric disequilibrium is. Furthermore, the Earth's atmosphere has only been oxidizing for approximately half of its history (the past 2.0–2.2 Ga). If we take this as a basis for designing extrasolar life detection, then the missions should equally look for a reducing atmosphere with traces of oxygen as indicative of life.

What Can Astrobiology Do for Earth System Analysis? (and vice versa)

Modeling of potential extrasolar planets is bringing insights to Earth system analysis and the general question of planetary habitability (e.g., the influence of plate tectonics on habitability). The boundaries of the habitable zone in the solar system are mainly influenced by geodynamics (spreading, subduction, continental growth). It may be that the tectonic style of our Earth is an exception and the typical tectonic style for terrestrial planets is that of Mars (stagnant lid tectonics). The sudden stop of plate tectonics on Earth would result in an interruption of the global volatile cycles on the 100-Ma timescale and in an earlier loss of habitability.

The importance of silicate weathering in the long-term regulation of carbon dioxide and climate (Walker et al. 1981) has become a key mechanism in studying both the Earth system and possible extrasolar planets. It relies on a hydrological cycle and, perhaps not surprisingly, models suggest that water worlds have more chances of being habitable than land worlds. This raised the question of whether water retention on Earth may be tied to atmospheric composition? In particular, is the existence of an effective cold trap and a "dry" upper atmosphere related to the presence of an ozone layer? This is discussed in the section on Gaia below.

Terraforming

While it has the technological capacity, one thing that intelligent life may decide to do is attempt to make other planets or moons habitable. Such activity is usually described as "terraforming," or more holistically as "ecopoiesis" (Fogg 1995). A favorite candidate planet is Mars. Indeed the seeding of other planetary bodies with life may already have started with incomplete sterilization of Mars landing space capsules (although it is questionable whether any life thus transported could survive on the surface of Mars). In the catalogue of questions generated before the workshop, the possibility of genetically engineering an organism that can be an agent of litho-panspermia was raised (this involves seeding by transport in meteorite fragments). Clearly this must rely on genes from terrestrial organisms, but a novel combination could be produced. One must ask: In which system do you wish to put the organism? Such activities raise considerable ethical questions and, encouragingly in our view, there are already treaties and planetary protection methodologies in place that protect any Mars life and protect the Earth from it (when returning samples).

Why Is Space Mute?

Thus far, the search for extraterrestrial intelligence has failed. At face value, the absence of evidence seems to indicate that intelligence is an extremely rare phenomenon in the Universe. Does this in turn suggest that intelligence is extremely difficult to evolve? Or does it mean that intelligence is very short-lived in the cases where it does evolve? The puzzle is compounded by the recent estimation that most Earth-like planets are on average ~2 Ga older than the Earth. If they underwent a similar sequence of major transitions to the Earth at a similar rate, then they could have been emitting detectable signals for ~2 Ga. When we equate time with distance across space, this gives a large volume of the Universe from which we might expect to detect a signal.

Frank Drake (pers. comm.) offers, however, a cautionary note: although the Earth has been potentially detectable for the last 50–70 years of radio transmission, we are currently getting less detectable and may soon become "invisible" in the radio frequency range. The reason is that bandwidth-sharing relies on low energy transmissions (a good example being a mobile phone network), and the resulting signals are not strong enough to be detectable across space (see also Westall and Drake, this volume).

A further intriguing possibility is that sending radio transmissions across the galaxy may not be such an intelligent thing to do. If a civilization makes itself known, it may risk "consumption" by a more aggressive civilization. Thus, the really intelligent thing to do would be to employ a sophisticated technique to hide messages and avoid detection! This science fiction scenario introduces the notion of a population of interacting planetary civilizations subject to some form of selection, in this case for radio-invisibility.

THE EVOLUTION OF INTELLIGENCE

Is intelligence rare on habitable planets? Or is it a common, indeed probable phenomenon? These questions may be addressed not only by SETI but also from what we have learned about the evolution of life on Earth, where it has taken over three billion years to arrive at intelligent life. This suggests that intelligence is extremely difficult to evolve, and that a planet is quite likely to reach the end of habitability before it emerges (Carter 1983). Human-like intelligence may only appear in social, multicellular creatures. This implies that a planet's life-forms must first pass through the earlier major transitions in complexity: from naked replicators, to prokaryotes, to something like eukaryotes, to something like metazoans. If the events leading to many or all of these transitions are rare and difficult, then we would expect that there are far fewer planets with eukaryote than prokaryote life, even fewer with large-bodied forms, and fewest with intelligence. In that sense, intelligence must be rare in the Universe, even on the habitable planets. However, *if* a planet has multicellular, social organisms, and *if* it still has a billion years of habitability left, is the transition to intelligence likely or unlikely? Here opinions differ sharply.

It seems that the initial transition to intelligence may have to cross an adaptive valley before it begins to matter. Intelligence has high costs in the energy of brain metabolism, in the vulnerability of the young, the demands of parental care, and the built-in danger of learning maladaptive behavior instead of efficiently following instincts, which have proved right in most cases in the past (Martin 1990). However, benefits are also very high to animals with a particular kind of life history, that is, a niche where environmental changes come often, but not too often. "Too often" means chaotic surroundings, where the experience of the previous generation offers little guidance. In that case, the best strategy is to be *r*-selected and hope that you can breed your way out of disaster by random survival. The benefits of intelligence also dwindle if environmental changes are so slow that it is possible to evolve innate behavior patterns to cope with the environment in a predictable way (Lumsden and Wilson 1983).

On the right timescale, learning helps (e.g., for innovative omnivores or for seminomadic creatures exploring new habitats). One of the strongest pressures for flexible learning may be competition within the species, each group keeping up with others like intraspecific Red Queens. Any species competes, but the kind of competition we call warfare — where social groups challenge others — means that one must learn rather arbitrary definitions of our-own-group as opposed to the enemy-group. It also exerts intense selection pressure. Thus, chimp type tribal war could have been a strong spur to intelligence (Darwin 1871).

Another predisposing factor is fission–fusion societies, as found in chimpanzees, humans, and some cetaceans. In these, individuals of a group have strong social bonds but often forage alone or in subgroups. This means that one individual may possess information which others do not. There is then a value in sharing (or deliberately withholding) information about situations that are distant in space and time. This kind of society can lead to a "theory of mind" (Premack and Woodruff 1978), that is, the realization that another individual may not know what you know, and thus needs to be told. In experimental situations, apes and human four-year-olds demonstrate "theory of mind," whereas monkeys and three-year-olds do not (Byrne and Whiten 1988, 1997; Whiten and Byrne 1997).

Finally, a tendency to eat embedded food may help. Some foods require a multistep process to open or find them, or even the use of tools. Again, flexibility matters. No primate uses tools as complex as a spiderweb, but the ability to learn food-obtaining behavior, such as termite-dipping, nut-cracking, or fish-net weaving marks both ape and human intelligence (Parker and Gibson 1977; Parker and McKinney 1999).

Our own intelligence depends on grammatical language, the *sine qua non*. It was once widely thought that other animals could not acquire words as symbols. It is now clear that common chimps and bonobos can acquire a 300-word vocabulary (with extensive training) and that dolphins can also learn and combine symbols (Savage-Rumbaugh and Lewin 1994). Laboratory chimpanzees spontaneously develop different food-calls to indicate specific foods (Hallberg et al. 2003).

The new rubicon is grammar. Although other animals combine words, they do not spontaneously produce agent–action–object phrases. Furthermore, attempts to have computer models or robots produce grammar seem to fail (Szathmáry, pers. comm.). This would suggest that grammar is a supremely difficult transition. On the other hand, it seems very odd to think that if animals do not have an agent–action–object view of the world. "You groom me" is different from "I groom you." "Alpha male thumps Beta" is normal; "Beta thumps Alpha" is headline news. If an animal got as far as regular communication with symbolic words, it should not seem that hard to move on to simple grammar.

Two technical problems make the transition to grammar appear more difficult than it actually might be. One is that we have only one example: our own sophisticated language. It is as though we had to infer the evolution of the eye, given only one species with complex eyes, and nothing with simpler eye-spots. If we still had living *Homo erectus,* or even Neanderthals, we would not find it so hard to understand how the gap was bridged.

The other problem involves the picture of *H. sapiens* having emerged suddenly, around 50,000 years ago in Europe (and Australia), all ready to draw murals on cave walls, presumably commanding an assistant "Pass me the thick paintbrush and some red ochre." The more Darwinian, gradualist view, precedes this with a 2 million year run-up, via *H. erectus,* grunting "My bone!" or "Handaxe, moron!" (invective enters most trained chimps' vocabulary), or delightedly reporting, "Mabel baby!" A geometrically scratched stone, a few harpoon-heads, some ochre lumps in Africa suggest 100 thousand years of representation before the cave paintings, but the evidence is still very thin for representation earlier than 40 ka B.P. Still it seems more reasonable than the sudden, simultaneous appearance of artists and linguists in Europe and Australia, with no precursors.

A third, more fundamental, problem concerns the desire to see humanity as separate and special. It is a fair bet that if we did have semi-linguistic species around, we would raise the criterion of intelligence to the ability to tell a good story — including scientific stories of cause and effect. This, in fact, may be a more relevant criterion than grammar for *Pan narrans,* the narrating ape (Pratchett et al. 2002).

In sum: intelligence is rare, taking planets as a whole. Intelligence in the few planets with large creatures may be relatively common, if language can be taken step-by-step, following the logic of the need for communication brought by an innovative, social, competitive lifestyle. If, on the other hand, language is so complex that it is extraordinarily difficult to achieve, we may not be alone in the Universe but we might not have anyone else to talk to.

The Origin of Language May Indeed Be Difficult

The fact that there cannot be too many genes involved in the ape-to-human transition by no means rules out the possibility that language was a genuinely difficult transition. Path-dependence of small populations is one of the key issues (Szathmáry 2004). Theoretical analysis shows that this depends on the relation between population size N and the rate of beneficial mutations u (Wahl and Krakauer 2000). If $Nu > 1$, then beneficial mutations occur together and can be simultaneously selected for (in sexual populations), whereas if $Nu < 1$ then replicate populations *cannot* evolve in parallel because each will accumulate a different set of mutations, and even for an overlapping subset the order of incorporation will be different. Note that more complex organisms tend to be bigger, and they typically have a smaller population size; hence adaptive evolution will be more path-dependent for them. This is especially true for the evolving human lineage.

According to one view (Szathmáry 2001), the largely novel faculty selected for was the ability of the brain networks to process syntactical information. The specific hypothesis is that linguistically competent areas of the human brain have a statistical connectivity pattern that renders them especially suitable for syntactical operations. It is thought that: (a) The origin of human language required genetic changes in the mechanism of the epigenesis in large parts of the brain. (b) This change affected statistical connectivity patterns of the neural networks involved. (c) Due to the selectionist plasticity of brain epigenesis, coevolution of language and the brain resulted in the genetic assimilation of syntactical processing ability as such.

If this is so, why is language not more common? It is hard to assess at the moment why language is unique. Even the "not enough time" case could apply, which would be amusing. But preemption, due to the subsequent cultural evolution that language has triggered, may render further trials very difficult indeed. There is, however, yet another consideration that indicates that language could be variation-limited in a deeper sense.

The habitat of the language amoeba is a large, appropriately connected neural network: most of the information processing within the network elaborates on information coming from other parts of the network. A special type of processing is required: that of hierarchically embedded syntactic structures. This leads to the following difficulties:

1. Neural networks contain a large number of cycles: syntactic structures of language are tree-like. It seems difficult not to process large trees without getting into loops.
2. Overproduction of initial synapses or decreased pruning, both implied in the origin of language, may easily lead to "solipsist" network dynamics, with two consequences: (a) the activity of the network is detached more than optimally from external sources of information; (b) exaggerated internal processing leads to too much "internal talking": linguistic processing for its own sake.

GAIA

Having reviewed the geosphere–biosphere coevolution on Earth and introduced astrobiology, we are in a position to address more general Gaia questions.

How Important Is Habitation to the Maintenance of Habitability?

Habitation is contingent on habitability; however, once life is established, can it influence the maintenance of habitability? We broadly define habitation to mean a first-order influence on the geochemical cycling of a planet. This is a more quantitative definition of life than "abundant life" (Lenton, Caldeira, and Szathmáry, this volume). Thus, life eking out a meager existence in a few localities does not amount to habitation.

The factors required for habitation include a free energy source, essential elements, liquid water, a tolerable temperature, and not too much UV. Recycling of the crust (i.e., plate tectonics) was put forward as a further necessity, and this warrants further theoretical research. Autocatalytic recycling of essential elements is also required for life to achieve a first-order influence on geochemical cycling. This in turn is mediated by life, introducing an element of "boot-strapping" to achieve habitability. To retain liquid water, a planet must avoid high rates of H atom loss to space. It was noted that the rate of H loss can be influenced by biology; for example, in a methane-rich atmosphere it will be increased (Catling et al. 2001). Furthermore, H loss depends critically on the thermal structure of the atmosphere, and the existence and effectiveness of a cold trap may ultimately depend on oxygen and thus on life.

Existing studies have mapped out the dynamics of the habitable zone in the future. One study suggests that habitation will prolong the habitability of the Earth system, at least for plants, by maintaining the Earth in cool conditions when otherwise it would be too hot for them (Lenton and von Bloh 2001).

How Globally Destructive Could Life Become?

The habitability question may be inverted to ask: Could the biota force a critical environmental variable to the level that would make the biota extinct? It was generally agreed that this may be possible if there is a long time lag in the system. However, if there are no time lags, the system should be stable, because detrimental effects of life on the environment will become self-limiting.

There is no convincing evolutionary model to argue either way. There are, however, models that could form the basis of future research. The best such model of global coevolution for application to Gaia is that of Stenseth and Maynard Smith (1984). This couples within-species interaction with multi-species dynamics, speciation, and extinction. The population genetic timescale within species is captured by an average lag load (L). The lag load of a species is defined as: $(W_{max} - W)/W_{max}$, where W is the average fitness and W_{max} is the maximum fitness attainable if the evolution of every other species is assumed to be halted. The species number (S) captures global biodiversity. The rate of change of each with time is a function of both (an autonomous system). The four possible solutions are: (a) spiraling into an attractor with a constant rate of speciation and extinction (known as the "Red Queen" solution), (b) extinction (of all species), (c) explosion, and (d) stasis (when the species number is constant and maximal, with zero speciation and extinction rates). Only the Red Queen solution and stability are feasible for Earth history. Paleontologists have shown an interest in the model, for addressing what happens after a major extinction event. The few million years it took for recovery in the aftermath of the K/T extinction event would tend to support the Red Queen solution. Extending this model to include an environment variable is a topic for future research.

The destruction question was made more specific in the form: Can an organism evolve to produce a compound that disrupts the biosphere and nothing can consume it? Looking back into Earth history, a striking example presents itself: the evolution of the production of lignin, causing rising oxygen. After a time delay, fungi evolved the capacity to biodegrade lignin. However, it is unclear whether it was this evolution or simple thermodynamics in the form of fire that prevented oxygen from rising. The "invisible hand" argument (Smith 1776) was put forward and states that for every new waste product there is sufficient variation and creativity in evolution that something will eventually evolve to biodegrade it. This deserves to be tested more thoroughly.

Autocatalytic Gaia

Our group was clearly inclined toward the view that life is an inevitable planetary phenomenon. A rather bolder proposition was also agreed upon: that with life, autocatalytic recycling is an almost inevitable planetary phenomenon. This is a natural extension of the autocatalytic nature of life from the internal environment to the external environment. It was suggested that there is a lock-in effect of recycling. Some of the processes in the recycling loop may be abiotic. However, once closed recycling of essential elements has emerged, it will maintain itself. Such a system is not vulnerable to evolutionary "cheats" because "You can cheat when you are talking about information, but you can't cheat when you are talking about matter and energy" (Szathmáry, pers. comm.).

Can Gaia Be a Role Model for Human Society?

It was noted with wry humor that humans are currently "terrafouling" Earth rather than terraforming elsewhere. In other words, we are making a mess of our home planet. Our group proposed that Gaia may provide a role model for human society in our quest for sustainability. One way to proceed would be to identify the negative feedbacks that can provide stability to the social system without central control, and the positive feedbacks that can amplify features that are desirable. This was termed a "sociocatalytic" theory. It is irrational, or at least unwise, to assume that humans are rational actors. Game theory was largely unsuccessful in economics because of the assumption of rationality, but successful in evolutionary biology because natural selection is a supremely rational process.

Broad lessons of sustainability can be gained from Gaia. First and foremost, the system has developed and flourished without extra inputs of matter and energy. It is both creative and variant. This undermines the common assumption that economic growth is necessary for the development of society. We do not need an ever-increasing energy usage to develop as a society! Some cautionary thoughts on globalization also emerged. In particular, one superorganism may be more vulnerable than a population of systems. The bigger the system is, the more vulnerable it is. If the nation state becomes less relevant in the future, we may revert to smaller units that are more stable (there are historical examples to support this). Regarding energy systems, it is clear that the same systems are not appropriate everywhere, and there needs to be diversity to achieve efficiency. This is likely to be a more stable solution than the current drive to make the same energy system work everywhere.

THE IMPORTANCE OF AVOIDING TIME LAGS

We note from our discussion of Gaia that to protect the global system, time lags must be avoided. The reason is that a long delay between cause and effect can allow the cause to reach disruptive proportions before the effect generates any counteracting negative feedback. The introduction of time delays can cause a system to move from a stable domain to a limit cycle (oscillation) and potentially to chaotic behavior. In the Daisyworld model, the original form of which is extremely stable, the introduction of a large heat capacity generates thermal inertia in the system and causes it to oscillate (Nevison et al. 1999), whereas the introduction of discrete generations can lead to deterministic chaos (Jascourt and Raymond 1992). In the real world, long time lags may be caused by the slow response of large reservoirs; for example, atmospheric oxygen at present has a response time of millions of years. One alarming example of the introduction of a potential time lag is the proposal to modify trees genetically to suppress photorespiration in the carbon-fixing enzyme RuBisCO (ribulose bisphosphate carboxylase/oxygenase). This has the potential to increase the oxygen source and affect the long-term oxygen balance of the planet. It would take millions of years for changes in atmospheric oxygen to trigger a response, and that response might be quite catastrophic, for example, involving global wildfires.

FUTURE RESEARCH AGENDA

Proposals for future research that emerged during our discussions can be summarized as follows:
1. Origins and evolution of life:
    a. Given the proposed model of life, the concept of major transitions in evolution should be extended to encompass transitions in all three subsystems of life:
        i. Boundary system.
        ii. Metabolic system.
        iii. Genetic system.

2. Coevolution of the geosphere biosphere.
   a. Narrow the error bars in the timing of:
      i. The major transitions in the evolution of life.
      ii. The major changes in the state of the Earth system.
   b. Improve understanding of the causality of coevolution (or is it the co-causality of evolution?):
      i. Conduct thought experiments *in silico* with generalized (Earth) system models.
      ii. Is there only one pathway for life through the redox couple sequence?
      iii. Is there only one pathway for a Gaia system from reducing to oxidizing conditions?
   c. On specific linkages:
      i. What was the role of life in the Great Oxidation?
      ii. What was the role of life in Neoproterozoic environmental changes?
      iii. Is there a link between environmental change and the evolution of natural language?
3. Astrobiology
   a. Generalize Earth system models and biological evolution models for use in thought experiments in the search for planetary life:
      i. If we could stop plate tectonics on Earth, what would happen?
      ii. If we sterilized the Earth, how long would it remain habitable (e.g., to prokaryotes)?
   b. Minimum requirements for a sustainable ecosystem on Earth and elsewhere:
      i. What is the minimum functional diversity required to set up a sustainable ecosystem and what can we learn from terrestrial ecosystems?
      ii. Are there spatial and temporal limits for the persistence of communities and species?
      iii. Where and how can we detect habitable and potentially inhabited sites beyond the Earth?
   c. Extension of human activities beyond the Earth:
      i. What are the investments required for human migration beyond the Earth (needs, costs, benefits, social and ethical consequences)?
4. Gaia
   a. Develop generic models to explore the evolutionary dynamics of Gaia systems
      i. Add dependence on the environment as a third variable to the model of Stenseth and Maynard Smith (1984).
   b. Gaia as a role model

  i. Develop a systems model of society synthesizing biological, cybernetic and institutional theory.
  ii. Explore the feasibility of a flexible investment framework for achieving the transition to sustainability (highly mobile and available capital ~1% of global GDP).

In the Dahlem spirit, we offer a humorous final thought on the future of coevolution (created by Alison Jolly during our discussions):

The next stage of Gaia?

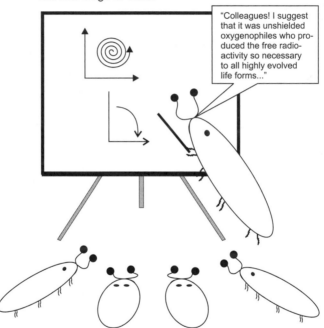

ACKNOWLEDGMENTS

We acknowledge the important contributions made to our report by Paul Falkowski and Andrew Watson, visiting members from Working Group 2.

REFERENCES

Altermann, W., and J.W. Schopf. 1995. Microfossils from the NeoArchean Campbell Group, Griqualand West sequence of the Transvaal Supergroup, and their paleoenvironmental and evolutionary implications. *Precambrian Res.* **75**:65–90.

Beaumont, V., and F. Robert. 1999. Nitrogen isotope ratios of kerogens in Precambrian cherts: A record of the evolution of atmospheric chemistry? *Precambrian Res.* **96**:63–82.

Brocks, J.J., G.A. Logan, R. Buick, and R.E. Summons. 1999. Archean molecular fossils and the early rise of eukaryotes. *Science* **285**:1033–1036.

Butterfield, N.J. 2000. *Bangiomorpha pubescens* n. gen., n. sp.: Implications for the evolution of sex, multicellularity, and the Mesoproterozoic/Neoproterozoic radiation of eukaryotes. *Paleobiology* **26**:386–404.

Butterfield, N.J., A.H. Knoll, and K. Swett. 1990. A bangiophyte red alga from the Proterozoic of arctic Canada. *Science* **250**:104–107.

Byerly, G.R., D.R. Lowe, and M.M. Walsh. 1986. Stromatolites from the 3300–3500 Ma Swaziland supergroup, Barberton Mountain Land, South Afrika. *Nature* **319**: 489–491.

Byerly, G.R., D.R. Lowe, J.L. Wooden, and X. Xie. 2002. An Archean impact layer from the Pilbara and Kaapvaal Cratons. *Science* **297**:1325–1327.

Byrne, R.W., and A. Whiten, eds. 1988. Machiavellian Intelligence. Oxford: Oxford Univ. Press.

Byrne, R.W., and A. Whiten. 1997. Machiavellian intelligence. In: Machiavellian Intelligence II: Extensions and Evaluations, ed. R.W. Byrne, pp. 1–24. Cambridge: Cambridge Univ. Press.

Carter, B. 1983. The anthropic principle and its implications for biological evolution. *Phil. Trans. Roy. Soc. Lond. A* **310**:347–363.

Catling, D.C., C.P. McKay, and K.J. Zahnle. 2001. Biogenic methane, hydrogen escape, and the irreversible oxidation of early Earth. *Science* **293**:839–843.

Cavalier-Smith, T. 2002. The phagotrophic origin of eukaryotes and phylogenetic classification of protozoa. *Intl. J. System. & Evol. Microbiol.* **52**:297–354.

Darwin, C. 1871. The Descent of Man and Selection in Relation to Sex. New York: The Modern Library, Random House.

Evans, D.A., N.J. Beukes, and J.L. Kirschvink. 1997. Low-latitude glaciation in the Palaeoproterozoic era. *Nature* **386**:262–266.

Fogg, M.J. 1995. Terraforming: Engineering Planetary Environments. Warrendale: Soc. Automotive Engineers.

Gánti, T. 1979. A Theory of Biochemical Supersystems. Budapest: Akademiai Kiado and Baltimore: Univ. Park Press.

Gánti, T. 2003. The Principles of Life. Oxford: Oxford Univ. Press.

Hallberg, K.L., D.A. Nelson, and S.T. Boysen. 2003. Representational vocal signalling in the chimpanzee. In: Animal Social Complexity: Intelligence, Culture, and Individualized Societies, ed. F.B.M. de Waal, and P.L. Tyack. Cambridge, MA: Harvard Univ. Press.

Han, T.-M., and B. Runnegar. 1992. Megascopic eukaryotic algae from the 2.1-billion-year-old Negaunee iron formation, Michigan. *Science* **257**:232–235.

Hayes, J.M. 1994. Global methanotrophy at the Archean–Proterozoic transition. In: Early Life on Earth, ed. S. Bengtson, pp. 220–236. New York: Columbia Univ. Press.

Jascourt, S.D., and W.H. Raymond. 1992. Comments on "Chaos in Daisyworld" by X. Zeng et al. *Tellus* **44B**:243–246.

Kazmierczak, J., and W. Altermann 2002. Neoarchean biomineralization by benthic cyanobacteria. *Science* **298**:2351.

Knauth, L.P., and D.R. Lowe. 2003. High Archean climatic temperature inferred from oxygen isotope geochemistry of cherts in the 3.5 Ga Swaziland Supergroup, South Africa. *Geol. Soc. Am. Bull.* **115**:566–580.

Kyte, F.T., A. Shukolyukov, G.W. Lulair, D.R. Lowe, and G.R. Byerly 2003. Early Archaean spherule beds: Chromium isotopes confirm origin through multiple impacts of projectiles of carbonaceous chondrite type. *Geology* **31**:283–286.

Lenton, T.M., and W. von Bloh. 2001. Biotic feedback extends the life span of the biosphere. *Geophys. Res. Lett.* **28**:1715–1718.

Lindsay, J.F., and M.D. Brasier. 2002. Did global tectonics drive early biosphere evolution? Carbon isotope record from 2.6 to 1.9 Ga carbonates of Western Australian basins. *Precambrian Res.* **114**:1–34.

Logan, G.B., J.M. Hayes, C.B. Hieshima, and R.E. Summons. 1995. Terminal Proterozoic reorganization of biogeochemical cycles. *Nature* **376**:53–56

Lovelock, J.E. 1975. Thermodynamics and the recognition of alien biospheres. *Proc. Roy. Soc. Lond.* **189**:167–181.

Lowe, D.R. 1994. A biological origin of described stromatolites older than 3.2 Ga. *Geology* **22**:387–390

Lumsden, C.J., and E.O. Wilson. 1983. The dawn of intelligence. *The Sciences* **23**: 22–31.

Martin, R.D. 1990. Primate Origins and Evolution: A Phylogenetic Reconstruction. Princeton: Princeton Univ. Press.

Nevison, C., V. Gupta, and L. Klinger. 1999. Self-sustained temperature oscillations on Daisyworld. *Tellus* **51B**:806–814.

Nisbet, E.G., and N.H. Sleep. 2001. The habitat and nature of early life. *Nature* **409**:1083–1091.

Parker, S.T., and K.R. Gibson. 1977. Object manipulation, tool use, and sensorimotor intelligence as feeding adaptations in cebus monkeys and great apes. *J. Human Evol.* **6**: 623–664.

Parker, S.T., and M.L. McKinney. 1999. Origins of Intelligence. Baltimore: Johns Hopkins Press.

Pavlov, A.A., J.F. Kasting, L.L. Brown, K.A. Rages, and R. Freedman. 2000. Greenhouse warming by CH_4 in the atmosphere of early Earth. *J. Geophys. Res.* **105(E5)**:11,981–11,990.

Pavlov, A.A., J.F. Kasting, J.L. Eigenbrode, and K.H. Freeman. 2001. Organic haze in Earth's early atmosphere: Source of low-^{13}C Late Archean kerogens? *Geology* **29**:1003–1006.

Porter, S.M., and A.H. Knoll. 2000. Testate amoebae in the Neoproterozoic era: Evidence from vase-shaped microfossils in the Chuar Group, Grand Canyon. *Paleobiology* **26(3)**:360–385.

Pratchett, T., I. Stewart, and J. Cohen. 2002. The Science of Discworld II. London: Ebury Press.

Premack, D., and G. Woodruff. 1978. Does the chimpanzee have a theory of mind? *Behav. Brain Sci.* **1**:515–26.

Ryder, G. 2003. Bombardment of the Hadean Earth: Wholesome or deleterious? *Astrobiology* **3**:3–6

Rye, R., P.H. Kuo, and H.D. Holland. 1995. Atmospheric carbon dioxide concentrations before 2.2 billion years ago. *Nature* **378**:603–605.

Savage-Rumbaugh, S., and R. Lewin. 1994. Kanzi: The Ape at the Brink of the Human Mind. New York: Wiley.

Schopf, J.W. 1993. Microfossils of the Early Archean apex chert: New evidence of the antiquity of life. *Science* **260**:640–646.

Schrödinger, E. 1944. What Is Life? Cambridge: Cambridge Univ. Press.

Smith, A. 1776. An Inquiry into the Nature and Causes of the Wealth of Nations. London: W. Strahan and T. Cadell.

Stenseth, N.C., and J. Maynard Smith. 1984. Coevolution in ecosystems: Red Queen evolution or stasis? *Evolution* **38**:870–880.

Summons, R.E., L.L. Jahnke, J.M. Hope, and G.A. Logan. 1999. 2-Methylhopanoids as biomarkers for cyanobacterial oxygenic photosynthesis. *Nature* **400**:554–557.

Szathmáry, E. 2001. Origin of the human language faculty. The language amoeba hypothesis. In: New Essays on the Origin of Language, J. Trabant and S. Ward, pp. 41–51. Berlin: Mouton.

Szathmáry, E. 2004. Path dependence and historical contingency in biology. In: Understanding Change: Models, Methodologies, and Metaphors. Basingstoke: Palgrave Macmillan, in press.

Szathmáry, E., and J. Maynard Smith. 1995. The major evolutionary transitions. *Nature* **374**:227–232.

Wahl, L.M., and D.C. Krakauer. 2000. Models of experimental evolution: The role of genetic chance and selective necessity. *Genetics* **156**:1437–1448.

Walker, J.C.G., P.B. Hays, and J.F. Kasting. 1981. A negative feedback mechanism for the long-term stabilisation of Earth's surface temperature. *J. Geophys. Res.* **86(C10)**:9776–9782.

Ward, P.D., and D. Brownlee. 2000. Rare Earth: Why Complex Life is Uncommon in the Universe. New York: Springer.

Westall, F., M.J. de Wit, J. Dann et al. 2001. Early Archean fossil bacteria and biofilms in hydrothermally influenced sediments from the Barberton greenstone belt, South Africa. *Precambrian Res.* **106**:93–116.

Whiten, A., and R.W. Byrne, eds. 1997. Machiavellian Intelligence II: Extensions and Evaluations. Cambridge: Cambridge Univ. Press.

7

What Do We Know about Potential Modes of Operation of the Quaternary Earth System?

M. CLAUSSEN[1,2], H. HELD[1], and D. P. SCHRAG[3]

[1]Potsdam Institute for Climate Impact Research (PIK), 14412 Potsdam, Germany
[2]Physics Institute, University of Potsdam, 14415 Potsdam, Germany
[3]Laboratory for Geochemical Oceanography, Department of Earth and Planetary Sciences, Harvard University, Cambridge, MA 02138, U.S.A.

ABSTRACT

The problem of Earth system stability during the Quaternary, approximately the last two million years, is addressed from a global perspective. Despite efforts over the last 160 years to obtain geological evidence of ice ages, we conclude that the question of which potential modes of operation impacted the Quaternary Earth system is as yet unsolved. However, there are some clues as to which elements should be included in a theory of the Quaternary Earth system. The search for direct paleo-analogues is unlikely to answer the question of potential modes in a physically meaningful manner. Assessment of a number of conceptual models — ranging from models in which forcing is necessary to yield observed climate variability to models of free climate oscillations — does not favor any particular model because of the difficulties of tuning each model to the time series of global ice volume. Hence, geographically explicit fully coupled climate system, or natural Earth system, models are required to analyze the system's response to geographically varying forcing and internal feedbacks. Evidence emerges that much of Quaternary climate variability was due to internal feedbacks involving ice sheets and biogeochemical cycles as critical elements and orbital forcing as the pacemaker.

PALEO-ANALOGUES VERSUS PALEO-SYSTEMS ANALYSIS

In 1795, the Scottish scientist James Hutton published his two-volume "Theory of the Earth," which established him as one of the founders of modern geologic thought. In brief, one could summarize his novel approach with the phrase "the present is the key to the past." Today, it has become popular to reverse this phrase by saying "the past is the key to the future" in order to highlight the value of paleoclimatology and possible paleo-analogues for the discussion of recent, present-day, and possible future climate change.

Numerous papers have put forth the idea of paleo-analogues. For example, Petit-Maire (1990) asked the question of whether the Sahara will become greener in a warmer climate, thereby proposing a paleo-analogue of potential modern climate change to the early and mid-Holocene wet phase in Northern Africa. In their model of the natural Earth system, Claussen et al. (2003) found indeed a possible greening of Northern Africa in a greenhouse gas climate warming scenario. However, they state that the early and mid-Holocene greening of the Sahara is not a proper paleo-analogue as the processes involved appear to have different weight.

The duration of the Holocene has been subject to controversial debate. From statistical interpretation of interglacial–glacial cycles, it has been suggested that the present interglacial is likely to end in only a few hundred years, whereas from the astronomical theory of glacial cycles, a rather long interglacial is expected (Berger and Loutre 1997). According to the astronomical theory, the interglacial associated with the Marine Isotope Stages 11 (MIS11) would be a dynamical paleo-analogue of the Holocene. However, it is unclear whether MIS 11 could serve as analogue of Anthropocene climate dynamics, that is, climate dynamics in the presence of strong perturbation of land cover and atmospheric chemistry.

There is general consensus that the current anthropogenically induced increase of atmospheric greenhouse gas concentrations is unprecedented in terms of the last 420 thousand years (ka), or perhaps even the last 20 million years (Ma) (Prentice et al. 2001). Hence, no direct paleo-analogue exists of present-day climate change when considering similar tectonic forcing, that is, land–sea distribution and natural outgassing of CO_2, water, and other substances from the Earth's interior, and total carbon of the fast pools (atmosphere, ocean, and biosphere). Perhaps we could construct paleo-analogues in a more general sense, that is, paleo-analogues for the identification and testing of functional mechanisms determining the potential modes of operation of the Quaternary Earth system. A recurrent theme directly relevant to concerns about future climate change is whether the Earth system is robust or highly sensitive to small forcings. In this respect, time periods prior to the Quaternary would also appear useful for study. For example, the Late Paleocene Thermal Maximum (LPTM), some 55 Ma ago, could be such a time period. During the LPTM, a dramatic increase in (natural) CO_2 emission occurred and was presumably of comparable magnitude as the anthropogenic greenhouse gas forcing (Schrag and McCarthy 2002; Watson et al., this volume). There is, however, one major caveat: some cases (e.g., the thermohaline ocean circulation discussed by Rahmstorf and Sirocko, this volume) indicate that the stability of the system depends on the state of the system and, hence, on the boundary conditions.

Therefore, we surmise that the search for paleo-analogues is unlikely to answer the question of potential modes of operation of the Quaternary Earth system and suggest that exploration of the phase space, through the implementation of numerical climate system models, would yield an answer. Accordingly, in this chapter, we focus discussion on dynamical models of the natural Earth system, or climate system; we use the terms "natural Earth system" and "climate system" interchangeably. We will summarize existing theories of Quaternary dynamics, from the perspective of conceptual models and comprehensive models, and explore possible conclusions to be drawn from modeling studies.

REVIEW OF THEORIES FOR QUATERNARY EARTH SYSTEM DYNAMICS

A Brief History of the Theories of Ice Ages

In 1842, shortly after L. Agassiz proposed the existence of ice ages on the grounds of geological evidence, J. A. Adhémar suggested the first astronomical theory of climate change based on the known precession of equinoxes. Over the following decades, glacial geology became strongly tied to the astronomical theory that was advanced by J. Croll in the 1860s. Since Croll's theory appeared to be increasingly at variance with emerging geological evidence, his theory was eventually refuted. In 1896, S. Arrhenius concluded that

> It seems that the great advantage which Croll's hypothesis promised to geologists, viz. of giving them a natural chronology, predisposed them in favor of its acceptance. But this circumstance, which at first appeared advantageous, seems with the advance of investigation rather to militate against the theory, because it becomes more and more impossible to reconcile the chronology demanded by Croll's hypothesis with the facts of observation (Arrhenius 1896, p. 274).

Perhaps it would be useful to cite this classic statement more often in the light of present-day orbital tuning of data.

Arrhenius was convinced that changes in atmospheric transparency (due to changes in atmospheric CO_2) would "prove useful in explaining some points in geological climatology which have hitherto proved most difficult to interpret" (Arrhenius 1896, p. 275). The astronomical theory was modified and advanced by R. Spitaler, M. Milankovitch, V. Köppen, and A. Wegner at the beginning of the last century; however, it was disputed again, in 1955, after C. Emiliani detected that there were more glaciations than the "traditional four" (Günz, Mindel, Riss, and Würm glaciations, which were discovered by A. Penck and E. Brückner in 1901–1909). The astronomical theory saw a strong revival after new geological evidence, presented by J. D. Hays, J. Imbrie, and N. J. Shackleton in 1976, appeared to corroborate many of the predictions advanced and refined by A. Berger in the late 1970s.

The theory of ice ages, in which geochemical reactions and CO_2 play a major role, is perhaps as old as astronomical theory. Early work dates back to J. J. Ebelman in the 1840s, J. Tyndall in 1861, and S. Arrhenius in 1896. In addition, there are present-day models (e.g., by G. Shaffer, developed in the 1990s) that could be described as biogeochemical oscillators in which ocean biogeochemistry is the key player.

Overview of Conceptual Models

Climate archives, which reveal climate variability over the last several million years, display four striking features (Figure 7.1):

1. Prior to 3 Ma ago, relatively small climate variations (with a period of approximately 20 ka) are observed.
2. Next, there appears to be a gradual transition to colder conditions until some 2 Ma ago.
3. Thereafter, until some 900 ka ago, that is, the so-called early Pleistocene, climate variations (with respect to temperature and ice volume) are seen with a dominant periodicity of approximately 40 ka.

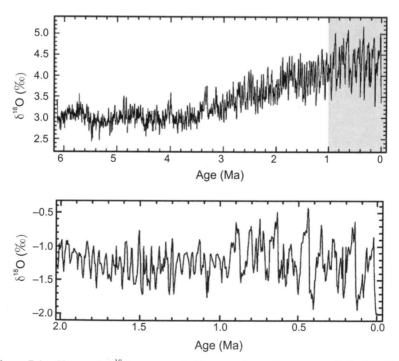

Figure 7.1   Changes in $\delta^{18}O$ reconstructed from marine sediment cores for the last 6 Ma and the last 2 Ma, respectively. These changes are interpreted as changes in global ice volume where increasing $\delta^{18}O$ values indicate increasing ice volume. The upper curve represents data from the tropical western Pacific sedimentary core ODP 806; the lower curve is from the same core, but using pelagic foraminifera. Figure from Saltzman (2002) and used with permission from Academic Press.

4. The last period, the late Pleistocene, is characterized by a predominant 100-ka periodicity with relatively large amplitude of temperature and ice volume. The time around 900 ka (or in other archives, around 700 ka) is often referred to Mid-Pleistocene Transition (MPT).

To explain these phenomena, a plethora of models has been developed over the past decades. These models can be categorized (see Saltzman 2002) into "forced" versus "free" models, that is, models in which astronomical forcing is necessary versus instability-driven, auto-oscillatory models.

Astronomical, or orbital, forcing consists of meridional and seasonal changes of insolation as a result of changes in the eccentricity of the Earth's orbit with dominant periods of 400 ka, 125 ka, and 95 ka, in the obliquity with a dominant period at 41 ka, and in the precession of equinoxes with a bimodal period at 23 ka and 19 ka. Models employing astronomical forcing need to resolve the "100-ka paradox": whereas the late Pleistocene time series displays a strong frequency component at 1/100 1/ka, astronomical forcing does not. Hence, within the astronomical paradigm, one has to conclude that the climate system responds in a strongly nonlinear way to the forcing. Muller and Macdonald (1997) proposed an alternative by suggesting that near-100-ka period fluctuations in the inclination of the orbital plane, in conjunction with the location of cosmic dust bands, might be able to generate the observed 100-ka oscillations. Again, the radiative forcing implied by this astronomical forcing appears to be rather small, and thus some additional amplifier is required to elevate the 100-ka period above the others.

From the viewpoint of systems theory, the forced models suggest different mechanisms in order to reproduce the observed features:

1. The simplest models are just a linear combination of the three astronomical tones.
2. Then, there are a number of more comprehensive models — energy-balance models, ice-sheet models, and atmospheric energy-balance/statistical dynamical models coupled to ice-sheet models — that are capable of reproducing the approximately 20-ka and 40-ka response to astronomical forcing, but do not show a 100-ka response.
3. A third class of models is based on the assumption that the near 20-ka precessional variations could produce 100-ka relaxation oscillations of ice sheets. Such behavior is well known to engineers and physicists from the paradigmatic van der Pol's oscillator or from aeroelastic fluttering (Guckenheimer and Holmes 1997). Relaxation oscillations are characterized by self-sustained periodic motion, where each period consists of sharp switching between small amplitude (energy gain) and large amplitude (energy loss) episodes.
4. Other forced models allow for internal oscillations, which are entrained by combination tones of precessional forcing (overtones within insolation) or produced by climate system's nonlinearity.

5. A more sophisticated class displays dynamics that result in multiple equilibria. If certain thresholds in forcing are reached, the system would jump from one equilibrium to another.
6. If, in addition, stochasticity is present, the periodic forcing can be less pronounced and yet can induce synchronized jumps as the result of stochastic resonance.

In general, one source of variability would be a random walk process. This idea was originally proposed by Hasselmann (1976), who assumed that the annual cycle of insolation generates variability in the fast climate components, which could be randomly accumulated by the more sluggish climate components. Recently, Wunsch (2003) pursued this idea by demonstrating that most long-term climate archives reveal a red-noise spectrum, that is, a spectrum with an amplitude of variance that decays with larger frequencies. Superimposed on the red spectrum are weak structures, which correspond to the frequency bands of orbital forcing (see Figure 7.2). To explain the dominant 100-ka climate variability, Wunsch suggests a stochastic forcing of a system with a collapse threshold, which yields a transition in the spectral domain from red to white, that is, a flat spectrum with amplitudes independent of frequency. This way, variability on

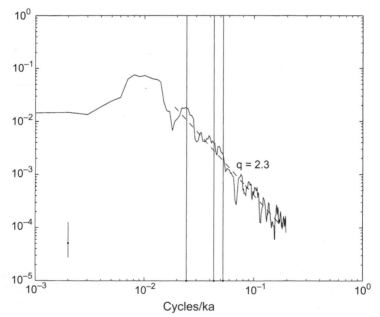

Figure 7.2 Power spectrum normalized to sum to unity, for the tropical (Panama Basin) core ODP677 (planktonic $\delta^{18}O$ of the last 1 Ma). Also shown are an approximate 95% confidence interval and a least-square fit to the high-frequency range of the power spectrum (dashed line). Vertical lines indicate periods of astronomical forcing, i.e., 41, 23, and 19 ka. This figure is taken with modification from Wunsch (2003) and used with permission by Springer-Verlag.

the 20- to 100-ka timescale, as well as on shorter timescales, can be explained in a combined way (Table 7.1).

If astronomical forcing is removed, forced models would not reproduce the major 100-ka period seen in most paleo-archives of the late Pleistocene. In contrast, free models do not rely on astronomical forcing at all; instead, the 100-ka period appears as an internal (free) oscillation once the system crosses critical boundaries in the parameter space. In other words, the 100-ka oscillation is the consequence of a bifurcation (i.e., a qualitative change of the system's state caused by a change in a control parameter, be it a key, internally generated slow variable or an external forcing) to a self-sustained oscillation (a so-called Hopf bifurcation) that is driven by an instationarity of the Earth system. Such free oscillators can be composed of ice masses interacting with another variable: ice-sheet location, bedrock depression, or thermohaline circulation. The free oscillator is forced by steady long-term variations, for example, by a tectonic forcing associated with a slow variation in CO_2 outgassing. The long-term forcing tunes the oscillator or switches it on or off in a Hopf bifurcation. An essential point is that the free models do not require amplification of small forcing and explain the relatively large climate changes over the Pleistocene as simply the

Table 7.1 Overview of models that try to explain Pleistocene variability, in particular the 20-, 40-, and 100-ka periods. *Any* of them performs well for the 20- or 40-ka frequency component.

---

1. Forced models: Orbital forcing is necessary
   a. Models with remarkable 100-ka response
      i. Astronomical forcing in conjunction with location of cosmic dust bands (Muller and Macdonald 1997)[1,2]
      ii. Linear combinations of the three astronomical tones; extreme necessity of 100-ka forcing[1,2]
      iii. Multiple equilibria; complicated threshold paths (Paillard 1998)[2]
      iv. Internal oscillations, entrained by combination tones of precessional forcing[1,2]
      v. Ice-sheet inertia[1,2]
      vi. 20-ka precession-induced relaxation oscillations[1]
      vii. Stochastic resonance and related effects; also produces higher-frequency variability (> 10 ka) in a red spectrum[1] (Wunsch 2003 expanded from Hasselmann 1976)
   b. Models without dominant 100-ka response:
      i. Some energy-balance models
      ii. Ice-sheet models
      iii. Coupled energy-balance/statistical dynamical models[1]
2. Free models: Orbital forcing is not necessary
   a. Free internal oscillation — limit cycle made up from a nonlinear combination of positive feedback loops[1,2]

---

[1] Examples for the model class are given in Saltzman (2002)
[2] Indicates a model that is included in the Roe and Allen (1999) intercomparison

response of relatively large internal oscillators that could in some cases even resist additional forcing over a variety of timescales.

Rial and Anaclerio (2000) demonstrated that the interplay of astronomical forcing with the Earth system's nonlinearity is not pure speculation. At a 95% level of statistical significance, by bispectral analysis, they showed that the 29-ka and 69-ka spectral peaks in the Vostok ice core time series are phase-coherent sidebands from the 41-ka obliquity and an overtone of the 413-ka eccentricity. This finding, based purely on time series analysis, without model assumptions, strongly supports nonlinear models.

The existence of switching behavior around 0.9–0.7 Ma B.P. provides further evidence for the system's nonlinearity, as the character of the astronomical forcing remained constant. Saltzman and his coworkers (cited in Saltzman 2002) assume tectonic CO_2 as a slow, steady external driving force, which causes their model to switch to the late Quaternary mode. According to their model, atmospheric CO_2 concentration decreases linearly in time over the last 5 Ma. Unfortunately, this assumption seems to be at variance with recent evidence from carbon isotopes of biomarkers by and from boron isotope reconstructions of seawater pH (see Pagani et al. and Pearson and Palmer, cited and referenced in Prentice et al. 2001). Paillard (1998) circumvents the problem of identifying a particular external forcing, but assumes a long-term linear shift in one of the thresholds defined in his model. An alternative hypothesis explains the mid-Pleistocene transition by the removal of regolith (i.e., more loosely aggregated soil till and sediment material) over several ice age cycles. According to that hypothesis, ice sheets would easily slide on regolith. After removal of soft sediments by early Pleistocene ice ages, ice sheets could eventually grow without the sliding effect of the soft sediment layer in the late Pleistocene. This hypothesis, however, seems to be at variance with Antarctic inland ice dynamics, where sliding can occur without significant disturbance of regolith mass.

Saltzman (2002) also suggests a switching behavior at the onset of the Quaternary ice age some 2.5 Ma ago, which, again, was triggered by steady reduction in tectonic CO_2 forcing according to his model. However, there is no unequivocal evidence for a switch of natural Earth system dynamics. At least some geological evidence (see Figure 7.1) suggests a more gradual transition into the Pleistocene. Although tectonic CO_2 forcing is disputed, the formation of the Isthmus of Panama (Haug and Tiedemann 1998), which lasted from 13 Ma to 1.9 Ma ago and which has led to a major reorganization of Atlantic ocean circulation, has been suggested as a candidate for triggering the onset of the Pleistocene.

The categorization of models of Quaternary climate system dynamics as "forced" and "free" can be complemented by ordering them by means of complexity in terms of physical processes involved. According to Saltzman (2002, p. 276) climate system models aim at "ever more complete representation of the full slow-response climate system"; hence the center manifold. Saltzman (2002,

p. 276) suggests that "ice sheets and their bedrock and basal properties, coupled with forced and free variations of carbon dioxide, operating on an Earth characterized by a high-inertia deep thermohaline ocean that can store carbon and heat ..." encompass the center manifold. (In the "vicinity" of a bifurcation (see above), one can identify slow and fast processes. Then the center manifold characterizes the interplay of the few slowest variables with the collective effect of all the other variables, and therefore allows the study of a complex system's long-time behavior by just a few degrees of freedom; hence by a conceptual model (see Guckenheimer and Holmes 1997). Behind this lurks the idea that few slow climate variables exist, which enslave the fast variables, and that this list of slow variables could be completed.

After several models had been suggested to explain the climatic time series, Roe and Allen (1999) investigated the performance of six representatives of these classes. They tuned each model to the time series (the last 900 ka) of global ice volume and also the time rate of change for the ice volume, while modeling the residuals as first- (and second-) order autoregressive process. They found that within 95% error bars one cannot favor any model over its competitors. This demonstrates that it is necessary to obtain further insight into the physical mechanisms underlying the paleo-time series and then to ask whether a hypothesis at hand would be consistent with the latter.

SOME LESSONS FROM COMPREHENSIVE MODELS OF THE EARTH SYSTEM

What Is the Appropriate Level of Model Complexity?

Thus far we have mainly considered conceptual models that dealt with relatively simple dynamic systems theory, in terms of the degrees of freedom. Saltzman (2002) called these models inductive models, that is, models based on the cross-understanding of the feedbacks that are likely to be involved. From these zero- or one-dimensional models, one gains valuable insight into possible feedback mechanisms. However, the predictive value of conceptual models is limited because of the large, in comparison with the degree of freedom, number of tunable parameters. Moreover, validation of conceptual models is difficult (see above). Some assumptions that form the basis for various conceptual models are oversimplified and, thus, could be misleading. For example, the assumption that orbital forcing is identified with summer insolation at high northern latitudes overemphasizes the ice-albedo feedback at high northern latitudes while neglecting the fact that other components of the climate system can react to varying insolation in a different way than Northern Hemisphere ice sheets do. Therefore, it seems sensible to explore the role of geographically varying forcing and feedback processes in geographically explicit models.

The degree of spatial and temporal resolution necessary for paleoclimate simulations is disputed. Comprehensive, "state-of-the-art" coupled models describing the general circulation of the atmosphere and the ocean (AOGCMs), in some cases including models of biosphere dynamics (referred to as "climate system models" or "Earth system models" — or quasi-deductive models, according to Saltzman) are supposed to be the most realistic laboratory of the natural Earth system. However, their applicability to long-term simulations is limited by high computational costs. Therefore, it has been proposed (Claussen et al. 2002) that Earth system models of intermediate complexity (EMICs) be used which operate at a higher level of spatial and temporal aggregation. In the case of CLIMBER-2 — an EMIC developed at the Potsdam Institute for Climate Impact Research — it was argued that the optimum degree of aggregation would be a spatial resolution finer than hemispheric scale, thereby resolving the distribution of continents and oceans, but larger than the correlation radius of the most energetic synoptic (weather) pattern, which is of the order of several 1000 km (Petoukhov et al. 2000). Temporal resolution and the degree of parameterization follow from that requirement. As a consequence of this aggregation, many processes resolved in AOGCMs have to be parameterized in EMICs. The advantage gained by this reduction is computational efficiency. Moreover, many EMICs are designed to include explicitly biogeophysical and biogeochemical processes, which makes them a useful tool for integrated paleoclimate modeling. Presumably, neither EMICs nor AOGCMs alone offer the best modeling tool, but rather it requires the appropriate use of the full spectrum of conceptual models, EMICs, and AOGCMs.

Some Results: Factors Contributing to Climate Differences between Glacial and Interglacial Climate

Most comprehensive model simulations have addressed the question of which processes could have contributed to the difference between glacial and interglacial states during the late Quaternary, more precisely between the Last Glacial Maximum (LGM) and the preindustrial, or even present-day, climate. For example, within the framework of the Paleoclimatic Modeling Intercomparison Project (PMIP), a number of atmospheric circulation model (AGCM) simulations were undertaken, in which oceanic characteristics were either prescribed on the basis of paleoclimatic reconstructions or simulated using a simple slab-ocean model with prescribed modern oceanic heat transport. It was shown that two major factors — the buildup of large ice sheets in the Northern Hemisphere and a lowering of CO_2 atmospheric concentration — could explain a global cooling at the LGM compared to the present in the range of $2°-6°C$. One observation that appears to be more difficult to reconcile with most model results is the large ($\sim 3°C$) cooling of the Western Pacific Warm Pool (e.g., Stott et al. 2002), as this region is relatively insensitive to ice-sheet emplacement and may suggest a somewhat higher sensitivity to atmospheric CO_2 or oceanic heat transport than most models currently employ.

Although ice-sheet cover and atmospheric CO_2 indeed may be the most important factors contributing to LGM–Holocene climate differences, several others may be relevant. In particular, paleo-data indicate that vegetation cover at the LGM was considerably different from the present one, with a much smaller forest area in Eurasia and in the tropics. Experiments with AGCMs have demonstrated that such changes in vegetation cover could have a pronounced regional impact, but show little global-scale influence (e.g., Kubatzki and Claussen 1998). This is not surprising, since it has been shown in several studies that models with fixed ocean characteristics considerably underestimate the impact of vegetation changes outside the regions where these changes occur.

Another important mechanism, which might have affected the glacial climate state, is related to changes in the ocean circulation. Data suggest a weakening and shallowing of the upper branch of the thermohaline circulation and northward penetration of the Antarctic bottom water. Model experiments, however, yield diverse results. Ganopolski et al. (1998) have demonstrated that a reorganization of the ocean circulation would have a profound impact on the climate. Other AGCM studies (e.g., Hewitt et al. 2001) show an intensification rather than weakening of the glacial thermohaline circulation. Hence the "thermohaline circulation riddle" is still not solved.

By using the Potsdam EMIC CLIMBER-2, Ganopolski (2003) made the first attempt to analyze geographically explicit factors that contribute to the difference between a fully glacial and an interglacial climate. (Actually, Ganopolski did not perform a complete factor analysis, because he did not differentiate between pure feedbacks and synergisms, i.e., feedbacks between feedbacks.) His results, shown in Figure 7.3, can be summarized as follows. Changes in ice-sheet distribution and elevation give the largest contribution to the glacial cooling, namely, about 3°C in globally averaged annual surface temperature. Regional cooling due to this process is strongest at high northern latitudes. Lowering of CO_2 by 80 ppm is a global process, which, according to CLIMBER-2, causes a global cooling by 1.2°C. These results are obtained in experiments with prescribed modern vegetation. The inclusion of vegetation dynamics leads to a drastically reduced forest area over Eurasia and some reduction of forest area in the tropics and subtropics. Hence, the strongest cooling associated with the biogeophysical feedback occurs at high northern latitudes (Figure 7.3c). In this region, the impact of vegetation changes is as large as the direct effect of CO_2. The global cooling due to changes in vegetation cover, however, is smaller than the CO_2 contribution, only some 0.7°C according to the model. The effect of a reorganization of the thermohaline ocean circulation is estimated as a difference between two equilibrium climate states corresponding to "cold" (stadial) and "warm" (interstadial) modes of the glacial thermohaline circulation discussed in Ganopolski and Rahmstorf (2001). Although reorganization of the thermohaline circulation does not have a pronounced global effect, due to the compensation between the Northern and the Southern Hemispheres, it has strong

regional impact (Figure 7.3d). The near-surface temperature is more than 5°C
lower over the Northern Atlantic in the cold mode than in the warm mode. To the
contrary, the Southern Ocean and Antarctica are warmer by 1°–2°C in the cold
(stadial) mode due to the seesaw effect.

Transient factor analyses have been undertaken by Berger (2001) using a
two-dimensional model of the Northern Hemisphere. Berger finds that factors
contributing to glacial–interglacial climate change are not constant in time. For
example, biogeophysical feedback — in particular the feedback between near-
surface temperature and expansion or retreat of highly reflecting snow-covered
tundra on the one hand, and darker, more insolation absorbing tundra on the
other hand — enhances the glacial inception, that is, the transition during MIS 5.
Once the ice sheets have grown, at the end of MIS 5, this feedback appears to
become negative.

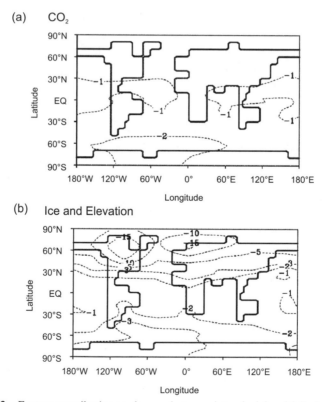

Figure 7.3   Factors contributing to changes between interglacial and full glacial equilibrium climate. Shown are changes in annual mean near-surface air temperature due to
(a) lowering atmospheric CO_2 concentration from 280 ppm to 200 ppm, (b) topographic
changes (i.e., changes in surface elevation, in surface albedo, and in land–ocean distribution) induced by prescribing ice sheets of the Last Glacial Maximum.

Another important lesson learned from EMICs is that changes in insolation are necessary to drive the climate system into a glacial mode, whereas changes in CO_2 operate as an internal amplifier. Keeping the geographic distribution of insolation, or more precisely the orbital parameters, constant at values typical for the Eemian interglacial, but prescribing CO_2 as some external forcing according to paleoclimatic reconstruction would yield no glacial inception (see Berger and Loutre 1997 and literature cited therein). In turn, keeping CO_2 constant at Eemian values does not prevent the last glacial inception. Moreover, experimentation with CLIMBER-2 reveals a rapid spread of ice sheets and a subsequent slower growth in ice volume once an insolation threshold is crossed. This suggests that the small ice cap instability seems to work in a geographically explicit model and that the glacial inception is presumably a bifurcation of the physical climate system (Calov et al. 2004).

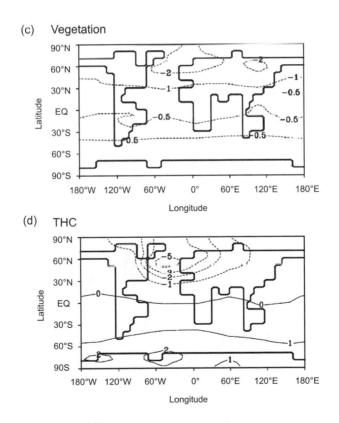

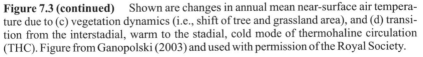

Figure 7.3 (continued)   Shown are changes in annual mean near-surface air temperature due to (c) vegetation dynamics (i.e., shift of tree and grassland area), and (d) transition from the interstadial, warm to the stadial, cold mode of thermohaline circulation (THC). Figure from Ganopolski (2003) and used with permission of the Royal Society.

The sensitivity of late Quaternary dynamics to atmospheric CO_2 as internal amplifier has been demonstrated by a number of numerical experiments (see Berger and Loutre 1997). Their studies suggest that the Earth system is more sensitive to CO_2 changes when insolation variations are small (e.g., during MIS 11 and in present-day climate as well as for the next 50 ka) than when they are large (e.g., during MIS 5e, the Eemian). It appears that there could be a natural threshold of CO_2 concentration of presumably 250 ppmv above which the Earth system is not able to sustain a 100-ka glacial cycle, but oscillated at higher frequencies with a much smaller amplitude of ice volume.

CONCLUSIONS

Our focus in this chapter has been on Earth system stability from a global perspective. We have not addressed the problem of "hot spots," that is, regions on Earth that appear to be highly sensitive to external forcing due to strong internal amplifiers and refer the reader to Rahmstorf et al. and Payne (both this volume). We have considered which potential modes of operation were implicit in the natural Earth system, or climate system, in the Quaternary. We conclude that even after some 160 years of geological research into the ice ages, this has not yet been solved. However, we have some clues on what elements a theory of Quaternary Earth system dynamics should consist of in terms of concepts and model structure.

Saltzman (2002) has proposed a unified theory of Quaternary Earth system dynamics. The term "unified" is used because it combines theories based on orbital forcing and on greenhouse gas forcing, respectively. Saltzman supposes that the center manifold of the climate system involves the ice sheets and their bedrock and basal properties, coupled with tectonically forced and free variations of CO_2, a high-inertia deep thermohaline ocean. Whereas it is very likely that slow variables exist, it is still not obvious that those can be identified with particular physical entities just mentioned. Quite the contrary, in highly resolved spatiotemporal dynamics, such slow variables may emerge as complex patterns across physical entities, which strongly supports the approach of spatially resolved climate models. Interestingly, validation of inductive models appears to be an almost futile task: assessment of a number of inductive models cannot favor any model over its competitor on the grounds of tuning each model to the time series of global ice volume.

In the range of more comprehensive, quasi deductive models, only EMICs have been used for long-term studies. These numerical experiments support the idea of orbital forcing acting as a pacemaker of glacial–interglacial cycles (Hays et al. 1976). However, the situation is rather complex. Obviously, the response of the Earth system to a given forcing is a function of the actual state of the Earth system as well as meridional and seasonal changes of the forcing. Presumably, changes in insolation associated with changes in orbital parameters trigger fast

internal feedbacks such as the water vapor–temperature feedback and the snow–albedo feedback, which then are further amplified by slower feedbacks, such as biogeochemical and biogeophysical feedback and the isostatic response of the lithosphere to ice-sheet loading. Some of these feedbacks even change sign during the course of an glacial–interglacial cycle.

The biggest mystery of all appears to be wrapped up with ice, whether the continental ice sheet at one pole or the extent of sea ice at the other. Evidence emerges that the ice sheets in the Pleistocene as well as biogeochemical cycles are critical components of both the free and forced conceptions of Pleistocene glacial cycles. Their response to a future of higher atmospheric CO_2 may well determine the gross dynamics of the climate system in the future.

ACKNOWLEDGMENTS

We wish to thank all members of the Dahlem Workshop, in particular Gilberto Gallopín, for constructive comments which helped improve the manuscript. We thank Ursula Werner, PIK, for technical assistance.

REFERENCES

Arrhenius, S. 1896. On the influence of carbonic acid in the air upon the temperature of the ground. *Philos. Mag. J. Sci.* **41**:237–276.

Berger, A. 2001. The role of CO_2, sea-level, and vegetation during the Milankovitch-forced glacial–interglacial cycles. In: Geosphere–Biosphere Interactions and Climate, ed. L.O. Bengtsson and C.U. Hammer, pp. 119–146. New York: Cambridge Univ. Press.

Berger, A., and M.-F. Loutre. 1997. Palaeoclimate sensitivity to CO_2 and insolation. *Ambio* **26**:32–37.

Calov, R., A. Ganopolski, V. Petoukhov, M. Claussen, and R. Geve. 2004. Transient simulation of the last glacial inception. Part I: Glacial inception as a bifurcation in the climate system. *Climate Dyn.*, in press.

Claussen, M., V. Brovkin, A. Ganopolski, C. Kubatzki, and V. Petoukhov. 2003. Climate change in northern Africa: The past is not the future. *Climatic Change* **57**:99–118.

Claussen, M., L.A. Mysak, A.J. Weaver et al. 2002. Earth system models of intermediate complexity: Closing the gap in the spectrum of climate system models. *Climate Dyn.* **18**:579–586.

Ganopolski, A. 2003. Glacial integrative modelling. *Phil. Trans. R Soc. Lond. A* **361**: 1871–1884.

Ganopolski, A., and S. Rahmstorf. 2001. Rapid changes of glacial climate simulated in a coupled climate model. *Nature* **409**:153–158.

Ganopolski, A., S. Rahmstorf, V. Petoukhov, and M. Claussen. 1998. Simulation of modern and glacial climates with a coupled global model of intermediate complexity. *Nature* **391**:351–356.

Guckenheimer, J., and P. Holmes. 1997. Nonlinear oscillations, dynamical systems, and bifurcations of vector fields, corr. 5th print. In: Applied Mathematical Sciences, vol. 42. New York: Springer.

Hasselmann, K. 1976. Stochastic models. I. Theory. *Tellus* **28**:473–485.

Haug, G.H., and R. Tiedemann. 1998. Effect of the formation of the Isthmus of Panama on Atlantic Ocean thermohaline circulation. *Nature* **393**:673–676.

Hays, J.D., J. Imbrie, J., and N.J. Shackleton. 1976. Variations in the Earth's orbit: Pacemaker of the ice ages. *Science* **194**:1121–1132.

Hewitt, C.D., A.J. Broccoli, J.F.B. Mitchell, and R.J. Stouffer. 2001. A coupled model study of the Last Glacial Maximum: Was part of the North Atlantic relatively warm? *Geophys. Res. Lett.* **28**:1571–1574.

Kubatzki, C., and M. Claussen. 1998. Simulation of the global bio-geophysical interactions during the Last Glacial Maximum. *Climate Dyn.* **14**:461–471.

Muller, R.A., and G.J. McDonald. 1997. Glacial cycles and astronomical forcing. *Science* **277**:215–218.

Paillard, D. 1998. The timing of Pleistocene glaciations from a simple multiple-state climate model. *Nature* **391**:378–381.

Petit-Maire, N. 1990. Will greenhouse green the Sahara? *Episodes* **13**:103–107.

Petoukhov, V., A. Ganopolski, V. Brovkin et al. 2000. CLIMBER-2: A climate system model of intermediate complexity. Part I: Model description and performance for present climate. *Climate Dyn.* **16**:1–17.

Prentice, I.C., G.D. Farquhar, M.J.R. Fasham et al. 2001. The carbon cycle and atmospheric carbon dioxide. In: Climate Change 2001: The Scientific Basis. Contribution of Working Group 1 to the 3rd Assessment Report of the IPCC, ed. J.T. Houghton, Y. Ding, D.J. Griggs et al., pp. 183–237. Cambridge: Cambridge Univ. Press.

Rial, J.A., and C.A. Anaclerio. 2000. Understanding nonlinear responses of the climate system to orbital forcing. *Quat. Sci. Rev.* **19**:1709–1722.

Roe, G.H., and M.R. Allen. 1999. A comparison of competing explanations for the 100,000-yr ice age cycle. *Geophys. Res. Lett.* **26**:2259–2262.

Saltzman, B. 2002. Dynamical Paleoclimatology. San Diego: Academic Press.

Schrag, D.P., and J.J. McCarthy. 2002. Biological-physical interactions and global climate change: Some lessons from Earth history. In: The Sea, ed. A.R. Robinson, J.J. McCarthy, and B.J. Rothschild, pp. 605–619. New York: Wiley.

Stott, L.D., C. Poulsen, S. Lund, and R. Thunell. 2002. Super ENSO and global climate oscillations at millennial timescales. *Science* **297**:222–226.

Wunsch, C. 2003. The spectral description of climate change including the 100 ky energy. *Climate Dyn.* **20**:353–363.

8

Modes of Oceanic and Atmospheric Circulation during the Quaternary

S. RAHMSTORF[1] and F. SIROCKO[2]

[1]Potsdam Institute for Climate Impact Research (PIK), 14412 Potsdam, Germany
[2]Institute for Geoscience, University of Mainz, 55099 Mainz, Germany

ABSTRACT

Paleoclimatic evidence shows that the ocean and atmosphere have undergone major changes during the Quaternary. For atmospheric mean circulation, data are consistent with changes in strength and shifts in position of major atmospheric circulation features (e.g., the westerly wind belt), whereas the structure of these main features appears to have persisted. For the ocean, evidence points to qualitative reorganizations in the thermohaline circulation. Modes of ocean–atmosphere variability, such as El Niño-Southern Oscillation (ENSO) and North Atlantic Oscillation (NAO), also appear to have changed or even been absent at times.

INTRODUCTION

In this chapter, we discuss the modes of operation of the oceanic and atmospheric circulation during the Quaternary. We begin by briefly highlighting key features of the oceanic and atmospheric circulation of the present-day Earth. Then we explore the range of possibilities for different behaviors of the planetary circulation system, as suggested by models, and review paleoclimatic evidence on the types of circulation that actually occurred during the Quaternary. We focus on dynamic changes in the ocean and atmosphere and less on the direct thermodynamic response of surface temperature to radiative forcings (Milankovitch, CO_2, etc.), although we do not completely ignore the latter.

Is the current state of the thermohaline ocean circulation, with deep water forming in the North Atlantic and around Antarctica, unique? Was the current three-cell (Hadley, Ferrel, and polar cells) structure of the atmospheric meridional circulation maintained in the Quaternary? Did the dominant ocean–atmosphere variability modes (ENSO and NAO) persist throughout this period? Answers to these questions will yield important clues as to what dynamic changes in ocean–atmosphere behavior might be expected during the Anthropocene.

THE PRESENT STRUCTURE OF OCEANIC AND ATMOSPHERIC CIRCULATIONS

Atmospheric and oceanic circulations are turbulent, complex flows with variability on all spatial and time scales. Despite this, some first-order, large-scale structures of the mean flow can be described, as can a few dominant coherent variability modes. These main structures and modes are our primary focus in this chapter.

Mean *atmospheric circulation* is organized in three main meridional cells in each hemisphere: (a) the Hadley cell, which extends across the tropics and subtropics, (b) the Ferrel cell in midlatitudes, and (c) the polar cell in high latitudes (Figure 8.1). This circulation is a response to solar heating in the tropics (causing rising air) and net radiative cooling in high latitudes. The most simple circulation driven by such heating and cooling would be one cell rising in the tropics and sinking near the poles, with warm air flowing aloft toward the pole and cold air flowing along the surface toward the equator. The Earth's rotation, however, does not permit such a simple flow since the Coriolis term dominates the equations and vorticity conservation inhibits north–south flow. As a result, air rising in the tropics descends in the subtropics, and the near-surface, equatorward winds of this Hadley cell take on a strong easterly component (i.e., the easterly trade winds). The seasonal cycle further modifies the circulation by causing a rising motion in the summer hemisphere and sinking in the winter hemisphere

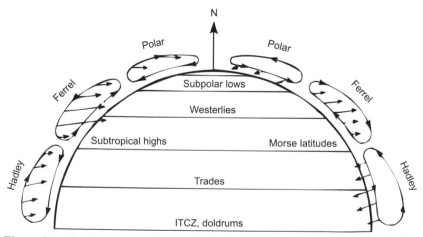

Figure 8.1   Basic structure of atmospheric circulation, showing the tropical Hadley cells, midlatitude Ferrel cells, and polar cells. Upper- and lower-level winds have directions suggested by the top and bottom arrows attached to the cells. From Apel (1987).

(i.e., one cross-hemispheric Hadley cell); only in the annual mean does the nearly symmetric picture, with one Hadley cell in each hemisphere, appear. As a result, the intertropical convergence zone (ITCZ) shifts north and south across the equator with the seasons. In midlatitudes, the near-surface winds, driven by the north–south pressure gradient, form the westerly wind belt. This westerly flow is dynamically unstable, forming synoptic eddies (cyclones and anticyclones) which contribute much to the meridional transport of heat and moisture in the atmosphere.

Wind-driven ocean circulation is predominantly organized in basin-scale horizontal gyres (i.e., subtropical and subpolar gyres) and is driven by the vorticity input from the prevailing surface winds (trade winds, westerly belt). As a consequence of the vorticity balance on a rotating sphere, flow on the western side of these gyres is concentrated in swift western boundary currents (e.g., the Gulf Stream in the North Atlantic, the Kuroshio in the North Pacific); the return flow, however, is more diffuse. These gyre circulations occupy the upper portion (several hundred meters) of the oceans.

By contrast, *thermohaline ocean circulation* covers the full depth range of the ocean. It is driven by heat and freshwater fluxes at the ocean surface and subsequent interior mixing of heat and salt. It can be described as a global-scale overturning of water masses, with sinking (i.e., deepwater formation) in a few locations balanced by a broader upwelling motion. There are deepwater masses from two main sources: (a) North Atlantic deep water (NADW), presently formed mainly in the Greenland–Norwegian and Labrador seas, and (b) Antarctic bottom water (AABW), presently formed primarily in the Weddell and Ross seas. The compensating upward motion in the ocean is not as localized and cannot be measured; tracer and modeling studies (Toggweiler and Samuels 1993) suggest that much of it may occur in the Southern Ocean aided by the wind (Ekman divergence).

Variability in the ocean–atmosphere system is not just random, uncorrelated noise but is partly organized in coherent large-scale variability modes with characteristic timescales, as can be shown by empirical orthogonal function analysis of global weather observations. The two leading variability modes on short timescales (decades or shorter) are ENSO and NAO or northern annular mode.

Below, we explore how stable and persistent these structures of the mean circulation and variability modes were in the Quaternary as well as what we know about the first-order changes in planetary circulation during this period.

THE LAST GLACIAL MAXIMUM

A good starting point for discussion is the Last Glacial Maximum (LGM), ~20 thousand years (ka) before present (B.P.). Quaternary climate is characterized by oscillations of global climate between two extreme states: cold glacials with large continental ice cover (covering much of North America and Eurasia) and interglacials with minimal ice cover (confined primarily to Antarctica and Greenland) (Figure 8.2). At least during the past four glacial cycles (420 ka), both the cold and warm limits of these oscillations appear to be remarkably constant, with similar values in global ice volume, atmospheric CO_2, and other parameters being reached every cycle. The present Holocene interglacial thus represents a warm extreme of the Quaternary climate range, whereas LGM represents the most recent cold extreme. There is a wealth of proxy data and model simulations about LGM climate which, taken together, allow many inferences about past ocean and atmosphere circulation.

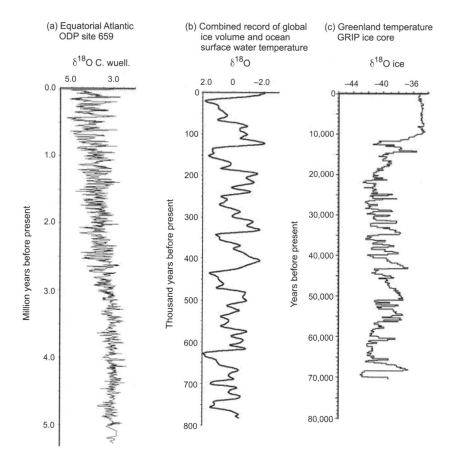

Figure 8.2   Timescales and time series of climate change during the Quaternary. (a) Oxygen isotope $\delta^{18}O$ composition of benthic foraminifera in a deep-sea core (ODP site 659) from the equatorial west Atlantic (Tiedemann et al. 1984). (b) The SPECMAP $\delta^{18}O$ stack of planktonic foraminifera from various deep-sea cores (Martinson et al. 1987). (c) Air temperature-related $\delta^{18}O$ record of the Greenland ice core GRIP (Johnsen et al. 1992).

Since the pioneering work of Milankovitch, insolation has been regarded as the primary forcing of the ice ages during the Quaternary. The climatic evolution during the past 2.5 million years shows clear cyclicities of about 19,000, 23,000, 41,000, and 100,000 years. These are attributable to rhythmic variations in the orbital constellation of the Earth to the Sun and are clearly seen in various paleoclimate records. According to the orbital theory of climate evolution, ice ages are triggered by insolation minima. Nevertheless, the LGM occurred during a time when the insolation patterns were almost identical to the present day, since the LGM occurred exactly one precessional cycle (of 20,000 years) ago. However, the Earth was very different in those days.

Proxy data from continental and marine geoarchives suggest a number of differences in the glacial atmospheric circulation compared to the present. The most obvious difference was the presence of ice sheets that were 3- to 4-km thick in northern Eurasia and North America, extending southward to about 52°–45°N (CLIMAP). The ice sheets must have acted as topographic barriers for the west wind zone, contributing to its southward shift.

Much of the North Atlantic was covered with sea ice, and tundra vegetation dominated northern America and Europe. Cold desert dust storms were frequent in all of northern Europe, Asia, and North America along a corridor several hundred kilometers south of the continental ice sheets. The Sahara expanded and the tropical belt of rain forests in Africa and southern America shrank under reduced precipitation. Summer monsoon winds in southern Asia vanished completely (Sirocko et al. 1996). The mean annual temperature of the globe was about 4°–8°C lower than at present; however, large regional variations occurred. Stute et al. (1995) measured a 5°C reduction of annual temperature from the noble gas composition of glacial groundwater aquifers in Brazil. Unfortunately, it is quite difficult to measure annual temperatures for the high latitudes because the seasonal temperature contrast must have been quite large near the ice-sheet margins.

This glacial world did not apparently result from insolation at the time (which was similar to the present) rather from its history, that is, from the amassment of ice sheets that began to form much earlier. Ocean records reveal a sharp sea-level drop at 118–115 ka, which marked the end of the last interglacial (128–118 ka). Thereafter, within a few thousand years, sea level dropped by about 50 m, or about 40% of the entire glacial–interglacial value (Figure 8.3). Modeling results indicate that the North American ice sheet formed during this time and that it was characterized by extremely low summer insolation over the Northern Hemisphere. Positive feedbacks since this glacial inception (near 117 ka) caused the ice sheets to grow and cover much of northern Europe, at least during stage 4 (the first major cooling phase of the last glaciation).

The major feedback mechanisms were the role of ice-sheet topography and albedo, atmospheric carbon dioxide and water vapor concentrations, ocean circulation, sea-level changes associated with the waxing and waning of ice sheets,

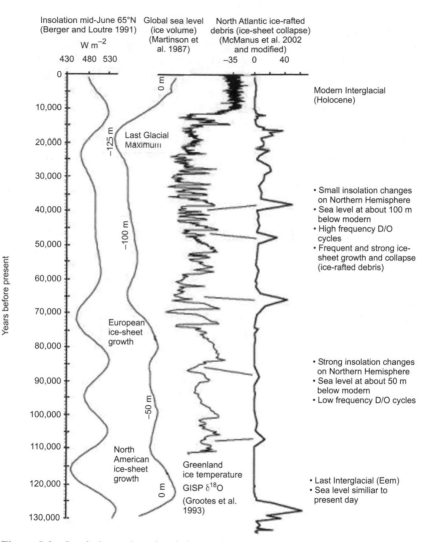

Figure 8.3 Insolation and sea-level change during the last 130,000 years in comparison to the structure of the Dansgaard/Oeschger (D/O) events (Greenland temperature) and surge of ice sheets during the Heinrich events, as recorded in North Atlantic deep-sea sediments by layers rich in drop stones from melting ice bergs (ice-rafted debris maxima after McManus et al. [1999, 2002] and Oppo et al. [2001, 2003]). For the latest version of global sea level, see Siddall et al. (2003).

as well as the continental albedo and vegetation. Each of these parameters was powerful enough to affect climate and environment on a large, even global scale.

Despite these large changes in surface climate, the limited evidence available is consistent with changes in *strength* of key circulation features (trade winds, monsoon circulation) and some *shifts in latitude* (e.g., ITCZ and the northern westerly belt). There is, however, no evidence to suggest that the basic structure of the atmospheric circulation (as described above) changed in a qualitative way. For example, coral data from the Huon peninsula, taken from different times of the Quaternary, consistently show a similar seasonal cycle (Tudhope et al. 2001), which under the present-day climate results from the seasonal passing of the ITCZ over this site. Apparently, this was a robust feature in the Quaternary. Proxy data for wet or dry conditions from a number of tropical South American sites suggest that the rainfall belt (i.e., the ITCZ) may have been further to the south in glacial times.

Turning now to ocean circulation, our discussion becomes more complex. Both quantitative and qualitative changes in ocean circulation have been discussed in the literature. Among the quantitative changes is a possibly enhanced subtropical gyre in the North Atlantic, which would be an expected result of stronger trade winds and westerlies. Oceanic evidence for an enhanced gyre is a regional warming in the Gulf Stream region.

For thermohaline circulation, earlier publications suggest that the LGM climate may be characterized by a complete absence of NADW formation. However, more recent evidence (Yu et al. 1996) suggests that NADW did form at a similar rate to that observed at present, though further south (to the south of Iceland) and reaching only lesser depths (1–2 km, as compared to 2–3 km in modern climate).

The second major deep water mass, AABW, was also present in glacial times. Proxy data suggest (Labeyrie et al. 1992) a larger volume of water masses of Antarctic origin in the Atlantic compared to the present. However, this does not necessarily mean an enhanced formation rate of AABW, as it could simply reflect a change in relative densities of northern and southern source waters.

Thus, our current understanding suggests that, just as in the atmosphere, the basic structure of ocean circulation resembled that of the present climate, with some adjustments in flow rates and in the latitude of northern deepwater formation. However, unlike during the Holocene (with the possible exception of the 8.2-k event), this basic circulation pattern appears to have been unstable and to have undergone a number of major changes during transient millennial events.

MILLENNIAL VARIABILITY DURING THE GLACIAL

The ice ages not only experienced generally colder climates than at present, they were also punctuated by abrupt climatic transitions. The best evidence for these transitions, known as Dansgaard/Oeschger (D/O) events, comes from the last glacial. D/O events typically start with an abrupt warming of around 10°C within a few decades or less, followed by gradual cooling over several hundred or thousand years. The cooling phase often ends with an abrupt final temperature drop back to cold ("stadial") conditions. Although they were first observed in the Greenland ice cores, D/O events are not a local feature of Greenland climate. Similar records have been found, for example, near Santa Barbara, California, in the Cariaco basin off Venezuela, and off the coast of India (for a compilation of evidence from 180 sites and references, see Voelker et al. 2002). D/O climate change is centered on the North Atlantic and regions with strong atmospheric response to changes in the North Atlantic, with a weak response in the Southern Ocean or Antarctica.

The "waiting time" between successive D/O events is most often around 1500 years or, with decreasing probability, 3000 or 4500 years (Alley et al. 2001). This suggests the existence of an as yet unexplained 1470-yr cycle that often (but not always) triggers a D/O event. This cycle has maintained its periodicity to within a few percent throughout at least the second half of the glacial (Rahmstorf 2003).

The second major type of climatic event in glacial times is the Heinrich event (Andrews 1998). Apparently, the large Eurasian and North American ice sheets surged every few thousand years and transported icebergs and ice-rafted debris (Figure 8.3) into the Atlantic (McManus et al. 2002). The nonlinear dynamics of ice sheets provide one plausible trigger mechanism. Ice sheets may grow for many thousands of years until their base melts due to trapped geothermal heat. Surges and strong calving result from a positive feedback: once the ice sheet moves, frictional heating causes further basal melting. This is an instability mechanism inherent to the continental ice sheets (see Payne, this volume).

The relationship of Heinrich events to sea-level change is not yet understood because we lack high-resolution sea-level reconstructions for the time of the ice-sheet decay. Records from the South Atlantic and parts of Antarctica show that the cold Heinrich events in the North Atlantic were associated with unusual warming there (Blunier et al. 1998), a fact sometimes referred to as "bipolar seesaw." A recent study from the Red Sea indicates that sea-level changes of up to 35 m may have occurred during times of the Antarctic warming (Siddall et al. 2003), that is, Heinrich layers likely occurred during a rise of global sea level by a few tens of meters. Initial sea-level rise, as a result of ice-sheet instability, could trigger further ice sheets by destabilizing the ice shelves at the margins of the continental ice.

Several climate models now incorporate continental ice sheets, and such models can reproduce irregular ice sheet surges. However, the interactions of the ice sheet with shelf ice and the ocean are currently only included in regional and process models, not yet in global climate models. Hence these models do not yet contain the mechanism for triggering and/or synchronizing ice-sheet surges via sea-level changes.

At the end of the last glacial, a particularly interesting abrupt climatic change took place, the so-called Younger Dryas event (12.8–11.5 ka B.P.). Conditions had already warmed to near-interglacial conditions and continental ice sheets were retreating when, within decades, the climate in the North Atlantic region switched back to glacial conditions for more than a thousand years. It has been argued that the cooling resulted from a sudden influx of fresh water into the North Atlantic through the St. Lawrence River, when an ice barrier holding back a huge meltwater lake on the North American continent broke (Fairbanks 1989). This could have shut down NADW formation. Alternatively, the Younger Dryas may simply have been the last cold stadial period of the glacial following a temporary D/O warming event (the Allerød event, the timing of which follows the 1500-yr pattern of previous D/O events), and ended by another D/O warming (1,500 years after the Allerød). In fact, both views may be correct, with the Allerød (and perhaps the Bølling 1,500 yr earlier) causing the meltwater influx.

For the D/O, Heinrich, and Younger Dryas events, the paleoclimatic data clearly point to a crucial role of Atlantic Ocean circulation changes. Modeling and analytical studies of the Atlantic thermohaline circulation show that there are two positive feedback mechanisms leading to threshold behavior. The first, called advective feedback, is caused by the large-scale northward transport of salt by the Atlantic currents, which in turn strengthens the circulation by increasing density in the northern latitudes. The second, called convective feedback, is caused by the fact that oceanic convection creates conditions favorable for further convection. These (interconnected) feedbacks make convection and large-scale thermohaline circulation self-sustaining within certain limits, with well-defined thresholds where the circulation changes to a qualitatively different mode.

Three main circulation modes have been identified both in sediment data and models (Figure 8.4): (a) a warm or interglacial mode with deep water forming in the Nordic seas and large oceanic heat transport to northern high latitudes, (b) a cold or stadial mode with deep water forming south of the shallow sill between Greenland, Iceland, and Scotland and with greatly reduced heat transport to high latitudes, and (c) a "switched off" or "Heinrich" mode with practically no deepwater formation in the North Atlantic (Sarnthein et al. 1994). In the last mode, the Atlantic deep circulation is dominated by inflow of AABW from the south.

Many features of abrupt glacial climate can be explained by switches between these three circulation modes. Model simulations suggest that the cold stadial mode is the only stable mode in a glacial climate; it prevails during the cold stadial periods of the last glacial. D/O events can be interpreted as temporary incursions of warm Atlantic waters into the Nordic seas and deepwater formation there, that is, a switch to the warm mode causing abrupt climatic warming in the North Atlantic region (Ganopolski and Rahmstorf 2001). As this mode is not stable in glacial conditions, the circulation starts to weaken gradually, and temperatures start to decline again immediately after the incursion until the threshold is reached where convection in the Nordic seas stops and the system reverts to the stable stadial mode. Heinrich events can be interpreted as a switch from the stadial mode to the Heinrich mode, that is, a shutdown of NADW. As this mode is probably also unstable in glacial conditions, the system spontaneously reverts to the stadial or warm mode after a waiting time of centuries, the timescale being determined partly by slow oceanic mixing processes.

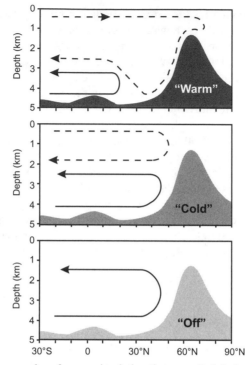

Figure 8.4 Three modes of ocean circulation that prevailed during different times of the last glacial period. A section along the Atlantic is depicted; the rise in bottom topography symbolizes the shallow sill between Greenland and Scotland. North Atlantic overturning is shown dashed, Antarctic bottom water solid. After Rahmstorf (2002).

This interpretation is consistent with the observed patterns of surface temperature change. Warming during D/O events is centered on the North Atlantic because this is where the change in oceanic heat transport occurs: the warm mode delivers heat to much higher latitudes than the cold mode. A switch to the Heinrich mode, on the other hand, strongly reduces the interhemispheric heat transport from the South Atlantic to the North Atlantic. This cools the Northern Hemisphere while warming the Southern Hemisphere, and explains the "bipolar seesaw" response in climate. It should be noted that the initial transient response can differ from the equilibrium response as the oceanic heat storage capacity is large. The patterns of these abrupt changes differ from the longer timescale (many thousands of years) response to the Milankovitch cycles because for the latter, the slow changes in atmospheric greenhouse gases (e.g., CO_2) and continental ice cover act to synchronize and amplify climatic change globally.

Although threshold behavior of the Atlantic can dramatically shape and amplify climatic change, the question remains: What triggers the mode switches? As mentioned above, D/O switches appear to be paced by an underlying

1,470-yr cyclicity that has as yet to be explained. This could result from an external (astronomical or solar) cycle or an internal oscillation of the climate system. However, the extreme regularity of the underlying 1,470-yr cycle strongly argues for an origin outside the Earth system (Rahmstorf 2003). That each cycle does not necessarily trigger a D/O event is probably the result of the presence of stochastic variability in the climate system as well as the presence of longer-term trends, such as the slow buildup of large continental ice sheets.

By contrast, ocean circulation change during Heinrich events can be explained by the amounts of fresh water entering the North Atlantic during these times in the form of icebergs. Simulations show that observed amounts of fresh water are sufficient to shut down deepwater formation in the North Atlantic.

SHORT-TERM VARIABILITY MODES

In present-day climate, ENSO is the strongest mode of natural climate variability. It is a coupled ocean–atmosphere mode that is centered on the tropical Pacific, with a variable period of 3–7 years, and it causes worldwide ecological and societal impacts due to its effect on the global atmospheric circulation. It is thus natural to ask whether this mode has also been in operation during different climatic states in the past. Annually banded corals provide a unique opportunity here, since they record climatic information at up to monthly rates of resolution based on the chemistry of their skeletons as they grow. As a result of tectonic uplift, fossil coral reefs from past climatic periods can be found on exposed terraces at some sites, such as the Huon peninsula of Papua, New Guinea.

Coral data from different time segments show convincingly that ENSO variability also prevailed under very different climates, including glacial times and the Eemian interglacial (Tudhope et al. 2001). The amplitude appears to have varied, however, with particularly weak ENSO variations during the mid-Holocene (6.5 ka B.P.) and the early glacial (112 ka B.P.), and the strongest ENSO during modern times. First attempts to simulate the effect of Milankovitch cycles on ENSO variations using a simple model suggest that the precession cycle could directly alter ENSO intensity by zonal asymmetric heating of the equatorial Pacific (Clement et al. 2000). This data shows, however, that this cannot be the only effect. Thus, more data and further simulations with more comprehensive models are needed to understand ENSO variations through time.

Less information is available on the past operation of the NAO. Appenzeller et al. (1998) have attempted to reconstruct NAO variability from a West Greenland ice core, a location where a correlation with the NAO index is found for the observational period. This reconstruction goes back 350 years and reveals that NAO variability is intermittent, that is, not always clearly present. However, based on such a single time series, it is impossible to distinguish conclusively a spatially coherent mode such as the NAO from ordinary uncorrelated stochastic variability.

CONCLUSIONS

The Quaternary underwent major climate changes, within consistent bounds, as a result of the insolation variations caused by the Milankovitch cycles. The response of the Earth system to this forcing is obviously highly nonlinear, since a simple linear relation of, for example, ice volume proxies and Milankovitch forcing does not exist. Much of the cause of this nonlinearity probably resides in the slow and nonlinear response of continental ice sheets to forcing (see Payne, this volume). Another part of this nonlinearity, however, resides in the response of atmosphere and ocean.

We have argued that the basic structure of the atmospheric circulation was probably not altered, although this is perhaps based more on the absence of evidence than on evidence for an absence of such a change. The limited paleoclimatic data for changes in atmospheric circulation is consistent with some quantitative changes and spatial shifts of key atmospheric circulation features, such as the westerlies or monsoons, but does not indicate, for example, qualitative reorganizations of the meridional circulation cells of the atmosphere. What might come closest to a qualitative change is the possibility of a shift in the location of tropical atmospheric convection, although evidence for this is thus far inconclusive. Such a shift could be important in that it could lead to large-amplitude global climate changes. It is intriguing to note that such a shift involves similar physics as in the case of the ocean: in both cases, the nonlinearity of the convection process is the deeper reason behind the large and nonlinear changes.

For ocean circulation, increasingly convincing evidence points to qualitative reorganizations of the thermohaline overturning circulation, and these have occurred repeatedly during the Quaternary. Some of these changes appear to have been in response to strong outside forcing, that is, a shutdown of NADW formation in response to ice-sheet surges (Heinrich events). For others (D/O events), no strong forcing mechanism is known. Data merely point to the timing being ruled by a 1,470-yr cycle of unknown origin, and simulations suggest that the ocean could have been close to a threshold where weak forcing was sufficient to trigger major circulation changes.

Concerning the major coherent short-term (decadal or shorter) variability patterns of the ocean–atmosphere system, ENSO, and NAO, the sketchy evidence available suggests that they may have operated only intermittently in the past. These patterns may thus be quite sensitive to forcing changes.

Which conclusions can be drawn for future anthropogenic climate change? Generally, the history of Quaternary climate suggests that the climate system is rather sensitive to forcing and responds in nonlinear ways, including the potential for strong amplifiers. Ocean circulation has responded with major changes in the past, not always due to strong forcing; thus the risk of future circulation changes must be taken seriously. Atmospheric variability patterns, like ENSO and NAO, may also have changed in the past and could be affected by future anthropogenic warming. Such nonlinear changes in the operation of ocean and atmosphere could have large and adverse impacts on human society. Although much progress has been made in modeling these mechanisms, it is difficult to predict the response of, for example, ENSO or the thermohaline circulation to anthropogenic increase in CO_2. Therefore, the possibility of nonlinear responses in ocean and atmosphere adds uncertainty to our assessment of future climate changes and is a cause for caution.

REFERENCES

Alley, R.B., S. Anandakrishnan, and P. Jung. 2001. Stochastic resonance in the North Atlantic. *Paleoceanography* **16**:90–198.

Andrews, J.T. 1998. Abrupt changes (Heinrich events) in late Quaternary North Atlantic marine environments: A history and review of data and concepts. *J. Quat. Sci.* **13**: 3–16.

Apel, J.R. 1987. Principles of Ocean Physics. London: Academic.

Appenzeller, C., T.F. Stocker, and M. Anklin. 1998. North Atlantic Oscillation in Greenland ice cores. *Science* **282**:46–449.

Berger, A., and M.F. Loutre. 1991. Insolation values for the climate of the last 10 million years. *Quat. Sci. Rev.* **10**:97–317.

Blunier, T., J. Chappellaz, J. Schwander et al. 1998. Asynchrony of Antarctic and Greenland climate change during the last glacial period. *Nature* **394**:739–743.

Clement, A., R. Seager, and M.A. Cane. 2000. Suppression of El Niño during the mid-Holocene by changes in the Earth's orbit. *Paleoceanography* **15**:731–737.

Fairbanks, R.G. 1989. A 17,000-year glacio-eustatic sea level record: Influence of glacial melting rates on the Younger Dryas event and deep-ocean circulation. *Nature* **342**: 637–642.

Ganopolski, A., and S. Rahmstorf. 2001. Rapid changes of glacial climate simulated in a coupled climate model. *Nature* **409**:153–158.

Labeyrie, L.D., J.C. Duplessy, J. Duprat et al. 1992. Changes in the vertical structure of the North Atlantic ocean between glacial and modern times. *Quat. Sci. Rev.* **11**: 401–413.

Johnsen, S.J., H.B. Clausen, W. Dansgaard et al. 1992. Irregular glacial interstadials recorded in a new Greenland ice core. *Nature* **359**:311–313,

Martinson, D.G., N.G. Pisias, J.D. Hays et al. 1987. Age dating and the orbital theory of the ice ages: Development of a high resolution 0 to 300,000-year chronostratigraphy. *Quat. Res.* **27**:1–29.

McManus, J.F., G.C. Bond, W.S. Broecker, et al. 1994. High-resolution climate records from the North Atlantic during the last interglacial. *Nature* **371**:326–329

McManus, J.F., D.W. Oppo, J.L. Cullen. 1999. A 0.5 million year record of millennial-scale climate variability in the North Atlantic. *Science* **283**:971–975.

McManus, J.F., D.W. Oppo, L.D. Keigwin, J.L. Cullen, and G.C. Bond. 2002. Thermohaline circulation and prolonged interglacial warmth in the North Atlantic. *Quat. Res.* **58**:17–21.

Oppo, D.W., L.D. Keigwin, J.F. McManus, and J.L. Cullen. 2001. Persistant suborbital climate variability in marine isotope stage 5 and Termination II. *Paleoceanography* **16**:280–292

Oppo, D.W., J.F. McManus, J.L. Cullen. 2003. Deepwater variability in the Holocene epoch. *Nature* **422**:277–278.

Rahmstorf, S. 2002. Ocean circulation and climate during the past 120,000 years. *Nature* **419**:207–214.

Rahmstorf, S. 2003. Timing of abrupt climate change: A precise clock. *Geophys. Res. Lett.* **30**:1510.

Sarnthein, M., K. Winn, S.J.A. Jung et al. 1994. Changes in east Atlantic deepwater circulation over the last 30,000 years: Eight time slice reconstructions. *Paleoceanography* **9**:209–267.

Siddall, M., E.J. Rohling, A. Almogi-Labin et al. 2003. Sea level fluctuations during the last glacial cycle. *Nature* **423**:853–858.

Sirocko, F., D. Garbe-Schönberg, A. McIntyre, and B. Molfino. 1996. Teleconnections between the subtropical monsoon and high latitude climates during the last deglaciation. *Science* **272**:526–529.

Stute, M., M. Forster, H. Frischkorn et al. 1995. Cooling of tropical Brazil (5°C) during the last glacial maximum. *Science* **269**:379–383.

Tiedemann, R., M. Sarnthein, and N.J. Shackleton. 1994. Astronomic timescale for the Pliocene Atlantic $\delta^{18}O$ and dust flux records of Ocean Drilling Program site 659. *Paleoceanography* **9**:619–638.

Toggweiler, J.R., and B. Samuels. 1993. New radiocarbon constraints on the upwelling of abyssal water to the ocean's surface. In: The Global Carbon Cycle, ed. M. Heimann, pp. 333–366. Berlin: Springer.

Tudhope, A.W., C.P. Chilcott, M.T. McCulloch et al. 2001. Variability in the El Niño-Southern Oscillation through a glacial–interglacial cycle. *Science* **291**: 1511–1517.

Voelker, A.H.L. et al. 2002. Global distribution of centennial-scale records for marine isotope stage (MIS) 3: A database. *Quat. Sci. Rev.* **21**:1185–1214.

Yu, E.-F., R. Francois, and M.P. Bacon. 1996. Similar rates of modern and last-glacial ocean thermohaline circulation inferred from radiochemical data. *Nature* **379**: 689–694.

9

What Is the Quaternary Phase-space Topology According to Cryosphere Dynamics?

A. J. PAYNE

School of Geographical Sciences, University of Bristol, Bristol BS8 1SS, U.K.

ABSTRACT

This chapter outlines the various mechanisms by which ice sheets can affect the evolution of the global climate system over Quaternary timescales. Discussion is divided into processes that contribute to the slow evolution of the system on the 100,000-year timescales of the ice ages and the rapid processes that could contribute to the variability observed within an individual glacial period. Particular attention is paid to the potential of ice sheets to act either as triggers of independent climate change or as amplifiers of changes initiated elsewhere.

QUATERNARY CLIMATE VARIABILITY

Growth and decay of ice sheets are, of course, the primary characteristics of the ice ages that dominate the Quaternary period. These approximately 100,000-year (100 ka) cycles are perhaps the most obvious feature of the Earth's climate record on timescales longer than the seasonal cycle. At their most recent maximum (the Last Glacial Maximum or LGM) 21 ka before present (B.P.), water stored in the terrestrial ice sheets accounted for approximately 120 m of sea-level depression relative to the present day. The locations and sizes of these former ice sheets, along with the present-day ice sheets of Greenland and Antarctica, are summarized in Table 9.1.

Ice sheets are, however, not merely passive recorders of the changing global environment. Their role as active components of the climate system has been highlighted in a string of modeling publications from the 1970s onwards. In this chapter, I will separate these interactions into two groups depending on the timescale over which they are thought to operate. The first set contains processes thought to contribute to the 100-ka cycles of the ice ages. Research in this area has a considerable pedigree dating back to the very early studies of the Quaternary. These processes will be referred to as "slow." A large literature,

Table 9.1 Estimates of present-day and maximum ice volumes (from Siegert 2002).

Region	Maximum volume (10^6 km^3)	Contribution to sea-level fall at LGM (m)	Present-day volume (10^6 km^3)	Potential contribution to sea-level rise (m)
North American (mainly Laurentide)	34.0	78–88	—	—
Eurasian Arctic (mainly Fennoscandian and Barents seas)	5.5	10–14	—	—
Greenland	4.0	2–3	2.9	7.2
Antarctica	37.0	14–18	25.7	61.1
All other ice masses	2.3	6	0.2	0.5
Total	82.8	110–129	28.8	68.8

including many useful reviews (e.g., Siegert 2001; Saltzman 2002), exists in this area and I will therefore concentrate on the second group of processes. These are processes that could generate the 6- to 10-ka cycles seen in many records of climate during the last glacial period (called the Weichselian, and separating the Eemian interglacial from the present-day Holocene), which will be referred to as "rapid." Particular attention has been paid to these processes because, in addition to having timescales more pertinent to society, they are characterized by phases of very intense, stepwise change.

The main theme of this chapter is to investigate the ways in which continental ice sheets can contribute to the variability of the global climate system over these timescales. Two possibilities exist: ice sheets could amplify changes generated elsewhere in the climate system or they could independently trigger change to which the rest of the climate system responds. These possibilities are not mutually incompatible, and both are thought to play contrasting roles over the slow and rapid timescales identified above.

I begin by briefly introducing terminology. Thereafter I discuss the ways in which ice sheets are thought to influence the evolution of climate on slow timescales. I then turn to the rapid timescales identified above and discuss evidence linking ice sheets with these processes, in particular the so-called Heinrich layers of the North Atlantic and the record of climate variability obtained from Greenland ice cores (for both glacial and interglacial periods). Finally, I review two sets of potential mechanisms by which ice sheets are thought to contribute to rapid climate variability. These are associated with the stability of the grounding line (which separates the floating and grounded portions of an ice mass) and with the potential for thermomechanical instability in the flow of ice sheets.

GLACIOLOGICAL BACKGROUND

We will focus exclusively on the continental ice sheets and their constitutive parts. Smaller terrestrial ice masses, such as glaciers and ice caps, will not be considered. Sea ice (frozen seawater) and permafrost (frozen land) will also be disregarded. This focus is primarily a function of the limited available space; however, the sheer size of the land-based ice sheets has led to a focus on their impacts within the literature. Vavrus and Harrison (2003) and Hibler and Flato (1992) offer useful reviews of the role of sea ice in the global climate system.

I start by introducing some basic definitions. Ice sheets are large masses of ice, typically between 1 and 4 km thick and having a horizontal extent of 10^6 to 10^7 km^2. They are therefore typified by extremely low aspect ratios of 10^{-3}. Ice sheets, in common with glaciers, gain mass primarily by snowfall. Snow is converted to firn and then to glacier ice on a timescale of ten to thirty years through various densification processes. The all-important feature of this ice is that it can deform, so that ice accumulated in one area can be lost in another. In the case of ice sheets, this horizontal transport occurs over thousands of kilometers.

There are three primary mechanisms of ice loss: First, ice can be lost via a variety of subaerial pathways, including direct sublimation, melt and subsequent runoff, and melt followed by water ponding and evaporation. The mass lost from the ice sheet by the combination of these processes is termed ablation. Much of the water generated by ablation makes its way into rivers and oceans on very short timescales, and the effect of this cold, fresh (fairly light) water on ocean circulation will be a recurring theme in this paper. In the Arctic, substantial volumes of meltwater are generated during summer only to be refrozen within the snow pack without contributing to ablation. The second method of loss is by submarine melt from the underside of floating ice shelves (see below), in which case there is a direct injection of light water at some depth (up to a kilometer) into the ocean circulation. Finally, ice may be lost by the calving of icebergs, either from floating ice shelves or directly from grounded ice. The frequency of calving events and the size of the icebergs produced vary enormously from daily events generating small icebergs (tens of meters) to decadal events generating the huge icebergs (10 km) found around Antarctica.

Ice sheets are typically separated into three distinct flow regimes, which will be referred to as inland ice sheet, ice streams (or outlet glaciers or fast flow features), and ice shelves. The interactions between these three components are crucial in determining the evolution of the whole ice-sheet system. Inland ice is very slow moving (1 to 10 m yr^{-1}) and flows primarily by the deformation of ice; however, it represents the bulk (> 90%) of ice within an ice sheet. Flow is controlled by a local stress balance between gravitational driving and basal traction, and largely depends on the magnitude of these stresses and on the temperature of the ice. Ice streams are distinct channels (~40 km wide) of relatively fast flow (1 km yr^{-1}) that drain the inland ice. They flow by slipping over the underlying substrate by either the deformation of the sediments on which they lie or decoupled sliding over the bed. The speed of ice streams is controlled by drag at the bed and at their lateral margins (where they interact with the surrounding inland ice). The type of substrate over which the ice is slipping and its water content largely control basal drag. Finally, ice shelves may form if local ice thickness is insufficient to prevent floatation in the surrounding oceans. In this case, basal drag is zero and rates of flow can be high ($1–10$ km yr^{-1}). Interaction between ice sheets and shelves across the grounding line will be discussed in more detail below.

Ice sheets are coupled in three main ways to the rest of the climate system. The first is associated with the high albedos of their snow and ice-covered surfaces. The albedo of clean snow ranges from 80–97%, whereas that of clean ice varies between 34–51% (Paterson 1994). This is many times greater than the albedos of other naturally occurring surfaces, such as water (3–10% at low zenith angles, and 10–100% at high angles), forest (10–25%), and bare soil (5–20%). Changes in ice-sheet extent, therefore, have a direct effect on the radiative balance of the Earth, with increased ice-sheet areas favoring increased reflectance of incoming shortwave radiation (insolation) and cooler temperatures. The second coupling arises because of the great depth of ice that is built up within ice sheets, which in this respect are equivalent to moving mountain ranges ~3 km high. The changes in Earth-surface topography associated with ice-sheet growth lead to cooler surface air temperatures ($\sim 6°–7°C$ km^{-1}) and the enhanced orographic blocking of atmospheric flow. This topographic effect is partially mitigated by the isostatic depression of bedrock beneath the growing ice sheet. Finally, the presence of an ice sheet dramatically alters the hydrological cycle. Ice sheets act as very long-term reservoirs of fresh water (residence times typically 1 to 10 ka) leading to sea-level changes. However, they may also release large quantities of fresh water (either as ice bergs or in meltwater floods) to the oceans. Several other more regional interactions also occur, for instance, the influence of strong catabatic winds generated over an ice sheet's surface on the thermal regime of the surrounding land and ocean.

SLOW PROCESSES

The results discussed in this section stem primarily from a class of models developed throughout the 1970s and 1980s. These models typically coupled a model of ice-sheet flow with simple energy-balance models (EBMs) of the atmosphere. The degree of complexity of these models varied considerably from simple, box-type models capable of analytical solution (Weertman 1976) to numerical models with a realistic (although coarse) land–sea distribution (e.g., Deblonde and Peltier 1991; Berger et al. 1990). These models were driven using seasonal variations in incoming shortwave solar radiation (insolation).

The well-known Milankovitch–Croll theory of the ice ages uses deviations in the seasonal distribution of incoming solar radiation to explain the growth and decay of the Northern Hemisphere ice sheets. These deviations are generated by variations in the properties of the Earth's orbit around the Sun (Figure 9.1). Three properties are of particularly interest. The first is precession, which varies with periods of 19 and 23 ka, and affects the time (in terms of season) at which the Earth reaches different points on its orbit (such as perihelion or closest approach to the Sun, and aphelion or furthest distance from the Sun). Precession controls the relative strengths of summer and winter insolation, with the expectation that "warm" Northern Hemisphere summers will occur when this hemisphere has its summer at a time of perihelion. The second property is that of obliquity, which controls the angle of tilt of the Earth relative to its orbital plane around the Sun (the ecliptic). This angle varies between 21° and 24° on a period of 41 ka, and again controls seasonality in that times of enhanced obliquity will be associated with a larger seasonal insolation cycle. The final property is eccentricity, a measure of the shape of the Earth's orbit, which can vary from nearly circular (where perihelion and aphelion are the same) to elliptical. Variations in eccentricity can change the total input of insolation to the Earth (in contrast to precession and obliquity, which only adjust its latitudinal and seasonal distribution) and occur on 100- and 400-ka cycles, as well as interacting with the precessional frequencies.

The comparison of the frequencies present in insolation predictions based on Milankovitch's calculations with long-term climate records (primarily the

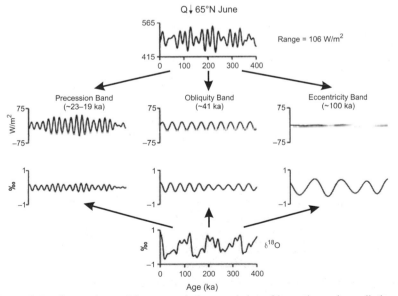

Figure 9.1 Comparison of the spectral characteristics of incoming solar radiation at 65°N (upper panel) and the deep-sea oxygen isotope record (lower panel) over the last 400 ka (after Imbrie et al. 1993).

oxygen isotope record available from deep-sea cores, which is taken as a close proxy of global land-ice volumes) reveals a very close correspondence with the 19-, 23-, 41-, and 100-ka periodicities occurring in both time series (Imbrie et al. 1993; Figure 9.1). However, two related issues remain. The first is the relative strength of the signals in that the 100-ka signal is dominant in the proxy record, whereas it accounts for very little of the variability in predicted insolation. Second, the changes in seasonal insolation generated by orbital variations are fairly small (e.g., at 9 ka B.P., Northern Hemisphere summers experienced 7% more insolation and 7% less winters; Saltzman 2002). A mechanism by which these small changes in forcing can be amplified within the Earth's climate system is clearly required.

Saltzman (2002) provides a very detailed account of the use of models in identifying the nature of this amplifier. The introduction of albedo coupling in models containing an EBM atmosphere and an ice-sheet model provides a mechanism for amplifying the small insolation changes discussed above (Pollard 1978; Oerlemans 1980a) — the mechanism being the survival of snow packs in Arctic Canada and Siberia through the relatively cool summers of an insolation minimum, which go on to initiate continental ice sheets. The long response times of these ice sheets then allow continued growth through subsequent insolation maxima. Although such models have demonstrated that the onset of ice ages is relatively easy to simulate, they often encounter difficulties in simulating the termination of an ice age.

A second theme in coupled EBM–ice-sheet modeling is the introduction of elevation–mass balance feedback. The coupling between surface elevation, air temperature, and mass balance implies that as an ice sheet thickens and experiences cooler surface air temperatures, its mass balance will become increasingly positive because ablation and the fraction of precipitation falling as snow are both strongly controlled by air temperature. In the presence of existing bedrock topography, this gives rise to the so-called "small ice cap" instability, which is a form of cusp catastrophe. In relatively mild climates, it is possible for small ice caps to exist on a mountain range because accumulation at higher altitudes can be balance by flow downhill and ablation in the lowlands (a set of small, stable equilibria). During a climate cooling, however, a point may be reached where the potential for ablation in the lowlands can no longer accommodate the ice accumulated above. A second set of equilibria therefore exists for very much larger (continental-scale) ice masses in which ablation and accumulation can only be balanced by flow over continental distances to lower latitudes or to the oceans (allowing calving to occur). One factor that opposes this type of behavior is the onset of an "elevation desert" effect over large ice sheets. This occurs at very cold temperatures ($< -20°$ to $-30°C$), where accumulation rates begin to fall with decreasing temperature because the amount of moisture that the atmosphere can hold begins to limit precipitation.

The isostatic response of bedrock to the load of the ice sheets introduces the potential for further complexity. This response also has a long timescale, and the interaction between this timescale and that of the ice sheet can potentially generate self-sustained oscillations (Oerlemans 1980b used a timescale of 30 ka). However, models that use a more realistic timescale (Peltier 1982 uses 3 ka) lose this property. Nevertheless, the lowering of ice-sheet surfaces because of isostatic depression is an important factor in counteracting the strong elevation–mass balance feedback and allows partial deglaciation to occur (Pollard 1982).

A common feature of many of the models discussed above is the need to introduce some additional physics in order to generate complete deglaciation at the end of an ice age. This may hint at the importance of rapid processes in controlling the transition from the glacial to the interglacial world. Pollard (1982) invoked iceberg calving into proglacial lakes as a mechanism, whereas Berger et al. (1990) employed a snow-aging mechanism. Finally, it should be mentioned that the atmosphere is treated in a very schematic way in most of these models, and the incorporation of more realistic atmospheric dynamics may remove the need for these deglaciation "fixes." An example is the recent work of Jackson (2000) and Roe (2002), who both simulate the interaction between the ice sheet's topography and atmospheric flow using a stationary wave analysis.

EVIDENCE IMPLICATING ICE SHEETS' INVOLVEMENT IN RAPID CLIMATE CHANGE

Over the last 15 years, a considerable body of evidence has developed suggesting that ice sheets play an active role in the climate system at timescales far shorter than those associated with the ice age cycles (considered above). This evidence stems principally from deep-sea sediment cores from the North Atlantic and from ice cores from the Greenland ice sheet. Most of the evidence relates to the last (Weichselian) glacial and to the preceding Eemian interglacial. Traditional glaciological thought suggests that ice sheets have very long response times (> 10 ka) because of their slow flow and immense size. Another factor that implies long response times is the fact that most (all) of the key processes affecting ice flow occur at or near the bed, while the effects of climate change will manifest themselves at the upper surface of the ice sheet. Any climate signal will therefore have to be propagated through 2 to 4 km of ice with the associated attenuation and delay. Clarke et al. (1999) estimate response times of 5 to 30 ka depending on the relative importance of diffusion (slower) and advection (faster) in transmitting the signal through the ice column.

If ice sheets do indeed play an active role at rapid (< 10 ka) timescales, then either they are triggering change independently or they are responding at timescales previously thought impossible. In support of the former possibility, there is a wealth of evidence suggesting that glaciers and smaller ice caps can exhibit cyclic behavior (surging) on decadal timescales. There is thus an expectation that ice sheets could also exhibit self-generated cyclic behavior on the assumption that glaciers are good scale models of their larger cousins.

The primary set of evidence implicating ice sheets in rapid climate change during the Weichselian is the existence of layers of centimeter-thick, terrigenous sand-sized sediment covering large areas of the North Atlantic in a spatially coherent pattern. The layers are known as the Heinrich layers, after their discoverer (Heinrich 1988), and there are six of them (named from youngest to oldest H1 to H6) occurring 6 to 16 ka apart (mean recurrence 11 ka). The Younger Dryas event (see below) has subsequently been included in this sequence as H0 (12.2 ka B.P.). A large body of literature exists on these layers, which can be summarized as follows. The only credible mechanism for their production is rafting on icebergs (because of the sediment's grain size). Their mineralogy strongly implies a source in the Hudson Bay region of Canada (Gwiazda et al. 1996). Maps of the thickness of each layer support this interpretation and show (Figure 9.2) a tongue of sediment reducing in thickness away from the Labrador Sea but covering virtually the whole zonal extent of the North Atlantic. The timing of the production of these layers has been compared to the climate record obtained from Greenland ice cores (Bond et al. 1992, 1993). The Heinrich layers occur at times of relative cold within the glacial period and are normally followed by a dramatic warming over Greenland of 3° to 6°C (see Figure 9.3). There is a hint that the other ice sheets fringing the glacial North Atlantic also generated significant layers of ice-rafted debris, and that the deposition of these layers was synchronous with that of the Heinrich layers (Bond and Lotti 1995).

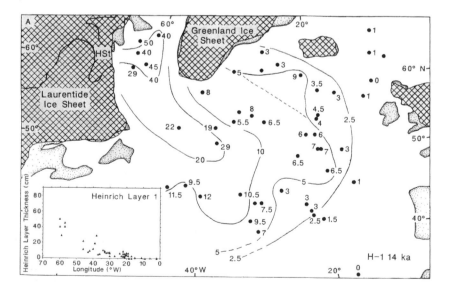

Figure 9.2 Thickness of Heinrich layer H1 across the North Atlantic. Figures are in centimeters and are based on magnetic susceptibility data from 51 deep-sea cores (from Dowdeswell et al. 1995; used with permission of the Geological Society of America).

I he ice-ratted origin ot these layers is irrefutable evidence linking their production to the ice sheets. It implies that the Laurentide ice sheet, in particular, experienced periods of dramatically increased iceberg production. These periods have been called Heinrich events and much of the rest of this chapter is concerned with potential mechanisms for their production. The release of huge numbers of icebergs into the North Atlantic is thought to affect global climate by freshening the surface water of the ocean dramatically. This then creates a lid of fresh, low-density water that prevents the density-driven formation of deep water in the northern North Atlantic and interrupts the operation of the global thermohaline circulation. The implications of these changes are discussed in greater detail elsewhere by Rahmstorf and Sirocko (this volume).

A similar mechanism is thought to have led to a very dramatic climatic reversal during the transition from glacial to interglacial conditions at the end of the Weichselian. The gradual retreat of the Laurentide and Eurasian ice sheets was punctuated by a 1.4-ka period in which temperatures fell back to full glacial conditions and the ice sheets either halted their retreat or advanced. This phenomenon (the Younger Dryas) is now reasonably well understood and is thought to have been caused by the release of massive amounts of meltwater from the retreating Laurentide ice sheet. In particular, the retreat of the ice sheet altered the topography of North America so that the primary route of drainage changed from flow down the Mississippi to the Gulf of Mexico, to flow along the Hudson and St. Lawrence rivers directly into the North Atlantic. This switch is thought

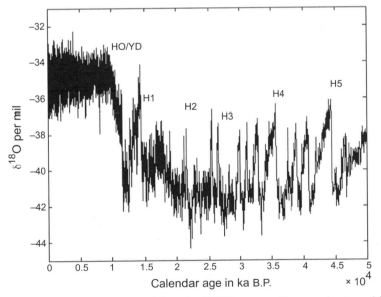

Figure 9.3 Oxygen isotope record from the GRIP ice core with approximate positions of the Heinrich layer shown (H0 to H5).

to have affected deepwater formation in the northern North Atlantic in the same way as described above.

Based on the importance of Heinrich events during the last glacial, it could be hypothesized that extensive Northern Hemisphere glaciation is a necessary prerequisite for rapid climate change because ice sheets provide a means of injecting very large quantities of fresh water on to the surface of the northern North Atlantic (thereby potentially interrupting deepwater formation). The climate records obtained from Greenland ice cores do, indeed, reveal the present Holocene period as being a period of great stability (Figure 9.3). A useful test of this hypothesis is to study the climate of the preceding Eemian interglacial for evidence of rapid climate change. The two Greenland ice cores (GRIP and GISP2) extend back into the Eemian. Their records for both the Holocene and Weichselian bear very close comparison, which implies that they reflect regional climate changes. Unfortunately, this similarity breaks down toward the bottom of the cores as ice from Eemian times is encountered. The down-core record in both cases implies very strong variability in the Eemian (Dansgaard et al. 1993). However, this variability is definitely attributable to ice folding and faulting at depth in the GISP2 core and is unlikely to have any climatic significance (Grootes et al. 1993). The origin of the Eemian-age variability in the GRIP core is less clear, although there is also evidence of disturbance in this core (Boulton 1993). Information from these ice cores cannot therefore be used to determine whether rapid climate variability in the North Atlantic occurs only during glacial periods. Ongoing drilling at a site north of the original GRIP site (N-GRIP) is aimed at obtaining an ice core that extends back to the Eemian but does not suffer from folding and faulting. The bedrock under this core is flat, unlike that at the GRIP and GISP2 sites.

There is, however, evidence from other sources that implies that the Eemian suffered at least one period of rapid climate change. Pollen records from central Europe (Field et al. 1994) and deep-sea cores in the eastern subtropical Atlantic (Maslin et al. 1998) support the existence of a single cold event during the middle of the Eemian about 122 ka B.P.

A second line of evidence that implies the potential for rapid climate change independent of the ice sheets are the Dansgaard/Oeschger events found in ice-core records of the Weichselian (Figure 9.3). Heinrich events are often associated with the cold phase of a Dansgaard/Oeschger event; however, not all Dansgaard/Oeschger events are associated with a Heinrich event. The two phenomena have been linked in so-called Bond cycles (Broecker 1994). Bond cycles have a saw-tooth shape with a gradual decrease in temperature occurring over a succession (4 to 5) of Dansgaard/Oeschger events, which is eventually terminated by a Heinrich event and the subsequent rise in temperature discussed above. The existence of Dansgaard/Oeschger events with no apparent direct association to a Heinrich event implies that there is a separate (probably oceanic) mechanism generating climate variability during a glacial, which intermittently interacts with the ice sheet based process that generates Heinrich events. Recent modeling (Rahmstorf and Ganopolski 2001; Rahmstorf 2003) implies that the ocean's thermohaline circulation may indeed be capable of abrupt switches in behavior without direct forcing by the ice sheets.

THERMOMECHANICAL INSTABILITY IN
THE FLOW OF ICE SHEETS

Two main mechanisms have been identified in the literature by which ice sheets could contribute to rapid climate change. They are associated with the behavior of the grounding line and with the interaction of ice flow and ice temperature.

Three types of thermomechanical instability are identified in ice-sheet models: (a) creep instability, (b) downstream transition from frozen to melting basal conditions, and (c) occurrence of warm-based ice encircled by frozen bed conditions. The first two processes are related to the temperature-dependence of the flow law for ice. The third relies on the geometry of the basal temperature field. These instabilities are internal in that they depend on flow-dependent thermal evolution of the ice sheet.

The flow of inland ice has the potential for unstable acceleration because the flow and temperature fields within an ice sheet are intimately linked. On one hand, the speed of ice flow depends sensitively on the temperature of ice (its viscosity obeys an Arrhenius-type exponential relation to temperature and can vary by three orders of magnitude over the natural temperature range of ice). On the other, the rate of ice flow partially controls temperature in that heat generated by frictional dissipation is a major component of the heat budget. Clarke et al. (1977) introduced the term "creep instability" for the process whereby an initial temperature anomaly leads to enhanced ice flow, increased dissipation, and further warming. This positive feedback process is only constrained by the onset of melt. Yuen et al. (1986) suggest that the process could lead to large-scale surging of the Antarctic ice sheet.

Initial numerical studies of creep instability employed vertical, one-dimensional models. However, Huybrechts and Oerlemans (1988) used a thermomechanical flowband model to assess the effects of changing ice-sheet geometry (affecting gravitational driving stress) and horizontal temperature advection. No indications of runaway warming were found and the modeled ice sheet responded to imposed climatic change in a well-behaved fashion.

Payne (1995) studied a related form of thermomechanical instability. In these models, instability arises because of an assumed abrupt increase in slip velocity with the onset of basal melting (i.e., ice stream flow). The sudden transition leads to a pronounced step in the ice-surface profile above the warm–cold ice transition. The steep surface slope, in turn, increases the gravitational driving stress and deformational velocity, and thus frictional dissipation also increases dramatically. The location of the warm–cold ice transition point can migrate

rapidly upstream as a consequence of this localized heating and associated enhanced flow, causing a surge. Eventually, reduced ice thickness and enhanced
cold-ice advection lead to stagnation. The validity of this instability mechanism
depends on the abruptness of the transition from frozen, immobile to
warm-based, sliding basal conditions. This is, in turn, determined by subglacial
hydrology and the deformation mechanism of subglacial sediments (Fowler and
Johnson 1996). A similar mechanism has been shown to operate in a three-dimensional model of the Laurentide ice sheet (Marshall and Clarke 1997; Calov
et al. 2002) and has generated Heinrich-type surges of the ice sheet. This process
is similar to the binge–purge mechanism of MacAyeal (1993), which was
largely schematic in nature and dealt with the transition between an ice sheet
dominated by inland-ice processes (slow flowing, cold based) and one that is
dominated by ice stream processes (fast flow, warm based).

Oerlemans (1983) and MacAyeal (1992) discuss a third form of internal instability. Both employ thermomechanical models that associate the presence of
basal meltwater with enhanced basal slip. The latter uses ice stream-specific
stress-balance equations but ignores horizontal temperature advection.
Oerlemans (1983) uses a constant climate forcing whereas MacAyeal (1992)
specifies a climate cycle according to the Vostok ice core record. In both cases,
the result is a cycle of slow ice-sheet growth and rapid discharge. The cyclic behavior relies on the development of melting in the interior of the ice sheet, where
ice is thick, while ice nearer the margins remains frozen to the bed. Eventually,
the pool of warm-based ice breaks through the encircling cold-based ice, leading to a large, rapid surge. The thin, post-surge ice sheet refreezes to its bed,
thickens over time, and the cycle repeats.

In summary, the coupling of ice flow and temperature introduces a number of
potential mechanisms for generating oscillatory behavior. Current glaciological
theory is therefore not inconsistent with the concept of ice sheets acting as triggers of the climate variability seen in the Quaternary.

INSTABILITIES ASSOCIATED WITH THE
GROUNDING LINE

Grounding lines separate the freely floating ice shelves from ice resting firmly
on the underlying bedrock. This transition has significance because the floating
ice experiences no basal drag and therefore flows very much faster than the
grounded ice. It is this observation that lies at the heart of "marine ice-sheet instability" hypothesis. Weertman (1957) derived an analytical expression for the
unidirectional spreading of confined floating ice, which implied that flow in the
shelf and, by assumption, at the grounding line was proportional to ice thickness
raised to power of three. This could then potentially generate instability because
flow at the grounding line should increase rapidly as the grounding line retreats
exposing progressively thicker grounded ice.

Subsequent modeling studies of grounding line migration and West Antarctic ice-sheet stability (Mercer 1978; Thomas and Bentley 1978) predicted rapid grounding line retreat into the inland-deepening basin. The crucial assumption in this analysis is that the stress regime at the grounding line can reasonably be approximated by that of the ice shelf. Hindmarsh (1993) argues that the transition zone between grounded ice sheet and floating ice shelf is of such limited longitudinal extent that it is unlikely for such transmission to take place. Numerical simulations (Herterich 1987) and an analytical analysis of stick-slip transitions (Barcilon and MacAyeal 1993) support this view. A consequence of this assumption is that the inland ice does not react to changes in the ice shelf. Hypotheses based on the disintegration of an ice shelf (and its subsequent effect on the flow of the inland ice) are therefore thought unlikely.

The Hindmarsh (1993) analysis suggests that the grounding-line system possesses neutral equilibria rather than the unstable equilibria proposed by Thomas and Bentley (1978). One of the many issues that this assertion raises is that numerical models of the system may not be sensitive enough to capture the dynamics of the underlying physical system. Predictions generated by numerical models may vary with the details of their numerical implementation and are therefore untrustworthy unless a detailed sensitivity analysis has been performed. This area of research remains extremely active and no general conclusions are yet possible, although the dire predictions of the marine instability hypothesis now seem unlikely. The original Hindmarsh (1993) work analyzed the junction between inland ice and an ice shelf; however, it is not clear whether its conclusions are also applicable to the junction between an ice stream and an ice shelf.

Other Fast-response Mechanisms

There are several other mechanisms that could allow ice sheets to respond to climate change at rates very much faster than those traditional accepted (> 1 ka). This possibility becomes important if the observation that several of the ice sheets fringing the Quaternary North Atlantic did indeed surge in synchrony. In order for this to happen, the climate signal imposed at the upper surface of the ice sheet must be translated to the ice-sheet bed at rates very much in excess of those associated with advection and diffusion.

Clarke et al. (1999) note that changes in the mass balance of an ice sheet will rapidly affect the geometry of the ice-sheet surface. This will alter the stress field within the ice sheet instantaneously. There is then the potential of changing the thermal regime of the ice bed because frictional dissipation is largely dependent on the stress field. A similar mechanism would be appropriate if the rheology of any deforming substrate was plastic and the changing stress regime crossed the yield stress of the substrate.

A recent observation by Zwally et al. (2002) prompts a further mechanism. Zwally et al. (2002) monitored surface velocity at a point near the equilibrium line of the Greenland ice sheet and found seasonal variations. The most logical interpretation of this finding is that meltwater generated at the ice-sheet surface is finding its way to the bed and affecting basal slip. This process is very well known in valley glaciers but was previously thought unlikely at ice-sheet scales (because of the great depths of very cold ice involved). This then provides a mechanism by which changing air temperatures can immediately generate changes in the flow of the ice sheet.

One further possibility is that of the weak coupling of nonlinear oscillators. This requires two sets of assumptions: (a) that we assume that the ice sheets fringing the North Atlantic were indeed surging independently by one of the mechanisms outlined above and (b) that we invoke some mechanism by which these separate ice sheets are weakly coupled. Examples of this coupling could be changes in precipitation rates associated with changing atmospheric and oceanic circulation patterns. One of the properties of this type of system is that behavior of the independent oscillators becomes synchronized through time (they become phase locked). Recently, Calov et al. (2002) found some evidence for this in a global climate model that incorporated surging ice sheets. However, phase locking usually only happens after many oscillations have occurred. It is not clear whether there were sufficient Heinrich events (a total of six recorded) to allow phase locking to occur.

SUMMARY

In this chapter I have highlighted evidence that implicates ice sheets as key players in the global climate system at both slow (100 ka) and rapid (< 10 ka) timescales. The role of ice sheets in the Milankovitch–Croll theory of the ice ages is well known. Albedo and elevation–mass balance feedbacks are capable of amplifying small variations in incoming insolation and can be used to explain the initiation phase of Northern Hemisphere glaciation; however, the termination of a glaciation is less well understood.

The role of ice sheets in generating climate variability within an ice-age cycle is less clearly understood. It appears that ice sheets (in particular the Laurentide) are intimately linked to the generation of Heinrich layers in the North Atlantic. It is likely that these layers are generated by large influxes of icebergs and that the meltwater released by these icebergs has a dramatic effect on the climate of the North Atlantic region. However, the origin of these surges in iceberg production is unclear. The literature supports two interpretations. First, that the surges are generated by internal instabilities of the Laurentide ice sheet (possibly of a thermomechanical origin). Several modeling studies indicate that this type of behavior is indeed possible. The complication associated with this mechanism is the possible synchronicity of surges from a number of separate ice sheets fringing the North Atlantic. The weak coupling of these separate oscillators could possibly be used to reconcile these observations with a surge-based theory. The second interpretation is that the ice sheets are amplifying changes generated elsewhere in the climate system (an obvious origin would be the postulated Dansgaard/Oeschger events of the oceans).

ACKNOWLEDGMENTS

I would like to thank Martin Claussen and Stefan Rahmstorf for their very helpful comments during the review of this paper.

REFERENCES

Barcilon, V., and D.R. MacAyeal. 1993. Steady flow of a viscous ice stream across a no-slip free-slip transition at the bed. *J. Glaciol.* **39**:167–185.

Berger, A., H. Gallee, T. Fichefet, I. Marsiat, and C. Tricot. 1990. Testing the astronomical theory with a coupled climate–ice sheet model. *Global Planet. Change* **89**: 125–141.

Bond, G., W. Broecker, S. Johnsen et al. 1993. Correlations between climate records from North Atlantic sediments and Greenland ice. *Nature* **365**:143–147.

Bond, G., H. Heinrich, W. Broecker et al. 1992. Evidence for massive discharges of icebergs into the North Atlantic ocean during the last glacial period. *Nature* **360**:245–249.

Bond, G., and R. Lotti. 1995. Iceberg discharges into the North Atlantic on millennial time scales during the last deglaciation. *Science* **267**:1005–1010.

Boulton, G.S. 1993. Two cores are better than one. *Nature* **366**:507–508.

Broecker, W.S. 1994. Massive iceberg discharges as triggers for global climate change. *Nature* **372**:421–424.

Calov, R., A. Ganopolski, V. Petoukhov, M. Claussen, and R. Greve. 2002. Large-scale instabilities of the Laurentide ice sheet simulated in a fully coupled climate-system model. *Geophys. Res. Lett.* **29**:2216.

Clarke, G.K.C., S.J. Marshall, C. Hillaire-Marcel, G. Bilodeau, and C. Veiga-Pires. 1999. A glaciological perspective on Heinrich events. In: Mechanisms of Global Climate Change at Millennial Time Scales, ed. P.U. Clark, R.S. Webb, and L.D. Keigwin. Geophys. Monograph 112, pp. 243–262. Washington, D.C.: Am. Geophys. Union.

Clarke, G.K.C., U. Nitsan, and W.S.B. Paterson. 1977. Strain heating and creep instability in glaciers and ice sheets. *Rev. Geophys. & Space Phys.* **15**:235–247.

Dansgaard, W., S.J. Johnsen, H.B. Clausen et al. 1993. Evidence for general instability of past climate from a 250-ka ice-core record. *Nature* **364**:218–220.

Deblonde, G., and W.R. Peltier. 1991. A one-dimensional model of continental ice volume fluctuations through the Pleistocene: Implications for the origin of the mid-Pleistocene climate transition. *J. Climate* **4**:318–344.

Dowdeswell, J.A., M.A. Maslin, J.T. Andrews, and I.N. McCave. 1995. Iceberg production, debris rafting, and the extent and thickness of Heinrich layers (H-1, H-2) in North Atlantic sediments. *Geology* **23**:301–304.

Field, M., B. Huntley, and H. Muller. 1994. Eemian climate fluctuations observed in a European pollen record. *Nature* **371**:779–782.

Fowler, A.C., and C. Johnson. 1996. Ice sheet surging and ice stream formation. *Ann. Glaciol.* **23**:68–73.

Ganopolski, A., and S. Rahmstorf. 2001. Rapid changes of glacial climate simulated in a coupled climate model. *Nature* **409**:153–158.

Grootes, P.M., M. Stuiver, J.W. White, S. Johnsen, and J. Jouzel. 1993. Comparison of oxygen isotope records from the GISP2 and GRIP Greenland ice cores. 1993. *Nature* **366**:552–554.

Gwiazda, R.H., S.R. Henning, and W.S. Broecker. 1996. Provenance of icebergs during Heinrich event 3 and the contrast to their sources during other Heinrich episodes. *Paleoceanography* **11**:371–378.

Hays, J.D., J. Imbrie, and N.J. Shackeleton. 1976. Variations in the Earth's orbit: Pacemaker of the ice ages. *Science* **194**:1121–1132.

Heinrich, H. 1988. Origin and consequences of cyclic ice rafting in the northeast Atlantic ocean during the past 130,000 years. *Quat. Res.* **29**:142–152.

Herterich, K. 1987. On the flow within the transition zone between ice sheet and ice shelf. In: Dynamics of the West Antarctic Ice Sheet, ed. C.J. van der Veen and J. Oerlemans, pp. 185–202. Dordrecht: D. Reidel.

Hibler III, W.D., and G.M. Flato. 1992. Sea ice modeling. In: Climate Systems Modeling, ed. K.E. Trenberth, pp. 741–757. Cambridge: Cambridge Univ. Press.

Hindmarsh, R.C.A. 1993. Qualitative dynamics of marine ice sheets. In: Ice in the Climate System, ed. W.R. Peltier, NATO ASI Series I, vol. 12, pp. 67–99. Heidelberg: Springer.

Huybrechts, P., and J. Oerlemans. 1988. Evolution of the East Antarctic ice sheet: A numerical study of thermomechanical response patterns with changing climate. *Ann. Glaciol.* **11**:52–59.

Imbrie, J., A. Berger, E.A. Boyle et al. 1993. On the structures and origin of major glaciation cycles: 2. The 100,000-year cycle. *Paleoceanography* **8**:699–735.

Jackson, C. 2000. Sensitivity of stationary wave amplitude to regional changes in Laurentide ice sheet topography in single-layer models of the atmosphere. *J. Geophys. Res.* **105(D19)**:24,443–24,454.

MacAyeal, D.R. 1992. Irregular oscillations of the West Antarctic ice sheet. *Nature* **359**: 29–32.

MacAyeal, D.R. 1993. Binge/purge oscillations of the Laurentide ice sheet as a cause of the North Atlantic's Heinrich events. *Paleoceanography* **8**:775–784.

Marshall, S.J., and G.K.C. Clarke. 1997. A continuum mixture model of ice stream thermomechanics in the Laurentide ice sheet. 2. Application to the Hudson Strait ice stream. *J. Geophys. Res.* **102(B9)**:20,615–20,637.

Maslin, M.A., M. Sarnthein, J.-J. Knaack, P. Grootes, and C. Tzedakis. 1998. Intra-interglacial cold events: An Eemian–Holocene comparison. In: Geological Evolution of the Ocean Basin, Ocean Drilling Program (ODP) Results, ed. A. Cramp, C. MacLeod, S.V. Lee, and E.J. Jones, vol. 131, pp. 91–99. College Station TX: ODP.

Mercer, J.H. 1978. West Antarctic ice sheet and CO_2 greenhouse effect: A threat of disaster. *Nature* **271**:321–325.

Oerlemans, J. 1980a. Continental ice sheets and the planetary radiation budget. *Quat. Res.* **14**:349–359.

Oerlemans, J. 1980b. Model experiments on the 100,00-yr glacial cycle. *Nature* **287**:430–432.

Oerlemans, J. 1983. A numerical study on cyclic behaviour of polar ice sheets. *Tellus* **35**:81–87.

Paterson, W.S.B. 1994. The Physics of Glaciers, 3rd ed. Oxford: Butterworth.

Payne, A.J. 1995. Limit-cycles in the basal thermal regime of ice sheets. *J. Geophys. Res.* **100(B3)**:4249–4263.

Peltier, W. 1982. Dynamics of the ice age Earth. *Adv. Geophys.* **24**:1–146.

Pollard, D. 1978. An investigation of the astronomical theory of the ice ages using a simple climate–ice sheet model. *Nature* **272**:233–234.

Pollard, D. 1982. A simple ice sheet model yields realistic 100 ky glacial cycles. *Nature* **296**:334–338.

Rahmstorf, S. 2003. Timing of abrupt climate change: A precise clock. *Geophys. Res. Lett.* **30**:1510.

Roe, G.H. 2002. Modeling precipitation over ice sheets: An assessment using Greenland. *J. Glaciol.* **48**:70–80.

Saltzman, B. 2002. Dynamical Paleoclimatology: Generalized Theory of Global Climate Change. San Diego: Academic.

Siegert, M.J. 2001. Ice Sheets and Late Quaternary Environmental Change. Chichester: Wiley.

Siegert, M.J. 2002. Role of ice sheets and glaciers in climate and the global water cycle. In: Encyclopedia of Hydrological Sciences, ed. M.G. Anderson et al., HSA 270. Chichester: Wiley.

Thomas, R.H., and C.R. Bentley. 1978. A model for Holocene retreat of the West Antarctic ice sheet. *Quat. Res.* **10**:150–170.

Vavrus, S., and S.P. Harrison. 2003. The impact of sea-ice dynamics on the Arctic climate system. *Climate Dyn.* **20**:7–8.

Weertman, J. 1957. The deformation of floating ice shelves. *J. Glaciol.* **3**:38–42.

Weertman, J. 1976. Milankovitch solar radiation variations and ice age ice sheet sizes. *Nature* **261**:17–20.

Yuen, D.A., M.R. Saari, and G. Schubert. 1986. Explosive growth of shear-heating instabilities in the down-slope creep of ice sheets. *J. Glaciol.* **32**:314–320.

Zwally, H.J., W. Abdalati, T. Herring et al. 2002. Surface melt-induced acceleration of Greenland ice sheet flow. *Science* **297**:218–222.

Back: Bob Scholes, Hermann Held, Tony Payne, Martin Claussen, and Stefan Rahmstorf
Front: Paul Falkowski, Quirin Schiermeier, Dan Schrag, Frank Sirocko, Victor Brovkin,
     and Andy Watson

10

Group Report: Possible States and Modes of Operation of the Quaternary Earth System

A. J. WATSON, Rapporteur

V. BROVKIN, M. CLAUSSEN, P. G. FALKOWSKI, H. HELD,
A. J. PAYNE, S. RAHMSTORF, R. J. SCHOLES,
D. P. SCHRAG, and F. SIROCKO

INTRODUCTION: WHY STUDY THE QUATERNARY?

The Quaternary, that is, the last approximately two million years of Earth's history, was a period that exhibited substantial variability and cyclicity in the climate of the Earth. For this period, we have by far the most extensive knowledge, as records of change preserved in recent sediments and in ice cores are the most detailed for the Quaternary. Only in the Quaternary do we have, for example, a record of changes in atmospheric composition from gas trapped in ice cores.

From this detailed knowledge, an extraordinary picture has emerged: the Quaternary climate behaved as a *system* of interlinked parts, changing in a complex but ordered way, in a sequence that recurs, with variations, like the themes of a symphony. The individual parts making up this composition were affected by changes in the physical circulation of the atmosphere and ocean, the coverage of vegetation, global ice sheets, marine biota, concentrations of carbon dioxide, methane, dust, and precipitation around the world. Unlike (most) pieces of music, however, this overall structure was not imposed by a single composer or conductor; rather, it was an emergent property of this nonlinear system of interlinked parts.

The study of the Quaternary can, therefore, allow us to probe the dynamics of the Earth climate system in all of its complexity. We can test our understanding of how we think this system works against the data that paleoclimate studies reveal. This should help us understand how this system will respond to the challenges of global change that we are now forcing on the planet. This may help us decide if we are pushing this system to the extent that we risk catastrophic, or at least unpleasant, outcomes.

THE QUATERNARY IN THE CONTEXT OF THE LONG-TERM EVOLUTION OF THE PLANET

Compared to most of Earth history, the Quaternary is characterized by comparatively low planetary temperatures. These gave rise to permanent ice caps at both poles and a cycle of climate in which the Northern ice sheets regularly expanded and contracted. The low temperatures were in part due to low concentrations of carbon dioxide in the atmosphere. At below 200 parts per million (ppm), CO_2 concentration during the cold phases of the Quaternary glacial cycles would have been a severely limiting factor for C3 carbon fixation, used by the great majority of plants on Earth. Models of atmospheric evolution suggest that CO_2 concentrations have very rarely, if ever, been so low in the more distant past (Berner 1994; Bergman et al. 2004).

Atmospheric CO_2 is believed to be influenced by the rate at which tectonic plate spreading occurs, this being the source of CO_2 to the atmosphere–ocean–biosphere system, and by the volume of sedimentary carbonate rocks. Over hundreds of millions of years, it is likely to be influenced by the Wilson cycle of continental rifting, ocean basin formation, subduction, and super-continent formation. According to this view, at times when there are extensive carbonate sedimentary rocks stored in ocean basins and margins, CO_2 concentration will be low because most of the carbon is locked up there. Subsequently, following subduction and metamorphism of these rocks, atmospheric CO_2 concentrations rise. Superimposed on this cycle may be an even longer-term trend toward low CO_2 values. With the decline in radiogenic heating of the interior, the Earth's mantle is now slowly cooling and the previously degassed volatiles, such as CO_2 and water, are slowly being re-incorporated into the planet. Ultimately, as the interior cools sufficiently, the oceans will dry up and CO_2 will disappear from whence it came, leaving a dry and lifeless planet.

What Do We Know, and What Do We Still Have to Learn about the Quaternary?

Table 10.1 provides a summary of some of the major features of the Quaternary climate, the areas in which we believe there is some consensus, and those on which there is none. Below, we address the points in the table in more detail.

Major Transitions and Orbital Forcing in the Quaternary

Figure 10.1 depicts the $\delta^{18}O$ record over the last 4 Ma. As recorded in benthic foraminifera, $\delta^{18}O$ responds to local sea temperature and to the total volume of ice on the planet in about equal measure; in both cases, negative values correspond to higher temperatures. These and similar records show that, over the entire period, variability in temperatures at periods ~20 ka^{-1}, typical of precessional forcing, can be discerned. Between 3 and 2 Ma B.P., the record begins a

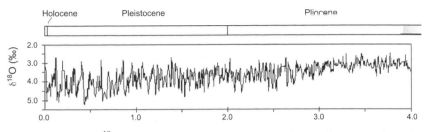

Figure 10.1 The $\delta^{18}O$ record over the last 4 Ma as shown in a sediment core from the Atlantic (after Tiedemann et al. 1994).

slow drift toward lower temperatures, and starting at about 2.6 Ma there is a transition to a dominant frequency ~40 ka^{-1}, associated with orbital obliquity forcing. At about 0.9–0.7 Ma, in the "mid-Pleistocene transition," the ~100-ka cycle appears and dominates thereafter.

Most researchers agree that these dominant frequencies are associated with the Milankovitch periods with which the orbital parameters of the Earth vary. The ~20-ka precessional and ~40-ka obliquity periods can be explained by moderately linear responses to the forcing, although there is no generally accepted explanation for why the dominant response should change with time at the transitions as described above (see Claussen et al., this volume). The 100-ka period remains a puzzle, because although orbital eccentricity does vary with a period close to 100 ka, there is very little power in the changes in insolation occurring at this frequency. No quasi-linear theory can therefore explain why this period is dominant in the late Quaternary. This is no real surprise, since we now appreciate that the components of the climate system are mostly very nonlinear. Following Hays et al. (1980), it is common to refer to orbital forcing as the "pacemaker" for climate change, since it is believed that it provides the timing or synchronization, but it is not clear whether it directly drives the change. The association of the dominant frequencies with orbital forcing could, for example, be due to phase-locking to the orbital periods of free oscillations that would in any case be present in the climate system (Saltzman and Verbitsky 1993).

Ice Sheets and Sea Level

The major feature distinguishing the long, cold glacial phases of the late Quaternary from the relatively brief warm phases is the buildup of ice on the continents (see Claussen et al., this volume). The major increases occurred in the Northern Hemisphere, on the North American continent (the Laurentide ice sheet), and on the Eurasian continent (the Fenno-Scandinavian ice sheet). The buildup of ice extracted fresh water from the oceans and resulted in a drop in sea level that exceeded 120 m compared to the present day. The ice sheets built up in incremental stages but tended to disintegrate relatively rapidly.

The presence of the ice sheets would have altered atmospheric and oceanic circulation, radiative balance, air, land, and ocean temperatures substantially.

Table 10.1   Points of agreement on the Quaternary.

Item	Points of agreement	Outstanding or unresolved issues
Ice and sea level	Buildup of N. American and Eurasian ice sheets as well as increases in existing ice sheets during glacial times. Ice buildup resulted in a maximum ~130 m reduction in sea level during glacial periods.	Mechanisms accounting for buildup and destabilization of ice sheets are not fully established.
Dominant periods observed in the records	Orbital forcing is the probable "pacemaker." 20- and 40-ka cycles may be reasonably linear responses to precessional and obliquity orbital forcing, respectively.	No agreed explanation for the strong response at the 100-ka orbital cycle.
Transition into the Quaternary	~2.6 Ma B.P. transition from smaller $\delta^{18}O$ variations at ~20 ka period to larger variations at period ~40 ka.	Cause of the transition is disputed.
Mid-Pleistocene transition	~900 ka B.P. transition to dominant period ~100 ka.	Cause of the transition is disputed.
CO_2, other greenhouse gases, and temperature are highly correlated on orbital timescales	Positive feedbacks link CO_2 and temperature. CO_2 oscillates between bounds ~190 ppm to 280 ppm.	Influence of CO_2 on temperature is well understood, but influence of temperature on CO_2 is still unresolved. A number of competing theories exist about why the ocean takes up more CO_2 during cold phases. What sets the bounds of CO_2 variation is not resolved.

Toward the end of the glacial periods, it is apparent that the ice sheets became unstable, leading to their relatively rapid collapse. The reasons for this instability are not properly understood, but they seem to involve the whole complex of feedback subsystems, described further below, that link the physical and biogeochemical climate systems.

Correlation between Carbon Dioxide, Methane, and Temperature on Orbital Timescales

Figure 10.2 shows the covariation between temperature, CO_2, CH_4, and local temperature (deduced from deuterium content of ice) in the Vostok core (data

Table 10.1 (continued)

Item	Points of agreement	Outstanding or unresolved issues
Dust and hydrological cycle	Glacial periods are drier and dustier than interglacial periods.	Influence of temperature on hydrological cycle broadly understood but details are not. Dust in Vostok core leads changes in temperature.
Terrestrial vegetation	Terrestrial biota have lower biomass in glacial periods.	Extent of the change is disputed.
Feedbacks	Positive feedbacks amplify climate sensitivity to insolation. Feedbacks include ice albedo, water vapor, land surface albedo, CH_4, and atmospheric CO_2.	Strength and importance of some of these are uncertain. Mechanisms for CO_2 feedback are uncertain.
Suborbital changes	Rapid warming events are recorded in Greenland ice cores during last glacial period (Dansgaard/Oeschger events). Dominant theory is that these are related to changes in ocean circulation in the N. Atlantic. Cold events are associated with ice-rafted debris layers (Heinrich events), indicating massive amounts of iceberg calving.	Unclear what triggers the changes in ocean circulation.

from Petit et al. 1999). It is well understood that increases in these greenhouse gases will tend to increase global temperatures, so increased concentrations during interglacials will contribute to the warm climate. However, the increased concentrations are not the sole cause of the climate change, and the timing at least is derived from astronomical variation. It follows that there must be a positive feedback in which the change to a warmer climate drives up the concentrations of the greenhouse gases; however, this linkage is not well understood.

In the case of CH_4, it is believed that a warmer, therefore wetter, climate will favor CH_4 release from wetlands and waterlogged soils. For CO_2, there is agreement that the increase with a warmer climate must involve re-partitioning of CO_2 from the oceans to the land and atmosphere. However, the mechanisms giving rise to this are still controversial. Many papers have been written on this issue and ca. 10 hypotheses in the literature attempt to explain this transition; still, no single mechanism seems to match all the data (for a review see, e.g., Sigman and Boyle 2000; Archer 1999). Thus, almost certainly, the true causes require several of these mechanisms acting in concert.

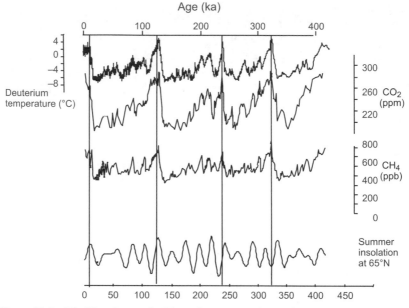

Figure 10.2   The Vostok record of deuterium content (proxy for local temperature), atmospheric CO_2, and CH_4. Redrawn from Petit et al. (1999).

Atmospheric CO_2 oscillates regularly between a low of ~190 ppm in peak glacial time and a high of 280 ppm during the interglacials. There appears also to be a third quasi-stable level, at about 210 ppm, that is frequently reached during glacial times (Ridgwell and Watson 2003). Just as there is no agreed mechanism for CO_2 variations, there is no consensus as to the reasons for these boundary values. It is notable that the lower value seriously limits net production of C3 plants. If terrestrial vegetation is an important factor in controlling the long-term rate of consumption of CO_2 by weathering of silicate minerals (Lovelock and Watson 1982; Berner 1994), this may help to determine the lower limit.

Hydrological Cycle and Aeolian Dust

It is clear that the glacial periods were much drier than the interglacials. The fundamental cause of this is well understood: it is the rapidly decreasing water vapor content of the atmosphere at lower temperatures, due to the lower vapor pressure of water evaporating over the oceans. However, the distribution in space of water vapor and precipitation in glacial times is important to a proper understanding of the buildup and decay of ice sheets. Our knowledge is inadequate in detail here. We need a better understanding of both the ocean and atmosphere interactions and how modes of variability, such as the El Niño Southern Oscillations (ENSO) and North Atlantic Oscillations (NAO), were affected by glacial climate.

During glacial times, there was much more windblown dust in the atmosphere as a result of the drier, more desert-like conditions that are associated with low vegetation cover (see below), and high winds probably prevailed. The precise controls on dust are quite subtle, however. Aeolian dust scatters short-wave radiation in the atmosphere, absorbs long-wave radiation, and can exert either a cooling or warming effect. It is also a major source of the limiting nutrient iron to the oceans. As such, it is an important player in the climate system and will be discussed more extensively below (see Table 10.2).

Terrestrial Vegetation

There is agreement that terrestrial biomass was lower during glacial time than interglacials. Evidence for this comes from several sources. Interpretation of deep-sea $\delta^{13}C$ records from foraminifera suggest that the glacial "terminations," or the global reservoir of organic carbon, increased substantially. Since

Table 10.2 Critical processes in the Quaternary system and likely changes in the future.

Process	Timescale of change (years)	Probability of major change	Uncertainty of our knowledge	Impact of these potential changes on society: An educated guess
Denitrification and phosphorus recycling	100–1000	Moderate	High	Low
Thermohaline circulation thresholds	10–100	High	Moderate	Moderate
Inland ice	100–10,000	Low	Low	High
Biogeophysical feedback	10–100	High	High	Low
CH_4 release from wetlands	10–100	High	Moderate	Low
Dust feedback	10–100	Moderate	High	Moderate
CO_2 release from land biosphere	10–1000	Moderate	High	High
$CO_2 + CH_4$ release from permafrost	10–1000	Moderate	High	Low
Sea ice	10–100	High	Moderate	Moderate
Gas hydrates	10–100	Low	High	High
Atmospheric circulation (ENSO, NAO, monsoon)	1–100	High	High	Moderate

terrestrial vegetation and soils are by far the largest reservoir of organic carbon on the planet, this indicates an increase in the biomass in these reservoirs between glacial and preindustrial time. Inventories of glacial vegetation cover made from sedimentological estimates of vegetation type suggest that the increase was substantially larger (Prentice et al. 2000). Finally, global models of vegetation type, when run for glacial conditions of precipitation, temperature, and atmospheric CO_2 concentration, also indicate that land biomass at this time was much less.

There is agreement, therefore, on the direction of the change, but not on how large the change was. Estimates range from 400 Pg C to 1500 Pg C by different methods.

Suborbital Changes

A major focus of research over the last decade has been on changes occurring at faster rates than can be explained by Milankovitch-type mechanisms. In particular, in $\delta^{18}O$ from ice cores in Greenland and North Atlantic sediment records, rapid warming events are seen in the later part of the last glacial period. These Dansgaard/Oeschger (D/O) events, of which some twenty have been identified, begin with a warming that may be as much as 8°C, occurring in a few decades. The warm phase may last a few hundred years or longer and is terminated by a less rapid cooling.

There is near-universal agreement that these events are associated with a change in the thermohaline circulation (THC) in the North Atlantic, with THC switching rapidly between two states, one of which transports more heat to the north than the other (Rahmstorf 2002). Such a switching behavior has long been known to occur in highly idealized conceptual models of the THC (Stommel 1961). However, what causes THC to switch between modes is unknown, and there is no consensus as to the details of the modes.

In addition, cold events associated with the launching of large numbers of icebergs into the North Atlantic have also been identified. These are termed Heinrich events and are marked in North Atlantic sediments by layers rich in ice-rafted debris, which appears to originate from the Hudson Bay area of Canada. Heinrich events and D/O events sometimes seem to be synchronized, so that Heinrich events occur immediately following a D/O event.

MAJOR FEEDBACK SYSTEMS, THRESHOLDS, AND CRITICAL POINTS

As discussed above, the major feature of the Quaternary climate system that makes it so interesting to study, and yet so difficult to understand, is the multiplicity of subsystems. These are mostly very nonlinear and subject to thresholds, multiple regions of stability, and transitions between regions of phase space.

Below, we discuss several of the subsystems and possible critical transitions which appear to be most important, and assess whether they may pose a threat as a result of human-induced climate change.

Denitrification and Phosphorus Recycling

It is known that the concentration of CO_2 in the atmosphere is dependent on the concentration of the limiting nutrient in the oceans, and in many parts of the ocean this is nitrogen. The residence time of nitrate in the ocean has been shown to be quite short, $<10^4$ years, and it is believed that the main terms in the budget are a balance between N_2 fixation in tropical and subtropical surface oceans and denitrification on continental shelves and in suboxic parts of the ocean. Therefore, it is apparent that either an increase in rates of nitrogen fixation or a decrease in rates of denitrification could increase the drawdown of CO_2. Nitrogen fixation rates may be sensitive to increases in the supply of aeolian dust to the tropics, whereas denitrification rates have tended to be less when there was less continental shelf area; therefore, we might expect more drawdown in the glacial ocean from this mechanism.

Phosphorus is close to co-limitation of productivity virtually everywhere in the oceans where nitrate is limiting. The efficiency of changes in the nitrogen balance of the oceans as a driver of CO_2 change depends on how the P:N requirement of plankton is assumed to change. In the limit that the Redfield P:N ratio is invariant, relatively small increases in CO_2 drawdown, ~20 ppm, could be realized before P limitation would set in, whereas if the ratio is plastic, substantially larger changes would be possible.

Humans have greatly increased the fixation of nitrogen and the rate at which it is washed to the sea through rivers. Fortunately, it appears that the majority of this excess nitrogen is denitrified in estuaries and the coastal ocean before it reaches the open sea. However, we expect that there will be an increase in the loading of the macronutrients P and N in the oceans during the Anthropocene.

Thermohaline Ocean Circulation Changes

Two known feedbacks render the THC a highly nonlinear system: an advective feedback and a convective feedback. Both are positive and act through salinity (Rahmstorf et al. 1996). As discussed above, we are moderately certain that during the Quaternary the THC underwent major and rapid transitions. These probably had some trigger that was external to the oceans (strong freshwater influx in the case of Heinrich events, a weak and so far unknown trigger in the case of D/O events), with the THC acting as a nonlinear amplifier. Based on our theoretical understanding of the feedbacks, our knowledge of the past circulation changes, and future scenario simulations with different models, we conclude that anthropogenic warming could trigger nonlinear changes in ocean circulation. These, in turn, could have major impacts, especially on marine eco-systems and on the climates of countries surrounding the northern Atlantic.

Instabilities of Inland Ice

There are two primary sources of climate sensitivity associated with the major ice sheets of the Quaternary: (a) a strong albedo effect of lying snow and ice, which was probably crucial in the onset of glaciation, and (b) the tendency for rapid melting, in particular dramatic iceberg production associated with Hein-rich layers, which had a major effect on North Atlantic climate (Bond and Lotti 1995). A surge-type mechanism operating over timescales of approximately 100 years is consistent with much of the evidence concerning the Heinrich events (Payne 1995; Calov et al. 2002). The precise cause of the deglaciation is uncer-tain. Models have been able to reproduce partial deglaciation using only insola-tion forcing; however, this is clearly insufficient to cause complete deglaciation.

In the Anthropocene, warming is likely to lead to the decay of much of the Greenland ice sheet over timescales of millennia (Huybrechts and de Wolde 1999). This may have an effect on the THC and may be irreversible. Much con-troversy surrounds the future of the West Antarctic ice sheet, which has been postulated to be prone to rapid decay via grounding line retreat (Mercer 1978; Thomas and Bentley 1978). A range of modeling work has failed to confirm this hypothesis (Hindmarsh and Le Meur 2001; Huybrechts 2002). However, con-tinued retreat may still be occurring as a response to deglaciation from the Last Glacial Maximum (LGM) (Bindschadler and Vornberger 1998; Conway et al. 1999), and there is considerable local variability in the flow of the ice sheet (Rignot 1998; Retzlaff and Bentley 1993; Joughin and Tulaczyk 2002).

The Greenland and Antarctic ice sheets are important because of their poten-tial contributions to future sea-level change. Greenland is likely to respond to lo-cal climate warming of 5°C over the next millennium, and recent predictions indicate that around half of the ice sheet may be lost (Huybrechts and de Wolde 1999). If the ice sheet falls below a critical threshold, it is unlikely to be able to recover to its present size under any climate other than one that is much colder than today. The amount of additional meltwater generated by Greenland is likely to freshen the surface of the North Atlantic and may have implications for the formation of deep water in the area as well as the THC of the world's oceans. Data generated by airborne and satellite altimetry suggest that the ice sheet is presently thinning, although there is much spatial variability in the signal. This thinning appears to be concentrated in many of the outlet glaciers draining the ice sheet (Abdalati et al. 2001). If this thinning is a direct result of recent climate change, then it hints at a rate of ice-sheet response not previously thought possi-ble. One potential explanation is that meltwater generated at the surface is reach-ing the bed (through 1 to 2 km of very cold ice) and increasing basal lubrication, thereby accelerating ice flow. This is supported by observations of a seasonal increase in ice velocity at a site near the equilibrium line in East Greenland (Zwally et al. 2002). This mechanism occurs regularly in valley glaciers but was previously thought inappropriate on the very much larger scale of ice sheets.

The Antarctic ice sheet differs from Greenland in that it does not experience significant mass loss though surface melting; therefore, all accumulated ice is lost via the calving of large icebergs. Modeling strongly suggests that a warming of 15°C is required before significant melt will occur (Huybrechts 1993). This is extremely unlikely, and it is fair to assume that the ice sheet will not respond directly to atmospheric warming in the manner of Greenland. Recent observations show that a number of West Antarctic ice streams are thinning dramatically and that this thinning is largest near the coast (Rignot 1998; Shepherd et al. 2001). One potential explanation is that these ice streams are experiencing reduced basal drag (and therefore acceleration of flow) as a consequence of enhanced oceanic melt. This implies a link between ice flow and coastal circulation around Antarctica, although it is unclear whether the ice streams' behavior reflects climate change or internal variability.

Land Surface–Vegetation–Albedo Feedbacks

Tundra–Taiga Feedback

Forests significantly reduce the albedo of snow-covered land surface. This is the basis for a positive feedback between forest and surface air temperature as follows: increased tree fraction → decreased surface albedo during snow season → increased air temperature and earlier snow melt → longer and warmer growing season → increased tree fraction. Paleosimulations for the mid-Holocene (e.g., Ganopolski et al. 1998), the LGM (e.g., Kubatzki and Claussen 1998; Levis et al. 1999), and the last glacial inception (deNoblet et al. 1996) highlighted the role of forest–temperature feedback as an amplifier of externally driven climate change. Sensitivity experiments with models of different complexity revealed that this biogeophysical feedback, although positive, is not strong enough to support multiple steady states for present-day climate and doubled CO_2 climate (Brovkin et al. 2003). Nonetheless, this feedback is important for climate–vegetation dynamics in the northern high latitudes as warming in the region will be substantially amplified due to a northward shift of the tree line.

The Green Sahara

According to paleobiological and geological evidence, the Sahara was much greener than today during the mid-Holocene, some 8000–6000 years ago (e.g., Prentice et al. 2000), during the early Holocene, and presumably even during the so-called Bølling–Allerød warm phase during the late deglaciation beginning some 16,000 years ago. The greening was tentatively attributed to changes in orbital forcing (e.g., Kutzbach and Guetter 1986) that amplified the North African summer monsoon. However, orbital forcing of the paleo-monsoon alone did not seem to be strong enough to explain widespread greening (Joussaume et al. 1999). Models which include atmosphere–vegetation interaction yielded a stronger, more realistic greening (Claussen and Gayler 1997; Doherty et al. 2000). With regard to potential future climate change, sensitivity tests (Claussen et al. 2003) suggest that some expansion of vegetation into today's Sahara is theoretically possible, if atmospheric CO_2 concentration increases well above preindustrial values and if vegetation growth is not disturbed, for example, by grazing. Stability analyses reveal that the atmosphere–biosphere system in northern Africa is able to attain multiple equilibrium states. For present-day as well as glacial climates, two solutions are possible: an arid Sahara similar to the present day and a green Sahara, resembling conditions during the Holocene. For the early and mid-Holocene, only one — the green — solution seems to exist (Brovkin et al. 1988). This transient behavior suggests that abrupt changes in the extension of the Saharan desert are possible and were indeed predicted to have occurred around 5,500 years ago (Claussen et al. 1999), in agreement with reconstruction models (deMenocal et al. 2000).

Methane Release from Wetlands

Wetlands and marshlands are sites of carbon-rich, anoxic muds in which methanogenesis (i.e., the production of CH_4 by bacteria) can occur. A proportion of this gas escapes to the atmosphere, so that today, wetlands such as rice paddies and salt marshes are recognized to be globally significant sources of CH_4. Since the worldwide prevalence of wetlands may be expected to increase in a warmer and wetter climate, it is reasonable to suggest that this is one major reason why atmospheric CH_4 concentrations, as recorded in ice cores, show an increase during interglacials. The CH_4 signal shows a relatively strong response at precessional frequencies (Figure 10.2), at which tropical monsoon signals are also heavily influenced. This may well be the response of tropical and subtropical wetland regions to increased precipitation resulting from monsoons. Recently, Lea et al. (2003) have shown there is synchronicity in the atmospheric CH_4 record during the Younger Dryas with tropical sea-surface temperature (SST) as recorded in a core from the Cariacao basin. They suggest this is due to sensitivity of both methane and local SSTs to the position of the intertropical convergence zone. If correct, this interpretation would indicate a strong role for tropical wetlands in modulating CH_4 concentrations.

Dust Feedback

Soil dust is an important contributor to aerosol loading and optical thickness. Mineral dust of grain size between 0.1–100 μm is entrained into the atmosphere from subtropical deserts in an amount of 1–5 Pg yr^{-1} (Duce 1995). Sand-sized dust faction (> 63 μm) usually creeps and saltates on the ground to form dunes, while the fine-grained faction is lifted up to ~5 km height by convective air movement over hot deserts, like the Sahara and Arabia. The entrainment of dust is restricted to regions where annual precipitation is below 200 mm per year (Pye 1988). Emission of mineral dust to the atmosphere generally exerts a cooling influence in the short-wave radiation band (reflection of solar radiation) but a warming influence in the long-wave radiation band (absorption of infrared); direct radiative effects are, however, grain-size dependent. Aerosols also cause an indirect radiative effect associated with changes in cloud properties (Penner et al. 2000).

Information on the dust load during the Quaternary comes from the accumulation of subtropical desert dust on the ocean floor and from dust in ice cores. Evidence from the subtropical Atlantic and Northern Indian Ocean reveal accumulation rates of dust that were about double during the LGM compared to the present (Mahowald et al. 1999; Kohfeld and Harrison 2001). Ice cores show a much larger relative change from glacial to modern conditions, with deposition rates fivefold higher than present at the LGM in Greenland, and ca. tenfold higher in Antarctica (Mahowald et al. 1999).

Possible climatic effects from mineral dust are a function of the total world area with precipitation less than 200 mm per year and the possibility of convective uplift of particles. Thus future climatic impact from aerosol dust depends on the amount of desertification in the low latitudes. If desert area would increase during the Anthropocene, as it did in the LGM, and convective processes are enhanced, we could expect that the atmosphere's mineral aerosol content will at least double. Agricultural intensification could either increase or decrease dust emissions; increased tillage may add to dust while increased dry-season irrigation and leaf area would inhibit dust production (Gregory et al. 2002).

Aeolian transport of mineral dust to the ocean presumably plays an important role in oceanic biogeochemistry. The supply of iron-rich dust to the Southern Ocean is increased during glacial periods (Mahowald et al. 1999). Fertilization of marine productivity by iron from this source could have contributed to a decrease in atmospheric CO_2 during the glacial period (Watson et al 2000). It is now well established that the Southern Ocean is limited in iron and that an increase in this supply might be expected to lead to drawdown in the region. Watson et al. (2000) suggest that a decrease in dust supply to this region at the glacial terminations was a principal cause of the initial rise in atmospheric CO_2. Ridgwell and Watson (2003) suggest that the following feedback process is responsible for observed instabilites in climate and CO_2 concentrations during glacial time: increased dust → Southern Ocean iron fertilization → decreased CO_2 → colder, drier climate → increased dust. Ridgwell et al. (2002) suggest that a possible effect of agricultural and forestry practices designed to sequester carbon in the terrestrial vegetation may be to reduce the flux of mineral dust to the ocean, which may in turn decrease the sink of CO_2 in the oceans due to decreased iron supply, partially offsetting any gain in carbon sink on the land.

Carbon Dioxide Release from the Land Biosphere

The first general circulation model (GCM) climate projection to include both the carbon cycle and vegetation as interactive elements showed an acceleration of global warming when these additional feedbacks were included (Cox et al. 2000). In their Hadley Centre model, the current land-carbon sink decreased, becoming a source of CO_2 from about the middle of the twenty-first century, adding to the anthropogenic emissions and driving the CO_2 concentration to about 1000 ppmv by 2100 under a "business-as-usual" emissions scenario (cf. 700 ppmv in the absence of carbon cycle feedbacks). The sink-to-source transition in the terrestrial biosphere was associated with the (usual) assumption that rate of decomposition (or "soil respiration") increases with temperature, whereas rate of photosynthesis saturates with CO_2 concentration. Consequently, although plant growth (and therefore litter production) increases with CO_2, this term is eventually outpaced by soil respiration, leading to a significant loss of soil carbon in the second-half of the twenty-first century. Results of this complex global model seem consistent with simple conceptual models (e.g., Scholes 1999), which show that the sink-to-source threshold in the terrestrial carbon cycle depends on uncertain quantities related to the climate sensitivity to CO_2, the half-saturation constant for photosynthesis as a function of CO_2, and the sensitivity of soil respiration to temperature (Cox et al. 2000). Since these quantities are all poorly known, it is not perhaps surprising that other coupled climate–carbon cycle GCMs do not see such strong feedbacks (Friedlingstein et al. 2001).

There is little evidence for such positive land-carbon cycle feedbacks in the Quaternary. Indeed, increases in land-carbon storage between the LGM and the Holocene (as inferred from ^{13}C records, Prentice et al. 2001) indicate that the land-carbon cycle largely acted as a negative feedback on the CO_2 increase during this period. However, this is not at odds with the Hadley GCM, which simulates a glacial-to-preindustrial increase in land-carbon storage of roughly the right magnitude (600 Pg C). Furthermore, the model produces a relatively modest reduction in the Amazon rainforest at the LGM, which is consistent with recent reconstructions from pollen records (Adams and Faure 1998). Based on these model validation exercises, it seems that the paleoclimate record of the Quaternary is consistent with significant positive feedbacks from the land-carbon cycle during the twenty-first century, which may be delayed but ultimately made more sudden by the decreasing cooling effects of sulfate aerosols (Jones et al. 2003). Key uncertainties associated with climate sensitivity to CO_2, regional climate change (e.g., drying in Amazonia), and long-term sensitivity of soil respiration to temperature need to be quantified before we can offer guidance on the probability of a land sink-to-source transition in the real Earth system.

Sea Ice

Measurements of the thickness of Arctic sea ice are available from periodic transits of submarines from the mid-1970s onwards. More recently they have been available from remote sensing by SAR and altimetry data mounted on satellites. Most of the data indicate that there is a thinning of the Arctic ice, which is on average today about 3 m thick, but which is thinning at a rate that suggests substantial retreat of the Arctic ice pack by 2050 and possible disappearance of sea ice in the Arctic summer by the end of the twenty-first century. If these predictions prove correct, there will be substantial consequences for the Arctic climate and for the Arctic ecosystem. A number of valuable species make the sea ice their principal habitat, and these will be directly threatened. The change in albedo of the region will be dramatic and will accelerate the warming of the region.

Gas Hydrates

Gas hydrates, or clathrates, are solids in which molecules of a gas are bound within open cavity-like cages formed by water molecules. Clathrates are stable at temperatures above those at which ice forms and at elevated pressures. Large reservoirs of methane clathrate are thought to exist at high latitude in the deep ocean, on continental margins, and in permafrost. For example, the reservoir in the Arctic permafrost region is estimated at 400 Pg C (Macdonald 1990) whereas the ocean reservoir may be $\sim 10^4$ Pg C. If this is correct, it would be the largest concentrated organic carbon reservoir on Earth, dwarfing the surface biosphere (~ 1500 Pg C living and dead matter) and twice the size of the estimated fossil fuel reservoir.

Methane hydrates are stable only where the sediment is cold enough and under sufficient pressure. Decreases in pressure, such as caused by a decrease in sea level, or increases in temperature may be expected to destabilize some of the hydrate and cause release of CH_4 gas, some of which may reach the atmosphere. Some of the CH_4 release associated with glacial terminations may therefore be due to the warming, and a positive feedback can be envisaged.

Limits can be put on the amount of hydrate release that could have occurred from glacial to interglacial time, because the ^{13}C content of CH_4 from methanogenesis is very distinct. Maslin and Thomas (2003) argue that a release of about 120 Pg C methane would be compatible with records of glacial–interglacial $\delta^{13}C$, and would indeed help to reconcile interpretations of the sedimentary $\delta^{13}C$ record with independent reconstuctions of the biosphere. Such a release is only a fraction of the available size of the reservoirs and would only contribute $\sim 30\%$ to the observed increase in atmospheric CH_4 if it occurred over a time period ~ 10 ka, so is consistent with the main increase being due to tropical wetland expansion.

Atmospheric Circulation (ENSO, NAO, Monsoon)

Atmospheric circulation may respond nonlinearly to greenhouse gas forcing. This applies to variability patterns such as ENSO and NAO as well as to the monsoon circulation. Paleodata suggest that changes in these patterns have occurred in the past, although the data on past atmospheric circulation are sketchy. In some cases, model scenarios for the future show a change toward stronger ENSO and a more positive phase of the NAO. Given the impacts that ENSO, NAO, and monsoon have on agriculture and food security, on extreme events and the economy, possible changes in atmospheric circulation patterns are of obvious concern (see Table 10.2).

USING THE PAST TO INFORM THE FUTURE

Although natural climate change was extensive and sometimes very rapid during the Quaternary, it does not provide a very close analogue to the present anthropogenically forced climate change. In looking for more direct analogues further back in the geological record, the Paleocene–Eocene Thermal Maximum (PETM) is of particular interest. PETM (previously called the Late Paleocene Thermal maximum) is a sharp rise (~7°C) in temperature that occurred about 55 million years ago, and is marked in both $\delta^{18}O$ and $\delta^{13}C$ records. The shift in carbon isotopes has such a sharp onset that it is difficult to resolve in the record, but recent evidence suggests that much of it occurs in two steps of less than 1000 years duration (Rohl et al. 2000). There is a subsequent, slower, exponential decay to former values which occurs over a period of order 200 ka. The most likely explanation for the huge $\delta^{13}C$ excursion is the release of a large quantity of biogenic CH_4 previously locked up in clathrates. Because the amount is so large and because the climate of the time was warm (contributing to a destabilization of deep ocean clathrates and reducing the amount stored in permafrost), it seems most likely that the deep sea is the source. Warming would occur because of the CH_4 increase in the atmosphere, but the amount of increase in CH_4 depends nonlinearly on how quickly the release is assumed to take place. Increases in CO_2 from the oxidation of the CH_4 plus changes in the water vapor content of the upper atmosphere would also exert warming influences on the planet. The PETM was a time of global extinctions of a number of important species of fossils, most likely associated with rapid climate change.

The PETM is of interest as an analogue of present conditions because it represents a period when global warming occurred at a rate normally never seen in the climate record; it may have been a warming rate similar to that happening today (but it may also have been about ten times slower). It tells us that there are potentially strong positive feedbacks "waiting in the wings" to amplify temperature increases that we have instigated, and that these are not only associated with the changing THC but also with CH_4 release. Furthermore, though the PETM was a very short event geologically speaking, in human terms it takes an immensely long time for the initial perturbation to dissipate. This should not be a surprise, as the PETM timescale of 200 ka is roughly that which we calculate for the carbon cycle to return to steady state after our present-day carbon emissions have ceased. However, it does provide a graphic illustration that the changes we are initiating are unlikely to be reversible over any short period. If we release all of the fossil-fuel carbon available to us, the climatic changes we initiate will last not just for centuries, but for tens or hundreds of millennia.

USING MODELS TO ADVANCE UNDERSTANDING

Because of the broad spectrum of typical timescales of the different components of the natural Earth system, simulation of the natural Earth system dynamics requires a spectrum of models of varying complexity. Comprehensive models of global atmospheric and oceanic circulation describe many details of the flow pattern, such as individual weather systems and regional currents in the ocean. Some models also include dynamic modules of biogeophysical and biogeochemical processes. These comprehensive models operate at the highest technically feasible spatial and temporal resolution, and they are supposed to be the most realistic laboratory of the natural Earth system (Grassl 2000). Because of their high computational cost, these models are applied to relatively short transient simulations of a few centuries or to so-called time slices during which the Earth system is assumed to be in a quasi-equilibrium state. At the other end of the spectrum of complexity are the conceptual or tutorial models. These are mainly inductive models based on a gross understanding of the feedbacks that are likely to be involved (Saltzman 1985). Conceptual models are designed to demonstrate the plausibility of processes. To bridge the gap between conceptual and comprehensive models, Earth system models of intermediate complexity (EMICs) have been proposed (Claussen et al. 2003). EMICs include most of the processes described in comprehensive models, albeit in a more reduced, that is, a more parameterized form. Most EMICs are designed for long-term simulations, and they include as many components of the natural Earth system as possible. Due to the implicit limitations in terms of their dynamics, application of EMICs and interpretation of results obtained by EMICs have to be done with greater care and caution than when using comprehensive models.

In conclusion, there is a clear advantage in making use of the full spectrum of models. EMICs, for example, can be used to construct large ensembles of simulations. Thus, they are more suitable for assessing uncertainty which comprehensive models can do to a significantly lesser extent. Moreover, EMICs can provide guidance to experimentation with comprehensive models by exploring new methodologies at comparatively low computational cost. Still, it is not sensible to apply an EMIC to studies that require high spatial resolution. High resolution would be needed, for example, to synthesize reconstructions from different paleoclimatic archives at different geographical locations or to interpret and forecast detailed patterns of climate processes. Finally, for developing understanding in terms of conceptual visualization and teaching, conceptual models are most useful.

FURTHER LESSONS FROM THE QUATERNARY: CAN WE "GEO-ENGINEER" THE CLIMATE SYSTEM?

Carbon Dioxide Sequestration in the Oceans: Iron fertilization

Shortly after the first evidence emerged in the 1980s that the Southern Ocean was probably iron limited, it was suggested, not entirely seriously, that the anthropogenic greenhouse effect could be "fixed" by massive fertilization of the (Southern) ocean with iron (Gribbin 1988). This idea generated intense media interest, but it was not long before relatively simple calculations were published to show that such a strategy could not work (Peng and Broecker 1991). The reason is simply that there is a dynamical limitation to the rate at which the ocean can take up excess atmospheric CO_2, due to the rate of ventilation of deep waters. This limit ensures that even if the Southern Ocean were completely fertilized so that all iron limitation in the region was alleviated, it could only take up an extra few tenths of a petagram of carbon per year. However, humans are currently emitting CO_2 at a rate of 7 Pg C yr^{-1}, about half of which remains in the atmosphere. Thus it would be necessary to increase the ocean uptake by ~3.5 Pg C yr^{-1} in order to stop the increase in the atmosphere — an order of magnitude more than is possible.

It remains true that, if carried out at a modest level, it is possible to imagine iron fertilization making a contribution to the global sequestration of carbon, of the same order as that which might be achievable by sustained increases in tree planting or agricultural reform, without too great an impact on the present state of the ocean ecosystem. Amounts of carbon on the order of 0.1 Pg C yr^{-1} might be sustainably sequestered in the ocean in this way (which is however much smaller than the effect claimed by entrepreneurs now actively promoting this idea). Any such effort would only be likely to be supported by marine scientists if it were accompanied by careful monitoring of the effects. In our present state of ignorance, we cannot rule out the possibility of undesirable side effects (e.g., generation of nitrous oxide and lowering the oxygen tension in subsurface waters). If these occurred it would be important to know about them and, if necessary, to stop the fertilization.

More generally, the study of the Quaternary climate system makes us aware of our own ignorance of this complex, nonlinear, and surprising system. We are only just beginning to learn about this organic entity, and our efforts at understanding the past, let alone forecasting the future, are still evolving rapidly. With our advances in knowledge, we may be tempted to believe that we have the understanding and the tools to engineer this system, but we would probably be deluding ourselves. How many of us have had the experience of taking a complex piece of equipment apart in the belief that we could fix some small fault, only to find that we could not put it back together? It would be a disaster if, collectively, we were to make the same mistake with the Earth system itself.

REFERENCES

Abdalati, W., W. Krabill, E. Frederick et al. 2001. Outlet glacier and margin elevation changes: Near-coastal thinning of the Greenland ice sheet. *J. Geophys. Res.* **106(D24)**:33729–33741.

Adams, J., and H. Faure. 1998: A new estimate of changing carbon storage on land since the last glacial maximum, based on global land ecosystem reconstruction. *Global Planet. Change* **16–17**:3–24.

Archer, D., A. Winguth, D. Lea, and N. Mahowald. 2000. What caused the glacial/interglacial atmospheric pCO(2) cycles? *Rev. Geophys.* **38**:159–189.

Bergman, N., T.M. Lenton, and A.J. Watson. 2004. A new biogeochemical Earth system model for the Phanerozoic (last 550 million years). In: Scientists on Gaia: The New Century, ed. S.J. Schneider et al. Cambridge, MA: MIT Press.

Berner, R.A. 1994. Geocarb-II: A revised model of atmospheric CO_2 over Phanerozoic time. *Am. J. Sci.* **294**:56–91.

Bindschadler, R., and P. Vornberger. 1998. Changes in the West Antarctic ice sheet since 1963 from declassified satellite photography. *Science* **279**:689–692.

Bond, G.C., and R. Lotti. 1995. Iceberg discharges into the North Atlantic on millennial timescales during the last glaciation. *Science* **267**:1005–1010.

Brovkin, V., M. Claussen, V. Petoukhov, and A. Ganopolski. 1998. On the stability of the atmosphere–vegetation system in the Sahara/Sahel region. *J. Geophys. Res.* **103(D24)**:31,613–31,624.

Brovkin, V., S. Levis, M.-F. Loutre et al. 2003. Stability analysis of the climate–vegetation system in the northern high latitudes. *Clim. Change* **57**:119–138.

Calov, R., A. Ganopolski, V. Petoukhov et al. 2002. Large-scale instabilities of the Laurentide ice sheet simulated in a fully coupled climate-system model. *Geophys. Res. Lett.* **29**:2216.

Claussen, M., V. Brovkin, A. Ganopolski et al. 2003. Climate change in Northern Africa: The past is not the future. *Clim. Change* **57**:99–118.

Claussen, M., and V. Gayler. 1997. The greening of Sahara during the mid-Holocene: Results of an interactive atmosphere–biome model. *Global Ecol. & Biogeog. Lett.* **6**:369–377.

Claussen, M., C. Kubatzki, V. Brovkin, A. Ganopolski, P. Hoelzmann, and H.J. Pachur. 1999. Simulation of an abrupt change in Saharan vegetation at the end of the mid-Holocene. *Geophys. Res. Lett.* **24**:2037–2040.

Claussen, M., L.A. Mysak, A.J. Weaver et al. 2002. Earth system models of intermediate complexity: Closing the gap in the spectrum of climate system models. *Clim. Dyn.* **18**:579–586.

Conway, H., B.L. Hall, G.H. Denton, A.M. Gades, and E.D. Waddington. 1999. Past and future grounding-line retreat of the West Antarctic Ice Sheet. *Science* **286**:280–283.

Cox, P.M., R.A. Betts, C.D. Jones et al. 2000. Acceleration of global warming due to carbon-cycle feedbacks in a coupled climate model. *Nature* **408**:184–187.

deMenocal, P.B, J. Ortiz, T. Guilderson et al. 2000. Abrupt onset and termination of the African Humid Period: Rapid climate response to gradual insolation forcing. *Quat. Sci. Rev.* **19**:347–361.

deNoblet, N.I., I.C. Prentice, S. Joussaume et al. 1996. Possible role of atmosphere–biosphere interactions in triggering the last glaciation. *Geophys. Res. Lett.* **23**: 3191–3194.

Doherty, R., J. Kutzbach, J. Foley, and D. Pollard. 2000. Fully coupled climate/dynamical vegetation model simulations over Northern Africa during the mid-Holocene. *Climate Dyn.* **16**:561–573.

Duce, R.A. 1995. Sources, distribution, and fluxes of mineral aerosols and their relationship to climate, In: Aerosol Forcing of Climate. Dahlem Workshop Report, ed. R.J. Charlson and J. Heintzenberg, pp. 43–72. Chichester: Wiley.

Friedlingstein, P., L. Bopp, P. Ciais et al. 2001. Positive feedback between future climate change and the carbon cycle. *Geophys. Res. Lett.* **28**:1543–1546.

Ganopolski, A., C. Kubatzki, M. Claussen et al. 1998. The influence of vegetation–atmosphere–ocean interaction on climate during the mid-Holocene. *Science* **280**: 1916–1919.

Grassl, H. 2000. Status and improvements of coupled general circulation models. *Science* **288**:1991–1997.

Gregory, P.J., J.S.I. Ingram, R. Andersson et al. 2002. Environmental consequences of alternative practices for intensifying crop production. *Agric. Ecosys. & Env.* **88**:279–290.

Gribbin, J. 1988. Any old iron. *Nature* **331**:570–570.

Hays, J.D., J. Imbrie, and N.J. Shackleton. 1980. Variations in the Earth's orbit: Pacemaker of the ice ages. *Science* **194**:1121–1132.

Hindmarsh, R.C.A., and E. Le Meur. 2001. Dynamical processes involved in the retreat of marine ice sheets. *J. Glaciol.* **47**:271–282.

Huybrechts, P. 1993. Glaciological modelling of the late Cenozoic East Antarctic ice sheet: Stability or dynamism? *Geografiska Annaler* **75A**:221–238.

Huybrechts, P. 2002. Sea-level changes at the LGM from ice-dynamic reconstructions of the Greenland and Antarctic ice sheets during the glacial cycles. *Quat. Sci. Rev.* **21**:203–231.

Huybrechts, P., and J. de Wolde. 1999. The dynamic response of the Greenland and Antarctic ice sheets to multiple-century climatic warming. *J. Climate* **12**:2169–2188.

Jolly, D., S.P. Harrison, B. Damnati, and R. Bonnefille. 1998. Simulated climate and biomes of Africa during the late Quaternary: Comparison with pollen and lake status data. *Quat. Sci. Rev.* **17**:629–657.

Jones, C.D., M. Collins, P.M. Cox, and, S.A. Spall. 2001. The carbon cycle response to ENSO: A coupled climate-carbon cycle model study. *J. Climate* **14**:4113–4129.

Jones, C.D., P.M. Cox, R.L.H. Essery et al. 2003. Strong carbon cycle feedbacks in a climate model with interactive CO_2 and sulphate aerosols. *Geophys. Res. Lett.* **30**:1479.

Joughin, I., and S. Tulaczyk. 2002. Positive mass balance of the Ross Ice Streams, West Antarctica. *Science* **295**:476–480.

Joussaume, S., K.E. Taylor, P. Braconnot et al. 1999. Monsoon changes for 6000 years ago: Results of 18 simulations from the Paleoclimate Modeling Intercomparison Project (PMIP). *Geophys. Res. Lett.* **26**:859–862.

Kohfeld, K.E., and S.P. Harrison. 2001. DIRTMAP: The geologic record of dust. *Earth Sci. Rev.* **54**:81–114.

Kubatzki, C., and M. Claussen. 1998. Simulation of the global biogeophysical interactions during the last glacial maximum. *Climate Dyn.* **14**:461–471.

Kutzbach, J.E., and P.J. Guetter. 1986. The influence of changing orbital parameters and surface boundary conditions on climate simulations for the past 18,000 years. *J. Atmos. Sci.* **43**:1726–1759.

Lea, D.W., D.K. Pak, L.C. Peterson, and K.A. Hughen. 2003. Synchroneity of tropical and high-latitude Atlantic temperatures over the last glacial termination. *Science* **301**:1361–1364.

Levis, S., J.A. Foley, and D. Pollard. 1999. CO₂, climate, and vegetation feedbacks at the last glacial maximum. *J. Geophys. Res.* **104**:31,191–31,198.

Lovelock, J.E., and A.J. Watson. 1982. The regulation of carbon dioxide and climate: Gaia or geochemistry? *Planet. Space Sci.* **30**:795–802.

MacDonald, G.J. 1990. Role of methane clathrates in past and future climates. *Clim. Change* **16**:247–281.

Mahowald, N., K.E. Kohfeld, M. Hansson et al. 1999. Dust sources and deposition during the last glacial maximum and current climate: A comparison of model results with palaeodata from ice cores and marine sediments. *J. Geophys. Res.* **104**:15,895–15,916.

Maslin, M.A., and E. Thomas. 2003. Balancing the deglacial global carbon budget: The hydrate factor. *Quat. Sci. Rev.* **22**:1729–1736.

Mercer, J.H. 1978. West Antarctic ice sheet and CO₂ greenhouse effect: A threat of disaster. *Nature* **271**:321–325.

Payne, A.J. 1995. Limit cycles and the basal thermal regime of ice sheets. *J. Geophys. Res.* **100**:4249–4263.

Peng, T.-H., and W.S. Broccker. 1991. Dynamic limitations on the Antarctic iron fertilization strategy. *Nature* **349**:227–229.

Penner, J.E., M. Andreae, H. Annegarn et al. 2000. Aerosols, their direct and indirect effects. In: Climate Change 2001: The Scientific Basis, ed. J.T. Houghton et al., pp. 289–348. New York: Cambridge Univ. Press.

Petit, J.R., J. Jouzel, D. Raynaud et al. 1999. Climate and atmospheric history of the past 420,000 years from the Vostok ice core, Antarctica. *Nature* **399**:429–436.

Prentice, I.C., D. Jolly, and BIOME 6000 members. 2000. Mid-Holocene and glacial-maximum vegetation geography of the northern continents and Africa. *J. Biogeog.* **27**:507–519.

Pye, J.M. 1988. Impact of ozone on the growth and yield of trees: A review. *J. Env. Qual.* **17**:347–360.

Rahmstorf, S. 2002: Ocean circulation and climate during the past 120,000 years. *Nature* **419**:207–214.

Rahmstorf, S., J. Marotzke, and J. Willebrand. 1996. Stability of the thermohaline circulation. In: The Warm Water Sphere of the North Atlantic Ocean, ed. W. Krauss, pp. 129–158. Stuttgart: Borntraeger.

Retzlaff, R., and C.R. Bentley. 1993. Timing of stagnation of ice stream-C, West Antarctica, from short-pulse radar studies of buried surface crevasses. *J. Glaciol.* **39**:553–561.

Ridgwell, A.J., M.A. Maslin, and A.J. Watson. 2002. Reduced effectiveness of terrestrial carbon sequestration due to an antagonistic response of ocean productivity. *Geophys. Res. Lett.* **29**:10.1029/2001GL014304.

Ridgwell, A.J., and A.J. Watson. 2002. Feedback between aeolian dust, climate, and atmospheric CO₂ in glacial time. *Paleoceanography* **17**:1059.

Rignot, E.J. 1998. Fast recession of a West Antarctic glacier. *Science* **281**:549–551.

Rohl, U., T.J. Bralower, R.D. Norris, and G. Wefer. 2000. New chronology for the late Paleocene thermal maximum and its environmental implications. *Geology* **28**: 927–930.

Rothrock, D.A., Y. Yu, and G.A. Maykut. 1999. Thinning of the Arctic sea-ice cover. *Geophys. Res. Lett.* **26**:3469–3472.

Saltzman, B. 1985. Paleoclimatic modeling. In: Paleoclimate Analysis and Modeling, ed. A.D. Hecht, pp. 341–396. New York: Wiley.

Saltzman, B., and M.Y. Verbitsky. 1993. Multiple instabilities and modes of glacial rhythmicity in the Plio-Pleistocene: A general theory of late Cenozoic climatic change. *Climate Dyn.* **9**:1–15.

Scholes, R. 1999. Will the terrestrial carbon sink saturate soon? *IGBP Newsl.* **37**:2–3. http://www.igbp.kva.se/uploads/nl_37.pdf.

Shepherd, A., D.J. Wingham, J.A.D. Mansley, and H.F.J. Corr. 2001. Inland thinning of Pine Island Glacier, West Antarctica. *Science* **291**:862–864.

Sigman, D.M., and E. Boyle. 2000. Glacial/interglacial variations in atmospheric carbon dioxide. *Nature* **407**:859–869.

Stommel, H. 1961. Thermohaline convection with two stable regimes of flow. *Tellus* **13**:131–149.

Tiedemann, R., M. Sarnthein, and N.J. Shackleton. 1994. Astronomic timescale for the Pliocene Atlantic $\delta^{18}O$ and dust flux records of ogram site-659. *Paleoceanography* **9**:619–638.

Thomas, R.H., and C.R. Bentley. 1978. A model for Holocene retreat of the West Antarctic Ice Sheet. *Quat. Res.* **10**:150–170.

Watson A.J., D.C.E. Bakker, A.J. Ridgwell et al. 2000. Effect of iron supply on Southern Ocean CO_2 uptake and implications for glacial atmospheric CO_2. *Nature* **407**:730–733.

Zwally, H. J., W. Abdalati, T. Herring et al. 2002. Surface melt-induced acceleration of Greenland ice-sheet flow. *Science* **297**:218–220.

11

Human Footprints in the Ecological Landscape

P. G. FALKOWSKI and D. TCHERNOV

Environmental Biophysics and Molecular Ecology Program,
Institute of Marine and Coastal Science, Department of Geological Sciences,
Rutgers, State University of New Jersey, New Brunswick, NJ 08901, U.S.A.

> "O Oysters," said the Carpenter,
> "You've had a pleasant run!
> Shall we be trotting home again?"
> But answer came there none —
> And this was scarcely odd, because
> They'd eaten every one.
>
> *Through the Looking-Glass and What Alice Found There*
> by Lewis Carroll (1872)

ABSTRACT

In this chapter we explore how the evolution of human behavior has led to the current condition, the quantitative impact of humans on ecological and biogeochemical processes, and potential strategies for developing a sustainable partnership between humans and the ecosystems in which they operate. Our basic thesis is that humans appear to have uniquely escaped from the Red Queen constraint of adaptive genetic selection. The subsequent consequences of that escape has led to rapid alterations of all ecosystems on Earth. The alterations were so rapid and so strong that they exerted selection pressures not unlike a mass extinction event.

INTRODUCTION

Human Evolution

The complete sequencing of the human genome has facilitated detailed molecular reconstructions of the evolution of *Homo sapiens*. A recent phylogenetic analysis suggests that our species arose approximately 200,000 years ago and is descended from a closely related, deeply rooted, but extinct species in the same genus (Figure 11.1), which itself arose approximately 5 million years (Ma) from another, extinct, primate lineage (Carroll 2003). The molecular reconstruction, in conjunction with incomplete physical anthropological evidence, indicates that for almost all of human history, subsistence was based on small nomadic groups cooperatively engaged in hunting and gathering. It is likely that these groups were closely related through family ties. Regardless, aided by geographic dispersion, restrictions in gene flow gave rise to several genetic clades or subpopulations, which can be identified by maternally inherited (i.e., non-Mendelian) information retained in mitochondrial genomes (Sykes 2001). Although genetic differences among individuals within subpopulations can be large, all human subpopulations remain capable of interbreeding. Competition between subpopulations has often led to violent outcomes and even extinction of some.

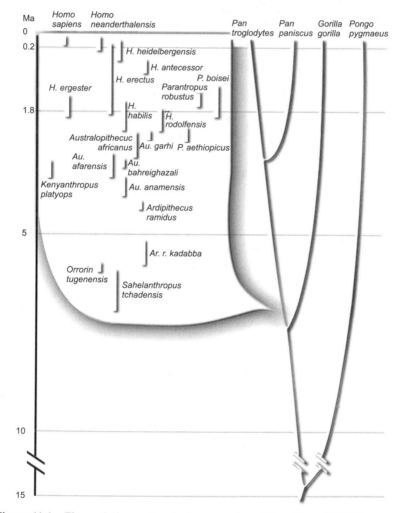

Figure 11.1 The evolutionary tree for humans adapted from Carroll (2003).

In the twentieth century of the common era, the ensemble of the subpopula
tions within the subpopulations comprising *H. sapiens* rapidly expanded. Over a
period of 100 years, the population grew from ca. 950 million to more than 6000
million. This unprecedented rate of expansion in population paralleled an unprecedented strain on Earth's natural resources. Humans presently consume or
exploit ca. 42% of the terrestrial net primary production (Vitousek et al. 1997).
Our species has displaced, extinguished, or impacted virtually every extant ver-
tebrate species (Jackson et al. 2001). With very few exceptions, humans have altered the flow and chemical form of all naturally occurring elements as well as
all of the fresh water on the planet (Falkowski et al. 2000) (Table 11.1, Figure
11.2). These activities, which further require inputs of energy, are claimed for
the food, fiber, and habitat of a single species. The continued growth of human

Table 11.1 Examples of human intervention in the global biogeochemical cycles of
carbon, nitrogen, phosphorus, sulfur, water, and sediments. Data are for the mid-1900s
(Falkowski et al. 2000).

Element	Flux	Magnitude of flux (millions of metric tons per year)		Change due to human activities (%)
		Natural	Anthropogenic	
C	• Terrestrial respiration and decay CO_2	61,000		
	• Fossil-fuel and land-use CO_2		8,000	+13
N	• Natural biological fixation	130		
	• Fixation owing to rice cultivation, combustion of fossil fuels, and production of fertilizer		140	+108
P	• Chemical weathering	3		
	• Mining		12	+400
S	• Natural emissions to atmosphere at Earth's surface	80		
	• Fossil-fuel and biomass-burning emissions		90	+113
O and H (as H_2O)	• Precipitation over land	111×10^{12}		
	• Global water usage		18×10^{12}	+16
Sediments	• Long-term prein-dustrial river suspended load	1×10^{10}		
	• Modern river suspended load		2×10^{10}	+200

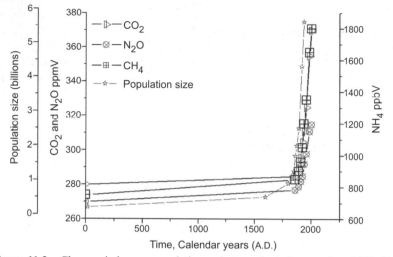

Figure 11.2   Changes in human population and atmospheric inventories of CO_2, N_2O, and CH_4 over the past 2000 years.

population through at least the first half of the twenty-first century will undoubtedly force an even greater exploitation of resources, with an inevitable increase in the human footprint on the ecological landscape. Clearly, such a condition is not sustainable, yet an off-ramp is not clearly visible in the trajectory of human domination of Earth's ecosystems.

SUSTAINABILITY

The concept of sustainability is frequently invoked but there is no consensus on a commonly accepted definition. We opt to define sustainability as a domain in space–time on a planet where the local time rate of change (i.e., the first derivative with respect to time) for all resources used by all organisms in the ecosystem is zero. This domain is not static. It changes with externally forced variables such as climate, tectonic activity, and planetary evolution in the broadest sense. On Earth these processes are ultimately linked to solar radiative output and the Earth's radiogenic heat flux. On long geological timescales, these processes change, yet Earth has remained habitable for over 3500 Ma out of the ca. 4600 Ma of its existence. Although sustainability on any planet is impossible on infinite timescales, it is clearly achievable for a relatively long period (Knoll 2003). Such sustainability must, therefore, be driven by internal feedbacks in biological processes that can adjust the output of metabolic processes to constrain broadly atmospheric gas composition and energy flow for long periods (Lovelock 1979).

   On a planetary scale, the chemistry of life involves sets of redox couples that are constrained by feedbacks (Nealson and Conrad 1999). The evolution of the feedbacks is itself not haphazard, yet we do not understand if there is a unique solution or there are multiple solutions that permit life to persist continuously as long as the planet remains within a zone of habitability (Kasting et al. 1988). The major metabolic sequences on Earth evolved over a period of ca. 2000 Ma and are based on coupled redox reactions that are far from thermodynamic equilibrium (Falkowski 2001). All organisms mediate phase-state transitions that add and remove gases from the atmosphere, and add to and remove solids from the lithosphere. For example, autotrophic carbon fixation converts gaseous CO_2 to a wide variety of organic carbon molecules, virtually all of which are solid or dissolved solids at physiological temperatures. Respiration accomplishes the reverse. Nitrogen fixation converts gaseous N_2 to ammonium and thence to organic molecules, whereas denitrification accomplishes the reverse. Calcification converts dissolved inorganic carbon and Ca to solid phase calcite and aragonite, whereas silicification converts soluble silicic acid to solid hydrated amorphous opal. Each of these biologically catalyzed processes is dependent upon specific metabolic sequences (i.e., gene families encoding a suite of enzymes) that evolved over hundreds of millions of years of Earth's history, and have, over corresponding periods, led to the massive accumulation of oxygen in Earth's atmosphere, and carbonate and organic matter in the lithosphere. Presumably, because of parallel evolution and lateral gene transfer, these metabolic sequences have frequently coevolved in several groups of organisms that, more often than not, are not closely related from a phylogenetic standpoint (Falkowski 1997). Based on their biogeochemical metabolism, these homologous sets of organisms are called functional groups or biogeochemical guilds, that is, organisms that are related through common biogeochemical processes rather than a common evolutionary ancestor.

Although functional groups form paired reaction pathways, their evolution was not initially so straightforward. The initial tempo of evolution of functional groups was almost certainly dictated to first order by selection based on redox gradients. Hence, depletion or production of substrates or products almost invariably leads to metabolic innovation, which couples the gradient through energy extraction or production. On Earth, the photobiological oxidation of water and subsequent oxidation of Earth's atmosphere liberated both the carbon reducers (oxygenic photoautotrophs) and carbon oxidizers (heterotrophs) from local electron sources and sinks, and thereby facilitated a pathway that ultimately led to colonization of land. However, although the innovation of water splitting altered the biogeochemical landscape of Earth forever after, the rate of evolution of that process has been extremely slow. All oxygenic photoautotrophs utilize a set of proteins that incorporate a Mn tetramer and pyrite-derived FeS clusters (Blankenship 2001). The evolutionary divergence of these proteins is extremely small, yet the biochemical turnover of the proteins is extremely fast. On average, all reaction centers in photoautotrophs are degraded and replaced every 30 minutes (Prasil et al. 1992), yet selection has not resulted in a solution to this seemingly incredible inefficiency. Similarly, nitrogen fixation, which evolved from a single common ancestral metabolic suite prior to the oxidation of Earth, remains an anaerobic process. In cyanobacteria, half of the nitrogenase appears to be inactivated by molecular oxygen at any moment in time. Although elaborate, and sometimes incredible, metabolic and structural innovations have evolved to protect the enzyme from oxidative damage, the core machinery has remained fundamentally unchanged (Berman-Frank et al. 2001).

Our point here is that the *realized* tempo of evolution of key metabolic pathways which characterize functional biogeochemical groups is very slow. Metabolic innovation was accomplished in the first 2500 Ma of Earth's history in microbes. In the succeeding 2100 Ma, *life-forms* changed, but *metabolic sequences* remained relatively unaltered. The changes in life-forms were driven by changes in substrate supplies and ecological opportunities. For example, in the oceans, there are eight divisions of eukaryotic oxygenic photoautotrophs comprised of 20,000 species. All these organisms obtained their metabolic machinery by pirating and enslaving a common cyanobacterial ancestral plastid (Delwiche 1999). Virtually no innovation occurred in this process; the host cells evolved specialized armor or nutrient acquisition strategies that facilitated ecological success, but even many of these processes were appropriated from prior evolutionary experimentation in prokaryotes.

The last grand experimentation in Earth's history occurred in the Paleozoic, with the invention of lignin. The rise of terrestrial plants and their rapid colonization of land resulted in resource plunder and may have accelerated weathering reactions that led to a decline in CO_2 and a (poorly documented) rise in O_2 (Berner et al. 2000). Feedbacks, driven by N_2 fixation as well as perhaps combustion, set Earth's O_2 concentration within narrow bounds (Lenton and Watson 2000; Falkowski 2002).

For the past 210 Ma, the carbonate isotope record shows an overall 1.5‰ increase in ^{13}C, suggesting a net burial of only 3000 Gt of organic C and an overall increase in O_2 of ca. 0.3%. Hence, despite several ocean anoxic events in the Cretaceous, the K/T extinction, and the Paleocene thermal maximum 55 Ma, from a biogeochemical perspective, the biologically mediated fluxes of gases and solids to Earth's atmosphere and lithosphere have been, to first order, close to steady state. Over this same period, however, it appears that a long-term depletion of CO_2 occurred and was accompanied by an increase in diversity of both photoautotrophs and aerobic metazoan heterotrophs (Rothman 2001). Changes in supply of substrates have been offset by changes in demands by consumers; the Earth system has been largely in the phase space we defined as sustained; however, one critical substrate, CO_2, has become increasingly scarce. In fact, over the past 15 Ma, the paucity of CO_2 is written in the carbonates, not by an increase in $\delta^{13}C$ but by a decrease. This apparently counterintuitive phenomenon reflects the invention and radiation of C4 photosynthesis, where the *core* metabolic machinery of oxygenic photosynthesis is preserved but several *previously invented* anapleurotic enzymes were summoned to extract CO_2 from an increasingly depleted pool — and to increase the supply of the substrate to the core machinery (Sage 1995). In so doing, the organic matter buried became isotopically heavier whereas the total amount buried barely changed. On the scale of biological innovation, the evolution of C4 plants was more akin to the transition from Baroque to Classical music than from Baroque to Hip Hop.

Although such innovations were small, they occurred on relatively long timescales compared with human evolution. For example, the evolution of C4 plants was itself a continuation of the evolution of photoautotrophs throughout the Cenozoic. Following the K/T extinction, the radiation of browsing mammals helped denude continental plateaus, which led to the evolution and radiation of (C3) grasses. The rise of grasses, which originated around the late Paleocene thermal maximum, was further accelerated by the evolution of grazing mammals, such as ungulates and horses. This feedback cycle accelerated silicon weathering (Conley 2002; Falkowski et al. 2003).

Grasses contain 6–10% dry weight of silicon (Epstein 1994). The deep roots of these organisms not only extract silicon from continental regolith but repackage the element in a form that is much more soluble. Natural burning of grasslands, rain on dead grass, and the mobilization of minerals in fecal material increased silicon fluxes to the coastal oceans and promoted a dramatic radiation of diatoms from the Eocene to present. Diatoms are the most efficient exporters of organic matter to marine sediments. The rise of diatoms appears to have further increased the drawdown of CO_2 and, by the Miocene, led to the rise of C4 grasses (Hayes et al. 1999). We suggest that these interconnected marine and terrestrial functional groups, connected by silicon, coevolved because of CO_2 depletion throughout the Cenozoic. The tempo of evolution is written in the genomic patterns of diatoms, grasses, and ungulates. Coevolution occurred through natural selection and while many species appeared and went extinct over this time, the overall biogeochemical pattern approached a sustainable condition because the time rate of change in the geochemical processes was slow or slower than the rates of evolutionary selection.

THE RED QUEEN HYPOTHESIS

The idea that coevolution increases stability by maintaining a constant rate of extinction and radiation over millions of years is called the Red Queen hypothesis (van Valen 1973). The gist of the idea is that, in tightly coevolved interactions, evolutionary change by one species (e.g., a prey or host) could lead to the extinction of other species (e.g., a predator or parasite), and that the probability of such changes might be reasonably independent of species' age. This idea, named after Lewis Carroll's character in Alice in Wonderland, proposes that evolution within a species must keep pace with environmental selection or the species will go extinct. In other words, each extant species has to "run" to stay in place. It is a useful heuristic device — which may or may not be correct — but serves as a starting point to examine how human evolution diverged from other species that inhabit Earth.

HUMAN ESCAPE FROM THE RED QUEEN

The evolution of *H. sapiens* rapidly changed Earth. Two major attributes of humans distinguish us from all other organisms (Table 11.2). These attributes have allowed humans to dominate the terrestrial landscape, but not without ecological costs, many of which are not yet recorded in the ledger of natural history.

A distinguishing feature of human evolution is clearly the evolution of complex language (Lieberman 2000). Human language permits communication of abstract thoughts through oral, visual, and written media. In the modern epoch, our communication skills are so honed that we can transfer, virtually instantaneously, vast bodies of knowledge across generational and geographic boundaries without changing a single gene within our gametes. Whereas other organisms, especially vertebrates, have limited communications skills, the quantum evolution that led to the extraordinary development of such attributes in *H. sapiens* appears unprecedented in the history of the planet. Language gave humans an incredible capacity to accommodate rapidly to, and indeed affect, the environment in ways no other organisms can.

The second attribute is the ability to create advanced tools. In this capacity, humans have excelled not only in fabricating instruments to acquire food and build shelters more efficiently — processes that clearly have parallels in other organisms — but in also altering natural materials to produce substances that otherwise never would have been found in nature. The examples of such massive alterations of materials are so enormous and so obvious to most of us that we tend to overlook their importance.

The result of the evolution of language and the ability to create advanced tools is, however, more subtle and dangerous. These two traits have permitted,

Table 11.2 Selected traits that distinguish humans from other apes (adapted from Carroll 2003).

Body shape and thorax	Elongated thumb and shortened fingers
Cranial properties (brain case and face)	Dimensions of the pelvis
Relative brain size	Presence of a chin
Relative limb length	S-shaped spine
Long ontogeny and lifespan	Language
Small canine teeth	Advanced tool making
Skull balanced upright on vertebral column	Brain topology
Reduced hair cover	Economic structure

and ultimately perhaps even required, a new form of knowledge, which call *distributed knowledge*. If we consider what each of us *individually* knows or knows how to do, we are hard pressed to recreate the world most of us know. For example, someone, somewhere knows how to make a light bulb, but very few of us individually have that knowledge. Moreover, we no longer go to a professional light-bulb maker and contract with him or her to make some specific light bulbs for us. Rather, a community of people has made machines that make and shape the glass for the bulbs, extracts, purifies, and fashions the tungsten elements, makes the metal base, pulls the vacuum during the manufacture, etc. Light bulbs are now made anonymously by groups of individuals, working with machines, made by other groups of people, each with specific individual knowledge. The knowledge is distributed.

The ensemble of human knowledge and skills is transmitted across geographical boundaries without need for genetic alteration. In so doing, skills are traded to create an *economy*. We assert that a fundamental *emergent property* of the evolution of speech and tool making is economic structure — a phenomenon unique to human society.

THE EVOLUTION OF CONSCIENCE AND SOCIAL BEHAVIOR IN HUMAN SOCIETY

One hallmark of human evolution is the concept of the distant future. Whereas experiential knowledge in many animals leads to both conditioned and learned behavior, anticipation of the distant future in humans is often intense. On an individual level, it can lead to anxiety, depression, or suicide. On a social level it can lead to war between subpopulations.

The ultimate anticipation in an individual's life is the concept of one's own inevitable mortality. The human fascination with death, the dead, and rituals surrounding mortality has led to many unique facets of human behavior, of which the emergence of irrational belief structures is but one of many manifestations (Becker 1997). Another facet is the striving to accumulate "wealth." Although it may be obvious that dogs do not pray (at least en masse), neither do they tend to hoard material resources.

The accumulation of "wealth" (an outcome of distributed knowledge and the evolution of economies) is not unique to humans, but the manifestation of "wealth" with regard to the amount of resources that can be acquired per individual human obeys no simple natural rules. The motivation is, however, usually quite obvious. Accumulation of wealth within a genetic lineage helps to ensure success of future generations regardless of physical or mental abilities (i.e., "skills"). Although humans without wealth have higher rates of fecundity, they also have higher rates of mortality. In human societies, wealth confers a greater probability of reproductive success among individuals who are physically or mentally deficient. Aided by tools in the contemporary epoch, otherwise infertile but wealthy human couples can even produce offspring by external fertilization.

Wealth is not a concept unique to humans — dogs bury bones and squirrels hoard nuts. However, dogs and squirrels do not use bones and nuts as a vehicle to entice other dogs and squirrels to build warmer and larger dog and squirrel houses, let alone breed more squirrels and dogs via external fertilization. The accumulation of wealth is understandable in a biological context as a form of competition; competition between humans and other species (e.g., for habitat), between subpopulations of humans (for material resources as well as habitat), and between individuals within the same subpopulations (e.g., for mates).

The anticipation of death and the accumulation of wealth within societies have been two forces motivating a second form of knowledge, which, for want of a better term, we will call *ethical knowledge*. Ethical knowledge is a code of conduct that is culturally derived, but transcends cultures. It provides a basis, not unlike game theory, for dynamic competition within subpopulations with clear biological outcomes. Over the course of human evolution, a sentimentalization of human life emerged, such that ethical knowledge forced increasingly altruistic behavior, ultimately leading to long-term desires to extend human life expectancy and decrease mortality. These desires, combined with language and tool-making capabilities, have greatly and increasingly led to resource plunder over the past several hundred years of human evolution.

INCREASING COMPLEXITY MAGNIFIES THE DEMAND FOR NEGENTROPY AND ENERGY

Although humans may (at least temporarily) escape the Red Queen constraints, no organism can escape thermodynamic constraints (Schroedinger 1945). One requirement of life is that an organism must maintain a state far from thermodynamic equilibrium with the environment. Hence, all forms of life increase the entropy of the surrounding environment, while simultaneously reducing their own entropy. This idea, formulated by Prigogine and Katchalsky as "nonequilibrium" thermodynamics (Prigogine 1980), established that the exchange of energy and matter between living organisms and the surrounding environment can only occur in an open system. An equation was developed to accommodate that change,

$$\delta S = \delta_e S + \delta_i S \quad \delta_i S > 0 \tag{11.1}$$

where δS is change in total entropy, $\delta_e S$ is change in entropy resulting from exchange with the external environment, and $\delta_i S$ is the change in entropy resulting from inefficiencies internal to the system. This equation requires the system to be in a lower than maximal entropy state as long as the system can utilize energy from the environment in exchange for high entropy matter. Hence, the more complex (ordered) a living system is, the larger is the difference in entropy between living organisms and the environment.

Increased demands for both energy and order (these are independent entities) result from enhanced exchange of materials with the environment, which can be achieved either by a single organism accelerating its metabolic rate or by an increase in the population of organisms (i.e., the growth of the system). The latter principle, developed by Ladsberg (1984) and Brooks and Wiley (1988), postulates that although the entropy of a system inevitably increases over time, it will always be lower than the maximal total entropy, that is, a living system can grow without violating the second law. Using a statistical mechanical approach, Brooks and Wiley defined the maximal possible entropy (or "phase space") as a function of information and suggested that genetic information sets the upper limits of entropy.

In the evolutionary trajectory, there is a persistent increase in the organizational complexity of metazoans and photosynthetic organisms (Stanley 1973). Although the increase in complexity can be biologically attributed to competition and selection, it physically reflects an increase in the potential level of entropy, as more and more instructional information is retained. Consequently, strictly from a thermodynamic perspective, the evolution of complexity has increased the consumption of energy and low entropy materials.

HOW DOES HUMAN EVOLUTION
AFFECT ENTROPY BALANCE?

The entry of *H. sapiens* onto the evolutionary stage introduced a new level of organization that simultaneously increased $\delta_e S$ and $\delta_i S$. Should humans be treated differently from the rest of evolution with respect to nonequilibrium thermodynamics, or is there simply an anthropocentric need to isolate human activities from the rest of the Universe? If the former, what feedbacks can be identified to produce sustainable human society given thermodynamic constraints?

During the course of human evolution, two major events marked the reorganization of human society and simultaneously affected planetary entropy balance. The first was the transition from a family-based hunter–gatherer existence to a tribally focused agricultural society. Examples of transitions can be found in the Natufian culture in the Levant and Jericho. The Neolithic revolution, which included domestication of plants and animals, was a turning point in the global entropy balance (Zohary and Hopf 2001), in which humans, by genetically selecting organisms, increased $\delta_i S$ at the expense of the rest of the system. This trend intensified as demands for more efficient food production for humans increased. The level of organization further increased with urbanization, which began in the early Bronze (Uruk) period.

Cities can be viewed as macro-sized dissipation mechanisms that are increasingly important in the contemporary world. They consume massive amounts of energy and low entropy matter for their maintenance, sustenance, and growth. Economy, technology, information, and social structures are an integral part of the flow of energy and matter that sustains the internal coherence of such organized superstructures. Since cities are open systems, there is no real limit to the amount of energy/matter that can be put through the system; therefore, there is no boundary on their level of complexity. Projections of supercities in the twenty-first century suggest populations within cities can grow to 25 to 50 million, requiring enormous increases in energy density. Indeed, the contemporary system demands huge amounts of energy. Global energy demand will have risen from 8.8 Gtoe (gigatons of oil equivalent) in 1990 to 11.3–17.2 Gtoe by 2020 (Birol and Argiri 1999). By the end of this century, the annual consumption is projected to reach 50 Gtoe or more.

As demand grows, there will be greater need for power plants, nuclear reactors, purification (detoxification) facilities, and depots for storage (Hoffert et al. 1998). These life-support systems are examples of yet further increases in complexity, requiring even more energy but simultaneously leading to more vulnerability to disturbance. (Witness the frequency of electrical blackouts in large sections of the United States caused by apparently trivial events.) The chance of a catastrophe en route increases in direct proportion to energy and material demands. At some point in this unfolding thermodynamic drama, the planet must plunge into a high-energy, low-entropy "trap," from which it will be increasingly difficult to escape. Clearly it is preferable to avoid the trap than to develop escape strategies *post facto*. One potential avoidance strategy employs altruism.

EVOLUTION AND ALTRUISM

Selection operates on individuals as manifested by phenotypic expression. Fitness, defined as the production of viable progeny, operates at the level of individuals and, barring a catastrophic mass extinction, determines the evolutionary trajectory of a species. As such, *intraspecific altruism*, which we define as a *reduction in reproductive output* by one individual in order to enhance the reproductive success of another, has no net effect on population growth but does affect phenotypic expression. Although there are many examples of altruistic behavior in vertebrates, it is a relatively rare behavioral characteristic outside of humans and other primates. In humans, however, altruism flourishes and is promoted culturally as an outgrowth of ethical knowledge. Altruism is not manifested by all individuals and does not appear to be genetically inherited. Therefore, we suggest that there is a third level of human knowledge, operating at the level of the individual, which we will simply call *individual knowledge*.

We propose that individual knowledge is obtained experientially as part of the developmental process, through parents, societal interactions (e.g., early childhood play, storytelling, and interactions with elders), and through cultural (tribal) rituals (e.g., religion, political structure). Because altruism is learned, it can be taught — and hence can rapidly pervade human society — but it can be lost just as rapidly.

In the evolution of modern, (especially) Western, societies, the confluence of uneven accumulation of wealth and ethical knowledge gave rise to increased pressure for altruism. Altruistic behavior, manifested, for example, in the redistribution of wealth or its application to large segments of society (e.g., the creation or endowment of a university, library, or hospital) can have a positive feedback. Ultimately, however, the accumulation of wealth, coupled with altruism and distributed knowledge, perpetuated a group of individuals who specialize in science and technology. These latter individuals have provided human societies with the tools and abilities to plunder resources more rapidly and are often looked to by other members of those societies for solutions to problems that innately require individual altruistic behavior. Indeed, scientists and technicians can manipulate genetic information and, in so doing, rapidly alter the phase space of the planet.

NONLINEAR OUTCOMES FROM ALTRUISTIC BEHAVIOR

Altruistic behavior can temporarily increase the evolutionary fitness of humans, but it can also have unintended negative impacts on the environment. We use, as an example, U.S. government policy at the end of World War II. To help returning veterans readjust to civilian life (and to help the economy), laws were passed that were ostensibly altruistic (by redistributing wealth), promoting the purchase of single-family houses outside of urban centers. Suburbia was born. The rapid expansion of single-family houses adjacent to cities (a phenomenon almost unique to the U.S.A.) fostered increased resource plunder. The development and expansion of suburbia destroyed farms and open spaces, replacing these not only with houses and schools and shopping centers, but with multilane highways, degraded lands, reduced habitat for other species, and gross overutilization of groundwater. This is an illusory reduction in entropy.

In suburbia, distributed transportation is virtually impossible. Americans became enamored with cars. Within a single generation, two energy-consuming (i.e., high energy, low entropy) pathways — a single-family house and one or more automobiles — became the norm for millions of Americans and remains so to this day. The net result is that government policy (or lack thereof) coupled with ignorance and arrogance (a lack of ethical knowledge) and only small numbers of altruistic consumers have produced the largest fossil-fuel demand ever seen on this planet. These types of nonlinearities in development can have profound negative effects on resource use and habitat. Songbirds, butterflies, and wildflower populations are decimated throughout much of the U.S. — victims of herbicides and insecticides designed to maintain perfect, "weed-free" lawns. Chlorine evasion to the atmosphere increases dramatically in the summer months as individual houses have individual swimming pools which consume annually millions of kilograms of chlorine.

The nonlinearities also affect societal structure. During World War II, the U.S. army segregated African-American soldiers from white soldiers. There were relatively few African-American draftees — most African-American males worked in factories in support of the war effort. These, nonmilitary personnel were not eligible for the housing subsidies and hence remained in the central cities. Those African-American veterans who tried to find a house in suburbia were often subject to discrimination. The government tax policies effectively (and inadvertently) promoted "white flight" and increased the racial segregation of America — a problem that remains and profoundly influences social structure and the distribution of wealth in the country to this day.

There are other, more dramatic examples, including the deforestation of Amazonia for short-term crop production, cattle ranches, and timber harvesting. In this case, government negligence and corruption rather than altruism are primary drivers. Similarly, strip mining in Africa, Australia, and China has led to massive alterations in surface topography which, in turn, has altered habitats for animals forever. Redistribution of water, for the production of energy and irrigation of crops, inevitably alters habitat as well. These and other activities, including expansion of grazing land and elimination of top predators, have massively altered the population structure and controls on most herbivorous mammals on the planet. Naiveté has played a role as well. The deliberate importation of mongooses to Hawaii and other islands, ostensibly to eat rats, another introduced species, led to a massive extinction of birds. Mongooses love bird eggs — they do not eat rats. Biological homogenization and direct exploitation (for food) has contributed to a rapid decrease in many species in a short time. Birds are very scarce in modern China — most have been eaten by humans. Codfish are almost extinct in the Northwest Atlantic — they have been efficiently harvested through the distributed technological knowledge of a more advanced species that seeks wealth (Jackson et al. 2001).

Social fixes to these complex problems are often hard to find. For example, a tax code could easily be written that would foster the return of people from suburbs to cities; however, there is no political motivation to do so. As long as energy and land are relatively affordable, suburban saturation is likely to persist — and that is hardly a path to sustainability. Extinction is forever — and although habitats can be restored, it does not mean that species will return. Is the sentimentalization of nature an important component of *ethical knowledge?*

THE ROLE OF SCIENCE

Over the past thirty years or so, scientists have increasingly documented the effects of humans in plundering Earth's resources. The documentation has had a relatively modest effect on societal responses. Sustainable development requires the mass expansion of individual altruistic behavior, a process that itself requires education and a reevaluation of how human economic structures can be used to preserve and conserve natural resources for future generations of humans. Education in developed countries can markedly alter patterns of resource use, but this must be coupled with intelligent investment of wealth in technologies that are inherently sustaining. For example, the photocatalyzed extraction of hydrogen from water would provide a potentially limitless, clean energy source, yet in the U.S. the investment in this process is less than $10 million per annum. A single breakthrough in catalysis could change the world forever. Similarly, the production of long-term human contraceptives, the development of N_2-fixing crops, or the replacement of relatively rare metals (such as titanium) in machines with easily produced alternative materials derived from renewable resources can alter the course of human impact on Earth.

Science and technology, however, are not the only solutions — human ingenuity must be coupled to human behavior. The concept that humans are partners in ecosystems is not new, but it does not pervade the human psyche, except in isolated, nomadic tribes, where there is a clearer, intuitive appreciation for habitat and a respect for it. We must leave the "documentation" stage of scientific enquiry and enter a social/technological stage, where realistic outcomes (both positive and negative) can be envisioned and integrated solutions explored. Nonlinearities in policy that can lead to dramatic changes in human behavior should be identified. Science does not simply serve as a knowledge base — it must also serve as a conscience of society — reminding wealth "creators" that sustainable resource management is the only viable option for future generations.

ACKNOWLEDGMENTS

We thank NSF and NASA for supporting our research as well as Andi Andreae, Andrew Irwin, Simon Levin, Peter Smouse, and Tamar Zohary for constructive comments on the manuscript.

REFERENCES

Becker, E. 1997. The Denial of Death. New York: Simon and Schuster.

Berman-Frank, I., P. Lundgren, Y.-B. Chen et al. 2001. Segregation of nitrogen fixation and oxygenic photosynthesis in the marine cyanobacterium *Trichodesmium*. *Science* **294**:1534–1537.

Berner, R.A., S.T. Petsch, J.A. Lake et al. 2000. Isotope fractionation and atmospheric oxygen: Implications for Phanerozoic O_2 evolution. *Science* **287**:1630–1633.

Birol, F., and M. Argiri. 1999. World energy prospects to 2020. *Energy* **24**:905–918.

Blankenship, R.E. 2001. Molecular evidence for the evolution of photosynthesis. *Trends in Plant Sci.* **6**:4–6.

Brooks, D.R., and E.O. Wiley. 1988. Evolution as Entropy: Toward a Unified Theory of Biology. Chicago: Univ. of Chicago Press.

Carroll, S.B. 2003. Genetics and the making of *Homo sapiens*. *Nature* **422**:849–857.

Conley, D.J. 2002. Terrestrial ecosystems and the global biogeochemical silica cycle. *Global Biogeochem. Cycles* **16(4)**1121.

Delwiche, C.F. 1999. Tracing the thread of plastid diversity through the tapestry of life. *Am. Naturalist* **154**:S164–S177.

Epstein, E. 1994. The anomaly of silicon in plant biology. *Proc. Natl. Acad. Sci. USA* **91**: 11–17.

Falkowski, P.G. 1997. Evolution of the nitrogen cycle and its influence on the biological sequestration of CO_2 in the ocean. *Nature* **387**:272–275.

Falkowski, P.G. 2001. Biogeochemical Cycles. In: Encyclopedia of Biodiversity, vol. 1, ed. S. Levin et al., pp. 437–453. San Diego: Academic.

Falkowski, P.G. 2002. On the evolution of the carbon cycle. In: Phytoplankton Productivity: Carbon Assimilation in Marine and Freshwater Ecosystems, ed. P. le B. Williams, D. Thomas, and C. Renyolds, pp. 318–349. Oxford: Blackwell.

Falkowski, P., M. Katz, B. van Schootenbrugge, O. Schofield, and A.H. Knoll. 2003. Why is the land green and the ocean red? In: Coccolithophores: From Molecular Processes to Global Impact, ed. J. Young and H.R. Thierstein, pp. 429–453. Berlin: Springer.

Falkowski, P., R.J. Scholes, E. Boyle et al. 2000. The global carbon cycle: A test of our knowledge of Earth as a system. *Science* **290**:291–296.

Hayes, J.M., H. Strauss, and A.J. Kaufman. 1999. The abundance of ^{13}C in marine organic matter and isotopic fractionation in the global biogeochemical cycle of carbon during the past 800 Ma. *Chem. Geol.* **161**:103–125.

Hoffert, M.I., K. Caldeira, A. Jain et al. 1998. Energy implications of future stabilization of atmospheric CO_2 content. *Nature* **395**:881–884.

Jackson, J.B.C., M.X. Kirby, W.H. Berger et al. 2001. Historical overfishing and the recent collapse of coastal ecosystems. *Science* **293**:629–637.

Kasting, J.F., O.B. Toon, and J.B. Pollack. 1988. How climate evolved on the terrestrial planets. *Sci. Am.* **258**:90–97.

Knoll, A.H. 2003. Life on a Young Planet. Princeton: Princeton Univ. Press.

Ladsberg, P.T. 1984. Can entropy and "order" increase together? *Physics Lett.* **78A**:219–220.

Lenton, T., and A. Watson. 2000. Redfield revisited, 2: What regulates the oxygen content of the atmosphere? *Global Biogeochem. Cycles* **14**:249–268.

Lieberman, P. 2000. Human Language and Our Reptilian Brain: The Subcortical Bases of Speech, Syntax, and Thought. Cambridge MA: Harvard Univ. Press.

Lovelock, J.E. 1979. Gaia: A New Look at Life on Earth. Oxford: Oxford Univ. Press.

Nealson, K.H., and P.G. Conrad. 1999. Life: Past, present and future. *Phil. Trans. R. Soc. Lond. B* **354**:1923–1939.

Prasil, O., N. Adir, and I. Ohad. 1992. Dynamics of photosystem II: Mechanism of photoinhibition and recovery processes. In: The Photosystems: Structure, Function, and Molecular Biology, ed. J.R. Barber, pp. 295–348. New York: Elsevier.

Prigogine, I. 1980. From Being to Becoming. San Francisco: W.H. Freeman,

Rothman, D. 2001. Global biodiversity and the ancient carbon cycle. *Proc. Natl. Acad. Sci. USA* **98**:4305–4310.

Sage, R.F. 1995. Was low atmospheric CO_2 during the Pleistocene a limiting factor for the origin of agriculture? *Global Change Biol.* **1**:93–106.

Schroedinger, E. 1945. What Is Life? Cambridge: Cambridge Univ. Press.

Stanley, S.M. 1973. An explanation for Cope's rule. *Evolution* **27**:1–26.

Sykes, B. 2001. The Seven Daughters of Eve. New York: W.W. Norton.

van Valen, L. 1973. A new evolutionary law. *Evol. Theory* **1**:1–30.

Vitousek, P., H. Mooney, J. Lubchenco, and J. Melillo. 1997. Human domination of Earth's ecosystems. *Science* **277**:494–499.

Zohary, D., and M. Hopf. 2001. Domestication of Plants in the Old World: The Origin and Spread of Cultivated Plants in West Asia, Europe, and the Nile Valley. Oxford: Oxford Univ. Press.

12

How Humankind Came to Rival Nature

A Brief History of the Human–Environment Condition and the Lessons Learned

B. L. TURNER II and S. R. MCCANDLESS

Graduate School of Geography and George Perkins Marsh Institute,
Clark University, Worcester, MA 01602, U.S.A.

ABSTRACT

Humankind is axiomatically tied to environmental change and was party to some forms of global-scale change reaching as far back as the Holocene. The capacity to induce environmental change, however, has increased throughout human history. This change follows a long-term, global-scale trajectory that tracks well with a logarithmic–logistic function describing human population growth. The model is applicable to technological and socioeconomic conditions as well. This review briefly describes the multiple steps of the resulting curve. Case-study examples illustrate the successive *or* stair-step changes in the human–environment condition. As these changes increase in variety, magnitude, pace, and spatial scale, they threaten the structure and function of the biosphere. Broad lessons are drawn from this history, with the caveat that the environmental changes underway today and the human–environment relationships precipitating them have no adequate analogues in the past.

THE LOGARITHMIC–LOGISTIC VIEW OF THE HUMAN–ENVIRONMENT CONDITION

Environmental change marks the history of our species, increasingly amplified as we have developed those attributes that designate us as social and reflexive beings. This history is relatively well documented at both local and global scales (e.g., Marsh 1864; McNeill 2000; Redman 1999; Thomas 1956; Turner et al. 1990; Williams 2003). From a global perspective, Deevey's (1960) multistep logarithmic–logistic function describing human population growth (Figure 12.1) serves as a basic template for human impact on the Earth and its biosphere.

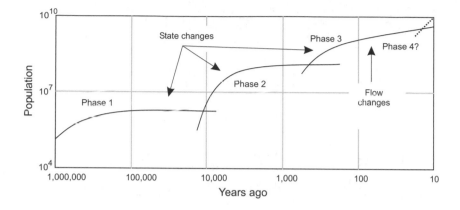

Figure 12.1 Logarithmic–logistic phases of the human–environment condition. See text for phase descriptions. From Deevey (1960).

It does so not because population necessarily acts as the primary driver of environmental change, but because the resulting curves track fundamental changes in the human–environment condition associated with different capacities to affect nature (Whitmore et al. 1990). For this reason, it warrants attention.

Three stair-step parts — from the emergence of our species to the present — comprise the logarithmic–logistic model (Figure 12.1). For Deevey, these parts constitute global-scale phases of ascending levels of population and technology capacity, presumably generated by the synergy between the two variables. Subsequent deliberations suggest that these phases track well with other characteristics of the human–environment condition. Primary among these are broad but dominant forms of political economy and social organization as well as the increasing per capita levels of consumption that they accommodate along with technological advances. Each phase step, therefore, represents fundamental changes in the sources and level of demand on nature's resources, and the techno-organizational capacities to fulfill those demands, including increasing "externalities" of waste from production and consumption. Thus, each phase step generates larger and more complex environmental consequences; by the third step, humankind matches or exceeds nature's contribution to many biogeochemical cycles that maintain the Earth system as we know it.

The first phase corresponds with human mastery of fire, advancement in toolmaking, and various pre-state social organizations. The second phase, beginning about 10,000 years ago, was triggered by the domestication of plants and animals which, in turn, supported the emergence of large, sedentary populations and state organization, culminating in the global transformation of biota. The third phase coincides with industrialization (fossil-fuel energy), the worldwide dominance of capitalism, and, to some degree, the democratization of the state. A fundamental question yet to be resolved is whether a new, fourth phase is emerging, characterized by globalization and advanced industrial processes.[1]

Each phase of the human–environment condition in this global accounting has not registered evenly or sequentially at subglobal scales. Some locales and regions experienced a particular phase long before or after other areas (e.g., the late arrival of agriculture to Australia). Others experienced long punctuations in occupation during which local and regional environments "recovered" from extensive human alteration (e.g., the southern Yucatán Peninsula and Petén, Guatemala, from the tenth to the twentieth centuries) (Turner and Butzer 1992). The global phasing of the human–environment condition, therefore, cannot be transferred uniformly to subglobal levels (Whitmore et al. 1990).

In the remainder of this essay, we trace the increasing capacities of humankind to alter the Earth and the biophysical processes that support the biosphere through these profound, if broad, phases of the human–environment condition. Again, each phase supports different kinds and rates of change with different consequences in terms of the magnitude and spatial scale experienced (Turner et al. 1990). These dimensions are not quantified but explored through examples illustrative of the phase.

The broad spatiotemporal dimensions of Deevey's phase steps favor such potential explanatory factors as change in population, technology, and political economy–societal organization (or the synergy among them) over those factors found in alternative explanations emphasizing institutions, values, and norms.[2] These last factors have yet to provide robust results in quantitative panel and cross-spatial tests of contemporary environmental change; they also prove problematic when population, technology, and time are controlled in historical

---

[1] Social sciences refer to a new, "post-modern" condition. This term focuses more on structural characteristics of capital and labor, which include their globally connected dimensions, than on the base technologies and their environmental consequences. The post-modern term, therefore, does not necessarily imply the advanced industrial dimension posed here. Whether this dimension is "real" (i.e., analytically useful) is the subject of future retrospective studies.

[2] Deevey's logarithmic–logistic function was developed to capture the global history of population growth. We expand it as a descriptive template of humankind's increasing capacity to rival the flux of nature. In doing so, our phase-step analogy is open to interpretative linkages to the IPAT identity and the contentiousness it invokes among many communities of social scientists, a literature that we do not review here (cf. Fischer-Kowalski and Amann 2001). This tension follows from interpretations that because P, A, and T are independent variables in the identity, they are explanatory. Such implications — predicated on an erroneous use of IPAT — are viewed by many researchers as insufficient, potentially "blaming the victim" (as in population growth in developing economies), and, more often than not, incommensurate with contemporary global economies in which the sources of environmental change are spatially incongruent with the location of the impacts. These and other critiques, however, do not hold for the long historical and global scale of the phase-step analogy.

analysis. Indeed, the only "tests" (historical comparative analysis) of which we are aware conclude that professed societal and religious values matter little in the degree and kind of environmental change societies have caused, controlling for the population and temporal dimensions of societies possessing similar technological capacity (see Whitmore et al. 1990). Such factors may prove more robust in explaining the *means* by which environmental change is enacted.

It is important to remember the ecological and individualistic fallacies: those factors giving rise to certain associations or patterns at one scale of analysis are not necessarily valid at levels ascending or descending scales (Gibson et al. 2000). Whereas population, technology, and broad characteristics of political economy–societal organization track at phase-step level of analysis, various case studies used to illustrate each phase step reveal the nuanced interactions among an array of factors that can be used to support various explanations of environmental change.

Finally, whether or not a new, fourth phase has emerged, the overall degree of human-induced change has entered a "no-analogue" state (Steffen et al. 2004). This state is matched by sociotechnological capacity to affect the very functioning of the biosphere for which there are also no analogues. For these reasons, lessons drawn from historical analysis must be viewed cautiously. Those advanced here with confidence are broad in scope and, for the most part, are not well matched spatiotemporally with the needs of modeling analysis, assessment, and decision-making communities.

LONG-TERM, GLOBAL PHASES OF THE HUMAN–ENVIRONMENT CONDITION

Phase 1: Fire-tools and the Reach of Ancient Cultures

The human imprint on the landscape was significant in the nascent development of society despite low levels of technology and population pressures. The control of fire provided our species with a powerful means to hunt, and its use for this and other purposes had serious environmental consequences (Pyne 2001; Sauer 1961). The imprints of historic and long-term burning can be found worldwide: from the distribution of certain grasslands and pine forests to bald mountain tops (e.g., Bird 1998; Denevan 1992). Observations of large-scale landscape burning by humans can be traced to antiquity, as in a Carthaginian reference to western Africa some 500 years B.C.:

> By day we saw nothing but woods, but by night we saw many fires burning ... we saw by the night the land full of flame and in the midst of lofty fire ... that seemed to touch the stars. (cited in Stewart 1956, p. 119)

Persistent burning, as observed in western and eastern Africa (in this case, surely for hunting and cultivation), changed states or faces of the Earth, contributing to some fire-climax ecosystems and landscapes understood to have evolved under persistent human burning (Bird 1998).

Mass Extinction of Megafauna

At least one unintentional consequence of the early uses of fire — extinction of biota — reached the global scale. Martin and Klein (1984) proposed that the extinction of Holocene megafauna at the end of the Pleistocene strongly corresponded to landscape disturbances by hunters. Recent assessments of the paleo-record for Australia interpret the loss of this biota as beginning 46,000 years ago with the arrival of fire-using Aborigines (Roberts et al. 2001), just as hunters, shortly after their arrival in New Zealand, eradicated the giant moa. Modeling exercises indicate that climate change alone was insufficient to trigger the extinction of Holocene megafauna in North America; human presence was necessary (Alroy 2001).

Phase 2: Domestication, Transportation, and States

Plant and animal domestication radically changed the human–environment condition and the scale of environmental consequences. Large, sedentary settlements developed worldwide; the growth in their size and number were complemented by the level and complexity of social organization. The rise of states and empires, in turn, concentrated power and resources. These were invariably employed to tackle the vagaries of nature, which constituted hazards to state and human well-being, especially with regard to food and fiber production. The expansion and intensification of domesticate production worldwide spurred changes in virtually every conceivable arable landscape, in many cases followed by significant environmental degradation (Redman 1999; Tainter 1988). Examples include soil salinization and erosion in desert environments (Adams 1965); deforestation and soil erosion in rugged mountain terrain; significant modification of surface hydrology, including wetlands; and large-scale deforestation across temperate and tropical landscapes (Denevan 1992; 2001; Doolittle 2000; Turner and Butzer 1992; Whitmore and Turner 2001). The vegetative consequences of some of these transformations would ultimately constitute criteria for the identification of certain physiographic-climatic provinces, such as the Mediterranean.

For the most part, the changes in question occurred at the local, landscape, and regional levels. These subglobal scales facilitate analysis of relationships in terms of coupled human–environment systems. They also primarily involved changes in the states or faces of the Earth, although the magnitude and spatial reach of certain activities, such as deforestation, affected the global carbon cycle long before the rise of fossil fuels through the process of cumulative change. The feedbacks of region-wide (cumulative) changes in vegetation may have affected local precipitation as well, a long-held, lay observation (Meyer 2000) that recently has gained scientific support. The magnitude and pace of these changes and their consequences increased throughout this second phase, although interpretations of human-induced environmental degradation and societal collapse can be traced back to antiquity (Redman 1999; Tainter 1988). Such interpretations, however, are hotly debated. Antagonists attribute environmental degradation and its social consequences to larger, political–economic conditions that weaken societal willingness or capacity to respond to change (Turner et al. 1990). These debates aside, various instances of environmental recovery have been well documented, as in the case of tropical forests in the Maya lowlands of Yucatán and Petén (Turner and Butzer 1992), and the extent of forest cover in New England, one of the three examples briefly outlined below.

Degrading Irrigated Lands

The rise of the early states in Mesopotamia was strongly linked to the development of irrigated cultivation in the plains of the Tigris and Euphrates rivers, beginning along the Euphrates as early as 4100 B.C. Reconstructions of settlement survey data (Adams 1965; Jacobsen and Adams 1958) display a regional population profile of long-term growth and decline between 4100 B.C. and A.D. 1250, punctuated by significant population declines about 450 B.C. and A.D. 1250. These profiles give rise to what Whitmore and colleagues (1990) label as two "millennial-long waves" of growth. Irrigation initially focused on the flood plains, but production was subsequently forced onto the higher portion of the immediate watershed via an elaborate canal system complemented by flood control devices. At no time were the entire plains cultivated, and the epicenter of activity appears to have shifted northward over time, perhaps partly in response to salinization.

As early as 1700 B.C., overzealous water use appears to have led to severe salinization, waterlogging, and marsh development at the tail ends of the distribution system, causing progressive settlement abandonment. Drainage canals constructed to alleviate these problems increased the scale of landscape transformation. Salinization, waterlogging, and accompanying agricultural abandonment were likely responsible, in part, for the concomitant declines in population at the time. Land degradation linked to irrigation is not restricted to this case, but has been repeated historically throughout the world. One estimate places cumulative soil loss worldwide (1700–1984) due to irrigation through inundation, waterlogging, and irreversible soil salinization at 1,000,000 km^2, or more than 37% of the total area (2,650,000 km^2) brought into irrigation over the past 300 years (Rozanov et al. 1990, p. 210).

Europeanization of Global Biota

The Columbian Encounter of 1492 is a good marker for the rise of European colonial states, their conquest of much of the world, and concomitant impacts on ecosystems, landscapes, and biota (Denevan 1992; Turner and Butzer 1992). Improvements in shipping that permitted long-distance, open sea voyages with relatively large cargo fostered this conquest and exchange. Plant domesticates from around the world were funneled to Europe, forever changing the region's culinary habits and, in some cases, cropping strategies related to the new cultivars. This impact, however, pales in comparison with that wrought by the distribution of the Europeans' favored biota and production strategies throughout the world (Crosby 1986). Nowhere were the impacts environmentally more significant than in the Western Hemisphere (Denevan 2001), where a combination of the plow, livestock, and, in some cases, the rifle reconfigured the land covers or keystone fauna that sustained them on a subcontinental scale (Turner and Butzer 1992).

In some cases, humans traumatized landscapes rapidly, as in the case of various islands in the Caribbean. In other cases, landscape and biota transformation played out in stages over centuries. The introduction of the horse to the Great Plains of North America, for example, gave rise to the Amerindian "horse culture." The subsequent eradication of this culture and the bison, the keystone species of the plains, made way for the introduction of longhorn cattle and ultimately, the large-scale cultivation of the American "breadbasket" (see below). The Europeanization process was as profound as it was replete throughout the landscapes of the Americas and Australia. Contrary to some arguments, however, it did not entail the first significant environmental change or degradation in the New World (Denevan 1992; Turner and Butzer 1992) or Australia (see Phase 1).

Deforesting New England

The seventeenth-century spread of European settlements across New England registered a "fundamental reorganization of plant and animal communities" (Cronon 1983, p. vii; Doolittle 2001) at a pace ten times faster than that in Europe (Merchant 1989, p. 2). Eventually, "old-world ecological relationships [were] reproduced in New England" (Cronon 1983, p. 155), complete with massive deforestation and the introduction of European species and associated diseases. Such introductions as Dutch elm disease, chestnut blight, hemlock woolly adelgid, gypsy moths, and beech bark disease continue to alter the long-term composition, structure, and function of New England's forests (O'Keefe and Foster 1998).

These and other changes in the New England landscape were, of course, the consequence of the invasion of new political economies and technological capacities, consistent with the thematic drivers operating historically at the global scale. Rich documentation and detailed studies of this case, however, reveal some of the mediating factors that shaped the specific consequences, especially the resulting landscape patterns. Cultural preferences valorized a visible confirmation of Anglo civilization, complete with well-tended farms, villages, and park-like forests, as captured by a Massachusetts colonist's record of the year 1642:

[T]his remote, rocky, barren, bushy, wild-woody wilderness, a receptacle for Lions, Wolves, Bears, Foxes, Rockoones, Bags, Bevers, Otters, and all kind of wild creatures, a place that never afforded the Natives better than the flesh of a few wild creatures and parch't Indian corn incht out with Chesnuts and bitter Acorns, [is] now through the mercy of Christ becom a second England for fertilness in so short a space, that it is indeed the wonder of the world. (Edward Johnson 1910 [1653; Book II, Ch. XXI], cited in Cronon 1983, p. 5)

A new export economy commanded commercially important species such as pine, oak, and chestnut, whose presence was repeatedly overestimated (Cogbill et al. 2002), spurring the expansion of economic activity across the region and promoting unsustainable offtake. Regardless, the deforested New England landscape would not last beyond the latter stages of the nineteenth century. Similar to the Maya case (above), significant tree cover regenerated throughout the twentieth century. Unlike the Maya case, this recovery took place in the face of increasing population and a transformation to industrial and service economies.

Phase 3: Industrialization and Increasing Lifestyle Expectations

The third phase, which began some 300 years ago but has escalated over the past 100 years, is commonly mislabeled the "Industrial Revolution." This phase encompasses the rise of fossil-fuel technologies, mass production–consumption, increased life expectancy and global population growth, as well as the hegemony of capitalism. These co-synchronic developments raised the magnitude and reach of state changes in the environment to unprecedented levels and have added a profoundly new kind of change: the ability to affect directly the biogeochemical cycles that function to maintain the biosphere (Turner et al. 1990; Steffen et al. 2004). By the late twentieth century, the consequences of these new human–environment conditions were so extensive that they generated worldwide concern and launched agendas regarding threats to the biosphere and the sustainability of the Earth systems. The documenting literature is large and so well known that it need not be reiterated here in detail (e.g., Tolba and El-Kholy 1992). It is noteworthy, however, that most types of human-induced changes in the states (e.g., forest to nonforest) and flows (e.g., carbon cycle) of the biosphere are occurring with unprecedented speed. These changes have reached a scale and magnitude, cumulatively and systemically, with serious consequences for the character and functioning of the biosphere — the so-called "no-analogue" human–environment condition (Steffen et al. 2004).

Fossil-fuel technology is fundamental to this condition. It possesses "liberating" capacities, rendered by the magnitude and quality of production with declining labor input. These capacities are tied to such socially desired consequences as improved nutrition and health, increased affluence, demographic transition, and enlarged flow of information and global connectivity (Grübler 1998). It is too simple, however, to reduce the human–environment condition of this phase to this technology, given the synergy between technology and the political economy. Critical elements of this phase have shaped societal decisions about the use of the environment. These elements include the solidification of state-sponsored capitalism, the distancing of consumption from production — literally across space as well as figuratively in the minds of the consumer — and rising economic expectations everywhere. The synergies involved have supported global population growth, the democratization of consumption and its related lifestyle expectations, an underlying belief in the power of technological solutions (substitutions) to deliver high quality goods and services *en masse*, and, until recently, an attenuated concern about environmental consequences. These consequences include changes in both states and flows of the biosphere, and are characterized by the large magnitude and rapid pace of individual activities triggering the changes in question, as the examples below illustrate.

The Southern Great Plains Dust Bowl

The southern Great Plains, stretching from northern Texas to southern Nebraska and central Oklahoma–Kansas to eastern Colorado, was labeled a desert by early American explorers who observed the landscape impacts of an extended dry spell. A period of favorable rainfall in the late nineteenth and early twentieth centuries, however, coincided with shifts in federal policy, which encouraged farmers to fill the region, and the arrival of rail transport, which opened markets to the east. Armed by technology, capitalism transformed the landscape and filled freight cars with nature's bounty bound for Chicago and points east. The paucity of farm-level conservation strategies, coupled with extensive drought and the Great Depression during the 1930s, desiccated the land and its crops and crippled its people. From 1933 to 1938, the southern Great Plains — the Dust Bowl — witnessed 301 dust storms that reduced visibility to less than a mile (Worster 1979, p. 15), effectively blowing away the topsoil. The economic and human impacts were immense, registered in multitudes of smallholders who lost their life-savings on the farm and were forced to migrate to California as farm laborers. Their numbers reached an estimated 300,000 in the last half of the 1930s alone (Worster 1979: 50). The southern Great Plains eventually recovered, at least in terms of land output, following a path of university-extension strategies informed by agroecological principles and aimed at reducing soil and water loss. Changing regional and national economies, however, have subsequently promoted substantial irrigation, drawing down the Ogallala aquifer and raising yet new concerns about the long-term sustainability of cultivation of the southern Great Plains (Brooks and Emel 1999).

Death of the Aral Sea

Beginning in the 1960s, until the demise of the Soviet Union, about 5 million ha of irrigated lands were added along the only two rivers feeding the inland Aral Sea (Perera 1993) in a series of projects implemented to generate hard currency for the state through the production of cotton. The state-sponsored and state-engineered projects employed inappropriately constructed irrigation canals with high rates of evaporation and infiltration. They were also controlled by water institutions that rewarded high use per unit of production (Kasperson et al. 1995; Micklin 1988). With the discharge of river water to the sea virtually stopped, the former fourth largest body of fresh water in the world lost 60% of its volume and 50% of its area, and more than doubled its salinity (Perera 1993). For all practical purposes the Aral Sea ecosystem is dead; resurrection, pending large-scale changes in adjacent land uses, will take a century or more. The consequences of its death have been far-ranging and include the collapse of the sea's former lucrative fishing industry, local climate changes with increased salt and dust storms, changes in surrounding land covers, and numerous human health problems, exacerbated by the collection of agricultural chemicals in canals and drinking water. The Aral Sea case represents one of the few collapses of a large coupled human–environment system during the industrial phase, illustrating the modern capacity to transform systems rapidly.

Identifying those conditions that make a system prone to collapse like the Aral Sea is a daunting task, one that has been marked as much by the ideology of the interpreter as the evidence. The Aral Sea case took place in a command economy and a Marxist-socialist political system. The "condition" that distinguishes the Aral Sea case from other, potential collapse cases appears to rest in the information and control structures, in which there were few or no checks on policy and decision making (Kasperson et al. 1995).

Mass Extinction of Species

The human modification and transformation of the Earth's surface is now so expansive in its reach and depth that the habitats lost, disconnected by landscape fragmentation, and invaded by exogenous species threaten a modern mass extinction of native species and a mega-mass extinction of populations (Hughes et al. 1997). Whereas the current rates (and magnitude) of extinction are difficult to calculate precisely, they appear to range from 100 to 1000 times higher than the background rate (Pimm et al. 1995). The pace of extinction is triggered by the high rates of species-rich tropical forest loss, beginning in the latter stages of the twentieth century and sustained to the present. Population extinctions offer a more sensitive metric of the effects of habitat loss than species extinctions (Pimm et al. 1995). By conservative estimates, tropical forest populations are disappearing at rates three to eight times those of species (Hughes et al. 1997). Currently, 90% of plant species threatened is highly restricted in their biogeography, endemic to one place or area, and thus highly prone to potential extinction. Land-use practices that transform critical habitats supporting threatened species can trigger species loss by making those ecosystems vulnerable to invasive species, which, in turn, outcompete the endemics. Fragmentation of landscapes disrupts pathways for species movement and recolonization of disturbed areas. Such impacts are not new, but they are magnified today by the magnitude and pace of land-cover changes taking place worldwide, especially in tropical forests.

Ecosystem Degradation and Restoration

The scale of land change underway worldwide is sufficiently large to threaten the provision of goods and services by a wide array of environments, including water filtration, flood and salinity control, air purification, pollination, sequestration of carbon, and genetic stocks, as well as "life-fulfilling" services: aesthetic beauty, spiritual renewal, recreation, and enjoyment (Daily et al. 2000; Daily and Ellison 2002). Heretofore, much of society has taken these goods and services for granted, with minimal attention paid to polluting, degrading, and other activities affecting ecosystem function. Historically, technological substitutes have countered these losses. The magnitudes of ecosystem changes underway and escalating human demand for the goods and services in question, however, have stretched the economic viability of many substitutes. For example, the costs of protecting watersheds may prove to be far less expensive for providing acceptable levels of water quality for urban users than the construction of new water treatment plants, exemplified in the case of New York City. Ninety percent of the city's water arrives, mechancially unfiltered and largely untreated, from the million-acre Catskill/Delaware basin. A major investment in a watershed protection program restored watershed functioning, saving the city several billion dollars (Daily and Ellison 2002).

Global Climate Change and Ozone

The role of humankind in inducing global climate change and the thinning of the ozone layer over the poles is so well known that the details need not be reiterated here. Both issues are illustrative of the newfound ability of humankind to affect directly the chemistry of the atmosphere and the biogeochemical flows that sustain the structure and function of the biosphere (Steffen et al. 2004). Changes of the states of the Earth (e.g., land cover) play a significant role in climate change; however, the overwhelming proximate cause of global climate change and the threats to the stratospheric ozone layer rest in the burning of fossil fuels and the release of synthetic compounds to the atmosphere. Either case constitutes systemic global change because a few point sources, if of sufficient magnitude, can trigger the impacts in question by moving through the fluid systems of biosphere (Turner et al. 1990). In this kind of global change, consequences everywhere follow from production–consumption anywhere, breaking the long-standing spatial linkage in the environmental consequences of production–consumption. Given the novel character and pace of this kind of change, its consequences are replete with uncertainties that render historical analogues suspicious. The connection of thinning stratospheric ozone and chlorofluorocarbon production constituted one of the first "surprises" with biosphere-level implications. A more recent surprise involves the possible degree to which climate change accounts for gains and losses in agricultural production in the United States (Lobell and Asner 2003).

Regional Air Pollution

The impacts of fossil-fuel burning extend well beyond global climate change and stratospheric ozone. Various human health problems and acid rain are long-standing recognized consequences; however, more recent work indicates large-scale regional feedbacks on food production systems from air pollution. About 60% of the world's food stock is produced in three large regions covering 23% of the Earth's land surface. These same three regions produce more than one-half of the global emissions in nitrogen oxides (NO_x), emitted largely through the burning of fossil fuels (Chameides et al. 1994). NO_x is critical to the creation of photochemical smog, including ozone (O_3), which is harmful to plant production. Enhanced by summertime high-pressure systems, repeated exposure to high levels of O_3 potentially reduces crop production by 5–10%, with the damage increasing with the magnitude and intensity of exposure (Chameides et al. 1994). Similarly, haze from regional air pollution in China, largely linked to energy production, is estimated to reduce about 70% of the crops grown by as much as 5–30% of their optimal yields (Chameides et al. 1999). Recognition and understanding of this feedback from large-scale urban-industrial complexes on plant productivity is relatively new, and illustrates the numerous and complex kinds of systemic changes underway in the functioning of the biosphere.

Phase 4: Advanced Industrialization and Globalization

The twenty-first century holds the promise of a new human–environment condition — globalization — characterized by the rapid and worldwide flow of capital, information, and people, increasingly driven by processes beyond the direct control of the state. Globalization is matched by science-driven industrialization, from information processing (e.g., nanotechnology) to molecular biology (e.g., transgenic crops), which portend not only major advances in production and consumption but also in the efficiency of energy and materials use (Ausubel 1996). These conditions coincide with a new level of global awareness about the human capacity to change Earth's environmental systems, with possible threats to the functioning of the biosphere. This last development attests to international attention given to climate change, loss in biodiversity, and threats to ecosystem goods and services, and the rise of Earth system, global change, and sustainability science (Kates et al. 2001; Schellnhuber and Wenzel 1998).

That humankind has actually entered a fourth phase of the human-environment condition or that the lenses through which we view human–environment conditions are moving toward a focus on sustainability are empirical questions that can only be answered in the future. Indeed, the history of human–environment relationships traced here would indicate that the last question warrants a cautious response.

THE ABSENCE OF ANALOGUES AND
THE VALUE OF HISTORY

History matters, according to an age-old adage. It does so in the sense that any human–environment condition is shaped by preceding conditions that reduced or enlarged future options, a process referred to as path-dependency. It does not in the sense that human–environment history appears to be punctuated by technological and political–economic innovations that fundamentally alter human-environment relationships. These punctuations, captured in Deevey's three stair-step phases, appear better suited to the metaphor of biological evolution and random gene mutation. The novelty and complexity of the interactions of these innovations with nature can lead to thresholds or tipping points, triggering punctuations in the functioning of the biosphere with consequences constituting surprises to humankind. This reality underlies the claims of global change science that the current global human–environment condition has no comparable analogues (Steffen et al. 2004) and thus human–environment relationships have entered the "Anthropocene" (Crutzen and Stoermer 2001). Likewise, contemporary society has little by way of analogues regarding its full capacity to affect the biosphere (e.g., nuclear war) or its potential to address human–environment problems at a global and large regional scale (e.g., the Montreal Protocol versus the Kyoto Convention). These concerns raise cautionary flags about the use of historical assessments to generate lessons or analogues, other than the most general in kind, to guide a "transition to sustainability" (Meyer et al. 1998). This caveat should be kept in mind when considering the list of broad principles and lessons that conclude this chapter.

Such caution notwithstanding, assessments of the kind offered here are useful because they help to reveal potential follies that may follow from more focused spatiotemporal analysis. A broader lens cautions the reader about extrapolations from studies with short time frames and small spatial extent that might identify capitalist economies or common property regimes as more environmentally damaging than other economic structures or land institutions, respectively; globalization as a special moment in regard to the spatial separation of production and consumption and their attendant environmental impacts; or worldwide environmental change as unique to the last century of human history. In addition, the details revealed in the more fine-tuned spatiotemporal scales of the example cases presented in this chapter reveal the variation and complexity of interacting factors that lead to certain human–environment consequences, many of which support explanations that differ significantly from those implied in the ultimate macroscale of the phase steps used to frame this review.

Broad Lessons

- *The human species is synonymous with environmental change.* This observation is as obvious as it is well established empirically. Its significance rests in those analytical traditions that find it useful to distinguish the environmental consequences of humankind from the ambient condition of biosphere absent our species.
- *Historically, societal needs and desires to generate more material goods and services, in quantity and quality, have driven environmental change,* often with unintended consequences.
- *Societal capacity and willingness to alter biophysical environments has increased throughout human history.*
- *Environmental change with global consequences is not new to Phase 3.*
- *Phase 1 and 2 changes appear to be linked to the high sensitivity and low resilience of certain biota and ecosystems to initial human contact or significant changes in human activity.*
- *The totality of changes* (i.e., number, pace, and magnitude*) and the systemic kind of change* (i.e., biogeochemical cycles) *underway is new to Phase 3.*
- *The global-scale history of sustained change is not necessarily replicable at subglobal scales.* Various sequences and outcomes of human disturbance exist, including regional abandonment, landscape (or partial landscape) recovery, and re-use and change.
- *In the past, environmental change posing various threats to coupled human–environment systems appears to be more common than the complete collapse of the coupled system.* Examples of coupled system collapse from human-induced degradation of the environment appear to be few or interpretations of them controversial.
- *Significant human-induced environmental change, including degradation, appears to transcend all forms of political economy and societal structure.*
- *The only documented collapse of a large coupled human–environment system during the third or industrial phase, the Aral Sea case, involved a human subsystem whose design and management had minimal societal participation and operated in a political system with few checks and balances.*
- *The capacity of a human subsystem to cope with environmental change is determined by the kind of changes underway and the dominant population-political economy-technology condition.*
- Given the no-analogue status of the environmental changes underway and the human subsystems involved, *historical assessments provide minimal principles or lessons of a specific kind for the assessment of present and future changes.*

REFERENCES

Adams, R.M. 1965. Land Behind Baghdad: A History of Settlement on the Diyala Plain. Chicago: Univ. of Chicago Press.

Alroy, J. 2001. A multispecies overkill simulation of the end-Pleistocene megafauna mass extinction. *Science* **292**:1893–1896.

Ausubel, J.H. 1996. Can technology spare the Earth? *Am. Sci.* **84**:166–178.

Bird, M.L. 1998. A million year record of fire in sub-Saharan Africa. *Nature* **394**:767–769.

Brooks, E., and J. Emel. 1999. The Llano Estacado of the U.S. Southern High Plains: The Rise and Decline of a Modern Irrigation Culture. Tokyo: United Nations Univ. Press.

Chameides, W.L., P.S. Kasibhatla, J. Yienger, and H. Levy II. 1994. Growth in continental-scale metro-agroplexes: Regional ozone pollution and world food production. *Science* **264**:74–77.

Chameides, W.L., H. Yu, C. Liu et al. 1999. Case study of the effect of atmospheric aerosols and regional haze on agriculture: An opportunity to enhance crop yields in China through emission controls? *Proc. Natl. Acad. Sci. USA* **96**:13,626–13,633.

Cogbill, C.V., J. Burk, and G. Motzkin. 2002. Fire on the New England landscape: Regional and temporal variation, cultural, and environmental controls. *Biogeography* **29**:1305–1318.

Cronon, W. 1983. Changes in the land: Indians, colonists, and the ecology of New England. New York: Hill and Wang.

Crosby, A.W. 1986. Ecological Imperialism: The Biological Expansion of Europe, 900–1900. Cambridge: Cambridge Univ. Press.

Crutzen, P.J., and E. Stoermer. 2001. The "Anthropocene." *Global Change Newsl.* **41**: 12–13.

Daily, G., and K. Ellison. 2002. The New Economy of Nature: The Quest to Make Conservation Profitable. Washington, D.C.: Island Press.

Daily, G.C., T. Söderqvist, S. Aniyar et al. 2000. The value of nature and the nature of value. *Science* **289**:395–396.

Deevey, E.S., Jr. 1960. The human population. *Sci. Am.* **203**:194–204.

Denevan, W.M. 1992. The pristine myth: The landscape of the Americas in 1492. *Ann. Assoc. Am. Geog.* **82**:369–385.

Denevan, W.M. 2001. Cultivated Landscapes of Native Amazonia and the Andes. Oxford: Oxford Univ. Press.

Doolittle, W.E. 2000. Cultivated Landscapes of Native North America. Oxford: Oxford Univ. Press.

Fischer-Kowalski, M., and C. Amann. 2001. Beyond IPAT and Kuznets Curves: Globalization as a vital factor in analysing the environmental impact of socio-economic metabolism. *Pop. & Envir.* **23**:7–47.

Gibson, C.C., E. Ostrom, and T.K. Ahn. 2000. The concept of scale and the human dimensions of global change: A survey. *Ecol. Econ.* **32**:217–241.

Grübler, A. 1998. Technology and Global Change. Cambridge: Cambridge Univ. Press.

Hughes, J.B., G.C. Daily, and P.R. Ehrlich. 1997. Population diversity: Its extent and extinction. *Science* **278**:689–692.

Jacobsen, T., and R.M. Adams. 1958. Salt and silt in ancient Mesopotamian agriculture. *Science* **128**:1251–1258.

Kasperson, J.X., R.E. Kasperson, and B.L. Turner II. 1995. Regions at Risk: Comparisons of Threatened Environments. Tokyo: United Nations Univ.

Kates, R.W., W.C. Clark, R. Corell et al. 2001. Sustainability science. *Science* **292**:641–642.

Lobell, D.B., and G.P. Asner. 2003. Climate and management contributions to recent trends in U.S. agricultural yields. *Science* **299**:1032.

Marsh, G.P. 1864. Man and Nature: Or, Physical Geography as Modified by Human Action. New York: Charles Scribner.

Martin, P.S., and R.G. Klein. 1984. Quaternary Extinctions: A Prehistoric Revolution. Tucson: Univ. of Arizona Press.

McNeill, J.R. 2000. Something New under the Sun: An Environmental History of the 20th-Century World. New York: Norton.

Merchant, C. 1989. Ecological Revolutions: Nature, Gender, and Science in New England. Chapel Hill: Univ. of North Carolina Press.

Meyer, W.B. 2000. Americans and Their Weather. Oxford: Oxford Univ. Press.

Meyer, W.B., K.W. Butzer, T.E. Downing et al. 1998. Reasoning by analogy. In: Human Choice and Climate Change, ed. S. Raynor and E.L. Malone, pp. 217–290. Columbus: Battelle Press.

Micklin, P. 1988. Desiccation of the Aral Sea: A water management disaster in the Soviet Union. *Science* **241**:1170–1176.

Myers, N., and A.H. Knoll. 2001. The biotic crisis and the future of evolution. *Proc. Natl. Acad. Sci. USA* **98**:5389–5392.

O'Keefe, J., and D.R. Foster. 1998. An ecological history of Massachusetts forests. *Arnoldia* **58**:2–31.

Perera, J. 1993. A sea turns to dust. *New Sci.* **140**:24–27.

Pimm, S.L., G.J. Russell, J.L. Gittleman, and T.M. Brooks. 1995. The future of biodiversity. *Science* **347**: 347–350.

Pyne, S.J. 2001. Fire: A Brief History. Seattle: Univ. of Washington Press.

Redman, C.L. 1999. Human Impact on Ancient Environments. Tucson: Univ. of Arizona Press.

Roberts, R.G., T.F. Flannery, L.K. Ayliffe et al. 2001. New ages for the last Australian megafauna: Continent-wide extinction about 46,000 yr ago. *Science* **292**:1888–1892.

Rozanov, B.G., V. Targulian, and D.S. Orlov. 1990. Soils. In: The Earth as Transformed by Human Action: Global and Regional Changes in the Biosphere over the Past 300 Years, ed. B.L. Turner II, W.C. Clark, R.W. Kates et al., pp. 203–214. Cambridge: Cambridge Univ. Press.

Sauer, C.O. 1961. Fire and early man. *Paideuma* **7**:399–407.

Schellnhuber, H.J., and V. Wenzel. 1998. Earth System Analysis: Integrating Science for Sustainability. Heidelberg: Springer.

Steffen, W., A. Sanderson, P.D. Tyson et al., eds. 2004. Global Change and the Earth System: A Planet Under Pressure. The IGBP Book Series. Berlin: Springer.

Stewart, X. 1956. Fire as the first great force employed by man. In: Man's Role in Changing the Face of the Earth, ed. W.L. Thomas Jr., pp. 115–133. Chicago: Univ. of Chicago Press.

Tainter, J.A. 1988. The Collapse of Complex Societies. Cambridge: Cambridge Univ. Press.

Thomas, W.M., Jr., ed. 1956. Man's Role in Changing the Face of the Earth. Chicago: Univ. of Chicago Press.

Tolba, M.K., and O.A. El-Kholy. 1992. The World Environment 1972–1992: Two Decades of Challenge. New York: Chapman & Hall for UNEP.

Turner, B.L., II, and K.W. Butzer 1992. The Columbian Encounter and land-use change. *Environment* **34**:16–44.

Turner, B.L., II, W.C. Clark, R.W. Kates et al., eds. 1990. The Earth as Transformed by Human Action: Global and Regional Changes in the Biosphere over the Past 300 Years. Cambridge: Cambridge Univ. Press.

Whitmore, T.M., B.L. Turner II, D.L. Johnson et al. 1990. Long-term population growth. In: The Earth as Transformed by Human Action: Global and Regional Changes in the Biosphere over the Past 300 Years, ed. B.L. Turner II, W.C. Clark, R.W. Kates et al., pp. 25–39. Cambridge: Cambridge Univ. Press.

Whitmore, T.M., and B.L. Turner II. 2001. Cultivated Landscapes of Middle America on the Eve of Conquest. Oxford: Oxford Univ. Press.

Williams, M. 2003. Deforesting the Earth: From Prehistory to Global Crisis. Chicago: Univ. of Chicago Press.

Worster, D. 1979. Dust Bowl: The Southern Great Plains in the 1930s. Oxford: Oxford Univ. Press.

13

Anthropogenic Modification of Land, Coastal, and Atmospheric Systems as Threats to the Functioning of the Earth System

M. O. ANDREAE[1], L. TALAUE-MCMANUS[2], and P. A. MATSON[3]

[1]Biogeochemistry Department, Max Planck Institute for Chemistry,
55020 Mainz, Germany
[2]Division of Marine Affairs and Policy, Rosenstiel School of Marine and
Atmospheric Science, University of Miami, Miami, FL 33149, U.S.A.
[3]School of Earth Sciences, Stanford University, Stanford, CA 94305–2210, U.S.A.

ABSTRACT

Over the past two centuries, the tremendous growth of the human population and the high resource demand of technologically developed societies has made humanity a geochemical and geophysical force that is able to compete with Nature's forces and to threaten Earth system functioning. Human activities are changing the composition of biosphere, atmosphere, and hydrosphere, affecting global climate, and may even perturb the main circulation patterns of the world's oceans. Human impact on the Earth system is illustrated by examining the effects of intensive agriculture, tropical deforestation, and excessive nutrient inputs into coastal ecosystems. Because of the numerous feedbacks and teleconnections in the Earth system, the change resulting from such perturbations is likely to be nonlinear and contain abrupt discontinuities. In this situation, the prudent course would be to maintain the Earth system as much as possible within the known parameter space and to pursue a course of sustainability using the knowledge gained from Earth system research.

INTRODUCTION

Throughout time, human species has exhibited a tremendous ability to transform its environment. With fire and spear, early humans changed the landscape as they spread about the globe, and at the end of the Pleistocene brought about the extinction of the Holocene megafauna on all continents except Africa (Turner and McCandless, this volume). As profound as these transformations were at the regional and even continental scale, they did not represent a threat to the functioning of the Earth system as a whole, as long as the human population was fairly small and its technology remained relatively simple.

This situation has changed dramatically over the last 200–300 years. The human population has "exploded" by an order of magnitude during this time period, to reach some 6000 million at present, and industrial development has sharply increased the per-capita environmental impact. Because of our tremendous resource use, humankind, which only accounts for about 0.001% of the Earth's biomass, has become a geochemical and geophysical force that is able to change the composition of biosphere, atmosphere, and hydrosphere, to affect global climate, and maybe even to perturb the main circulation patterns of the world's oceans.

Human energy mobilization equates to only about 0.01% of the solar energy available to the Earth system, a minute perturbation! This suggests that human activities perturb the Earth system not only by altering fluxes of energy and materials but by modifying processes in a catalytic manner, and by changing the way the system functions. It also points to the inherent fragility of the Earth system, at least from the perspective of humanity and the macrobial part of the biosphere: a change of 1–2 K in global temperature may represent a serious threat to human welfare and the survival of sensitive ecosystems.

In this chapter, we illustrate how human activities may be threatening key parts of the Earth system. The limited space available obviously does not permit a comprehensive discussion of human challenges to the Earth system, nor does it allow a detailed discussion of any particular threat, such as climate change. We will show how a complex of human activity results in systemic, not just quantitative, change. In other words, we want to look not just at the change in the magnitude of some system variable, such as temperature or rainfall, but at the change in how the system works. For this purpose, we have selected three activities that have major impacts on the Earth system: agriculture, tropical deforestation, and excessive nutrient inputs in coastal ecosystems. These activities are obviously connected in some of their causes and effects, highlighting the ultimate need to analyze global change through a systemic approach.

INTENSIVE AGRICULTURE AND THE EARTH SYSTEM

Over the past 35 years, development and utilization of Green Revolution technologies, including use of improved genetic materials and increased use of fertilizers, pesticides, mechanization, and irrigation, has led to major increases in grain yields worldwide. This pattern of "intensification" has complemented the ongoing expansion of food production into previously nonagricultural lands; indeed, intensification of agriculture has accounted for more of the global food production increase than expansion and is likely to continue to do so. To feed the human population of the next century, however, even more substantial increases in food production will be required. Given the consequences of deforestation and other land conversions for agriculture (see next section), the best hope for increasing food production while minimizing the impacts to the global system may be by increasing yields in both developed and developing regions through a more intensive yet efficient use of resources on lands currently under agriculture and through the recovery of degraded agricultural lands.

Intensification of land use has had important implications for global as well as regional and local Earth system processes (Matson et al. 1997; Ruttan 1999; Tilman et al. 2002). Soil loss, water pollution, hydrologic changes, as well as impacts on atmospheric chemistry and climate represent only a few of the concerns. The goal of sustainable agricultural production can only be achieved if these off-site consequences of agriculture are addressed and minimized. In this section, we identify important links between agriculture and the functioning of the Earth system and suggest alternatives for the future that may reduce the consequences of intensification.

Fertilizer Use

Intensive high-yield agriculture is dependent on the addition of fertilizers, especially those that are synthetic. Between 1960 and 1990, N fertilizer production and use increased fivefold, and this trend is expected to continue over the next 25 years (Cassman and Pingali 1995; Matson et al 1997; Tilman et al 2002). Consumption of inorganic P fertilizers likewise increased from less than 2 Mt yr^{-1} in 1950 to more than 13 today (Smil 2000). For both of these nutrient elements, agricultural fertilization represents a global-scale disruption in the natural element cycles (Falkowski and Tchernov, this volume). Fertilizer use alone has almost doubled the amount of available N coming into the global terrestrial system each year. Likewise, phosphorus fertilizer use has about doubled the flux of P in the global system.

Some of the consequences of fertilizer use are, of course, positive: without these fertilizers, world food production could not have increased at the rate it did. However, only about 30–50% of the N and P fertilizer applied is used by plants (Smil 1999, 2000; Tilman et al. 2002); significant proportions of it can be lost from agricultural fields in gaseous and solution forms and have numerous negative consequences for off-site ecosystems, water systems, and the atmosphere. Leaching and runoff of N and P to surface and groundwater have engendered concern at both local and regional scales. For example, Howarth et al. (1996) estimated that total riverine fluxes from most of the temperate zone land systems surrounding the North Atlantic have increased two- to twentyfold since preindustrial times. Nitrogen loading to estuaries and coastal waters is responsible for over-enrichment and eutrophication, often leading to the creation of low-oxygen conditions that endanger fisheries (see later section). The

proximate[1] causes of deforestation have remained the same: agricultural expansion, wood extraction, and infrastructure extension (Geist and Lambin 2001). In a meta-analysis of tropical deforestation worldwide, Geist and Lambin (2001) found that agricultural expansion, that is, the expansion of cropped land and pasture, is the most frequently reported proximate cause of deforestation, but that it almost always occurs in combination with one or both of the other two factors. Among the underlying forces, economic factors are dominant; however, here again we see that in most cases a combination of drivers is at work. Regional patterns vary: In Africa, agriculture-related drivers play a dominant role, whereas wood-related factors are more important in Southeast Asia. Poverty-related issues are more frequent drivers of deforestation in Africa, whereas large individual gains from land speculation are a significant factor in Latin America. Policy and institutional factors are the second most important underlying force of tropical deforestation and almost always act in concert with other drivers.

It is becoming clear that causes and drivers of deforestation cannot be reduced to a single or even a small number of parameters, but must be seen as a synergistic interplay of forces. Allocating blame to single causes or drivers (e.g., population "pressure" or shifting agriculture) is not scientifically valid and may lead to misguided policy. Moreover, the complex interplay of drivers and their regional variations makes it unlikely that universal policies to control deforestation can be found. Instead, region-specific policies based on analysis of specific driver interactions will be required. This is further complicated by the fact that some underlying socioeconomic causes may be far outside of the region being deforested. For example, demand for cheap beef in North America and Europe drives deforestation in Brazil and SE Asia, where soybeans and cassava are grown extensively as ingredients in cattle feed.

Extent of Deforestation

At the end of the twentieth century, ca. 3,500 million ha of forest remained on Earth, representing a decrease of about 40% from the time before the advent of sedentary agriculture about 8000 years ago (World Resources Institute 1994). As a fraction of land cover, forests have declined from ~45% to ~27%, and almost all of this change can be attributed to human activities. More than half of the remaining forest (~2,000 million ha) is located in developing countries, particularly in tropical regions, which are subject to intense deforestation pressure. Whereas there is a large uncertainty associated with the absolute magnitude of the rate of deforestation in developing countries, there can be no doubt that it has accelerated dramatically in the last four decades of the twentieth century,

---

[1] We define proximate causes as human activities that directly affect the environment and thereby bring about land-use and land-cover change. They reflect and implement underlying socioeconomic driving forces ("drivers": demographic, economic, technological, policy/institutional, and cultural or sociopolitical factors).

reaching about 16 million ha per year in the 1990s (Houghton 2003a). Lower deforestation rates for humid tropical forests have been proposed by Achard et al. (2002). Over the period 1980–1995, the world has lost some 200 million ha tropical forest, the equivalent of the total forest area of Mexico or Indonesia! Deforestation is particularly severe in the "top 10" countries of deforestation, which account for 50% of forests lost: Brazil, Indonesia, P.R. Congo, Bolivia, Mexico, Venezuela, Malaysia, Myanmar, Sudan, and Thailand.

There is no indication that this alarming rate of deforestation is declining significantly, if at all. Over the next 15 years, it has been estimated that the agriculture sector will require an additional 150–200 million ha of land to accommodate the expansion of commercial farming, subsistence cropping, pasture development, and range land (Roper and Roberts 1999). Given the lack of alternative sources of land, most of this will be at the expense of the remaining forest lands. Other developments are similarly discouraging: Brazil, the country that contains about 40% of the world's remaining rainforest and which already has the highest absolute rate of deforestation, has recently announced a new master plan for the region, "Avança Brazil." While not directly including any plans for deforestation, this plan is fast-tracking numerous infrastructure projects intended to accelerate the development of industrial agriculture, logging, and mining. This enhanced infrastructure is likely to lead to a sharp acceleration of deforestation over the next 20 years, to the point that few pristine areas will survive outside the western quarter of Amazonia (Laurance et al. 2001).

The deforestation rates cited above are based on assessments of areas fully converted from forest to other land uses, now usually determined using satellite surveys. However, as Nepstad et al. (1999) have suggested, this widely used measure of tropical land-use change may not tell the entire story. They show that logging severely damages 1.0–1.5 million ha of forest annually, and this is not included in the ~2 million ha deforestation rate reported by the deforestation mapping programs. Furthermore, surface fires during drought periods leave affected rainforest prone to ecological deterioration and loss of the rainforest vegetation. This process is substantially enhanced when the forest canopy is disturbed by selective logging. Human activity provides additional sources of ignition. Again, infrastructure development, such as promoted by "Avança Brazil," acts to aggravate this threat.

Consequences of Deforestation

Beyond the obvious ethical and aesthetic concerns about the destruction of one of the world's most majestic and species-rich ecosystems, why do we need to be concerned about the destruction of the tropical rainforests? The answer lies in the essential goods and services that the tropical region and its forests provide to humanity and the biosphere. At the most direct level, the forests produce a tremendous wealth of timber and non-timber products. Used sustainably, they can play an important role in the economies of developing nations. Unfortunately, it is currently more profitable to "cut-and-get-out" than to engage in sustainable practices (Myers 2002).

As important as the direct economic role of the tropical forests may be, more germane to this discussion are its functions in the Earth system: They are a key reservoir of biological species, play a crucial role in the carbon cycle, have essential functions in regulating climate and the water cycle, and have an important influence on the composition of the atmosphere.

Extinction of Species

The ultimately most disturbing threat to massive species extinction stems from tropical deforestation. Tropical forests are the homes of 50–90% of the world's plants and animals, including 70% of the world's vascular plants, 30% of bird species, and 90% of invertebrates (Roper and Roberts 1999). Given the high degree of endemism typical of tropical forests, the annual destruction of some 15 million ha means the extinction of thousands to tens of thousands of species, often before they have even been described scientifically. Estimates of the true rate of extinction are by necessity vague, given that we are not even sure about the number of existing species; however, values of the order of 50,000 species lost per year have been proposed (Roper and Roberts 1999).

The consequences of this loss are difficult to state in precise terms. There is a huge economic potential in medical drugs that could be derived from tropical forest species, as some classical examples like the rosy periwinkle have shown. The tropical forest may also hold large genetic resources for the development and improvement of crop species. Governments and industries have clearly become aware of this potential and compete with each other for its exploitation.

The argument often made — that diversity is important to stabilize ecosystems — is obviously of no great significance to those willing to destroy the rainforest ecosystem altogether. How does the loss of diversity affect the Earth system? At present, an answer to this question is not possible, as it requires a much deeper understanding of ecology and its role in the Earth system than we now possess. We are thus left to contemplate the ethical implications of eliminating a large fraction of our fellow passengers on Spaceship Earth, and the results of deleting large amounts of genetic information from the book of evolution. How many letters can we erase from a book or how many pages can we tear out before it loses its meaning?

Effects on the Water Cycle and Climate

The fundamental driving force of atmospheric and oceanic circulation stems from the contrast between a net excess energy input from solar radiation in the tropics and a net energy outflow of terrestrial longwave radiation at high latitudes. In the humid tropics, we see the massive conversion of solar radiative energy into sensible and then latent heat by evaporation of water from the sea and land surfaces. In a sense, the tropics play the role of the boiler in the huge heat engine that is the atmosphere. Intuition would suggest, therefore, that a change in the way the "boiler" operates, resulting from massive changes in land cover and land use as well as chemical composition of the atmosphere, would not only affect local climate but also the entire Earth climate system. The complexity and variability of climate, however, makes it difficult to provide scientific evidence for this conjecture.

Local changes in microclimate and regional climates are easy to prove. Deforested surfaces have different energy budgets, diel cycles of temperature and humidity, as well as different evaporation rates and runoff characteristics (e.g., Meher-Homji 1991; Silva Dias et al. 2002). At larger scales, observations suggest changes in rainfall patterns, rather than amounts, resulting in shifts in the intensity and timing of convective rainfall.

Deforestation may affect rainfall by a number of mechanisms, operating at scales ranging from the microscopic to feedbacks from large-scale circulation and climate dynamics. Conversion of rainforest to human-dominated forms of land use leads to a dramatic increase in the abundance of atmospheric aerosol particles, both from the smoke of deforestation and pasture maintenance fires, as well as from emissions associated with human activity in the settled regions after deforestation. These aerosol particles act as cloud condensation nuclei during cloud formation, and thereby have a strong influence on the optical properties of clouds and on the mechanisms of rain production (Rosenfeld 1999). Among the consequences are more intense updrafts and enhanced abundance of ice particles in convective clouds, increased lightning frequency, and less, but more intense, rainfall out of a given cloud. While this mechanism acts on a very small scale, it affects large regions, as the smoke from fires and the aerosols from human activities travel over distances of hundreds to thousands of kilometers.

Land-cover conversion — from forest to pasture, agricultural lands, degraded lands, human infrastructure (e.g., roads, buildings) — changes the radiative properties and evaporation behavior. Surface albedo is increased, and thereby the energy uptake by the Earth's surface is reduced. At the same time, the partitioning of the energy flux from the ground to the atmosphere shifts from latent to sensible heat, also reducing the moisture flux. Furthermore, increased runoff from deforested lands leaves less water for aquifer recharge and evaporation. These factors change the amount of water vapor available for rainfall formation, as well as the convective behavior in the atmosphere.

Large-scale effects from such changes of the functioning of the atmospheric heat engine are difficult to detect through observations because of the complexity of the climate system and the difficulty of attributing observed changes to specific causes. Therefore, climate models are normally used to explore the effects of such perturbations. For such simulations to be valid, however, convection and cloud processes must be realistically modeled. Unfortunately, because of the limitations of present-day computers and our incomplete understanding of some of the relevant processes, this is still beyond the state of the art in climate models. Despite these known shortcomings, some model experiments on the consequences of tropical deforestation have been conducted. These suggest significant effects on large-scale climate to result both from the land-cover change and from the injection of large amounts of smoke and other aerosols in the tropics (Nober et al. 2002; Werth and Avissar 2002). The strongest effects are predicted to occur in the tropics, but significant climate perturbations are also seen in some temperate regions. Both the Hadley and the Walker circulations are affected, and resonances with El Niño phenomena are possible. Since these, in turn, influence the flammability and stability of tropical forests, positive feedbacks leading to large-scale destabilization of tropical forests are conceivable. There is an urgent need to develop improved modeling tools for exploring these possibilities.

Effects on the Carbon Cycle

Tropical forests contain about 430 Pg (petagram = 10^{15}g) of carbon, with about equal amounts occurring in aboveground biomass and in the soil and root zone. This represents almost as much carbon as was present as CO_2 in the preindustrial atmosphere. The annual amount of carbon released to the atmosphere from tropical deforestation remains highly uncertain, despite considerable efforts to constrain this important component of the carbon cycle (Prentice et al. 2001). The most recent estimate of the *net* annual carbon flux from land-use change for the 1990s is 2.2 Pg yr^{-1}, dominated by tropical deforestation (1.6 Pg yr^{-1}) and timber harvesting in the tropics (Houghton 2003a). This estimate is based on a bottom-up approach using land-cover/use statistics, and the net flux is defined as the sum total of all the land-use changes that result in carbon losses (deforestation, etc.) minus those that result in carbon gains (afforestation, etc.) by the terrestrial biota. Top-down approaches, based on inversion of atmospheric CO_2 concentrations and isotope data, suggest that carbon sources and sinks in the tropics are roughly in balance, that is, that the land-use-related losses of CO_2 are approximately balanced by increases in the biomass of the intact, standing vegetation. Present data, however, do not allow an unequivocal choice between two alternatives: large carbon emissions from deforestation offset by large sinks in undisturbed forest, or moderate emissions from land-use change with only small changes in the carbon balance of the undisturbed forest (Houghton 2003b).

Wherever the truth may be between these two extremes, it is obvious that any increase in deforestation, or any decrease in the ability of the intact forest to take up carbon, will make the tropics into a significant net carbon source and add to the atmospheric CO_2 burden. This is particularly worrisome as we do not know what the reasons are for the net carbon uptake by the remaining tropical forest:

CO_2 fertilization and climate change are possible explanations, but scientific evidence is tenuous at best. As a consequence, we also do not know how reliable this carbon sink will be in the future. Is it already close to saturation, or will it grow with further increases of CO_2? Will it reverse, as temperature and rainfall are affected by climate change? Will changes in forest cover due to destabilization from drought cycles, accidental surface fires, or even large-scale instability of the Amazon rainforest ecosystem lead to significant carbon releases from the tropics, with potent feedbacks through climate mechanisms (Cox et al. 2000)?

Deforestation and Atmospheric Chemistry

Huge amounts of methane and other volatile organic compounds (VOC) are being released into the atmosphere from biogenic and anthropogenic sources, with a combined flux of over 1000 million tons per year (Prather et al. 2001). Tropical forests are the most prolific emitters of biogenic VOC, resulting in high VOC loadings in the boundary layer over the rainforest. The most important first reaction step for their removal and the self-cleansing of the atmosphere is the reaction with the hydroxyl radical, OH. This short-lived, very reactive molecule is formed from the photodissociation of ozone (O_3) to dioxygen (O_2) and an energetic oxygen atom (O), and the subsequent reaction of this atom with water vapor (H_2O). OH concentrations are highest in the tropics because of the high levels of ultraviolet radiation and water vapor, and most of the oxidation of CH_4, CO, and several other trace gases occurs in the "Great Tropical Reactor," the region of high OH concentrations in the tropical troposphere (Andreae and Crutzen 1997).

In addition to the presence of water and ultraviolet light, the abundance of O_3 and the relative amounts of hydrocarbons and nitrogen oxides ($NO_x = NO + NO_2$) also play crucial and interrelated roles. At very low levels of NO_x, hydrocarbon oxidation removes O_3 and consumes OH, while at higher NO_x levels, more O_3 and reactive radicals are produced. Under pristine conditions, the rainforest and soils are the dominant source of both hydrocarbons and NO_x in the tropics, and their relative amounts emitted are such that NO_x concentrations remain low and the troposphere is in a low ozone state. NO_x fluxes from the rainforest are kept low by a tight interaction of biological, chemical, and physical processes, which allow efficient internal turnover of fixed nitrogen, preventing it from escaping easily into the atmosphere. NO is produced during the breakdown of biological matter in soils, and part of this gas can escape into the air layers over the soil. Here, it can react with ozone to form NO_2, which is efficiently deposited on forest vegetation and made available for plant growth. Only a modest fraction of the NO emitted from the soil can therefore escape into the atmosphere and contribute to ozone formation.

The removal of the tree canopy as a result of deforestation breaks open this tight NO_x recycling system. In the shallow canopy of pastures and crops, there is much less chance for oxidation of NO to NO_2, and for the deposition of NO_2 to leaf surfaces. At the same time as more NO_x can escape, biogenic hydrocarbon emissions are reduced because of the change of vegetation from trees to grass or crops. Biomass burning for deforestation and land management supplies additional NO_x and hydrocarbons to the regional atmosphere. The consequence is a transition from a low-ozone state of the Great Tropical Reactor to a high-ozone photochemical smog situation, as is now observed in the deforested regions of the tropics (Andreae et al. 2002).

This change in gas phase chemistry also has consequences for aerosol production. Under natural conditions, the aerosol yield from the photooxidation of terpenes is quite low, because the prevailing reaction chains lead to relatively volatile compounds that do not readily condense into particles. However, at higher O_3 concentrations more low-volatility compounds are produced, which can then form aerosol particles. Tropical deforestation and land-use change thus enhance aerosol loading in three ways: through biomass burning, emissions from fossil-fuel combustion in vehicles, power plants, etc., and by increasing the aerosol yield from the oxidation of biogenic hydrocarbons. As we have discussed above, elevated aerosol levels change cloud dynamics and enhance lightning frequency. This in turn increases the NO_x formation by lightning, which further drives up ozone production.

Because of the highly nonlinear character of the chemical processes in the atmosphere, mixing and transport play a key role in regulating chemical transformations. Especially in the humid tropics, the water cycle strongly affects the dynamics of physical transport, in particular through the regulation of soil water content. Water availability also influences the emissions of NO_x and VOC from the soil–plant system. Consequently, there is a close linkage between hydrological, biological, physical, and chemical processes, which has the potential for complex feedbacks and interactions and is far from being adequately understood.

NUTRIENT INPUTS AND EUTROPHICATION IN COASTAL ECOSYSTEMS

Coasts and their associated ecosystems are among the most biogeochemically reactive, productive, and diverse interface domains in the Earth system. Stretching a total length of 2 million km surrounding continents and islands, they function as a complex biologically mediated filter to sort and process organic matter, inorganic nutrients, contaminants, and sediments delivered from land, sea, and air. N and P tightly cycle between the seafloor and shallow waters, fueling 20% of annual marine net primary production. This in turn supports 90% of global marine fisheries. The biomass produced in coastal systems is distributed over a vast array of diversity from microbial communities to coral reefs. Coastal areas are currently home to 2.4 of the 6.1 billion people worldwide.

Anthropogenic Drivers

Under pristine conditions, some coastal systems are naturally eutrophic, a condition where an increase in nutrient load causes elevated primary production that results in an accumulation of organic matter. Additions of decomposable organic matter, like sewage, further exacerbate this (Cloern 2001; Nixon 1995; Smith et al. 2003). For some sites in Europe, the onset of coastal eutrophication has been established from sediment cores to coincide with the flourishing of the Industrial Revolution in the mid-1800s. In the New World, similar events occurred about a century later.

Since then, the human imprint on coastal systems has grown. Population increase and consequent migration to coastal urban centers, growing affluence, and many human activities have profoundly altered these coastal systems (Curran et al. 2002). Eutrophication resulting from human-dominated waste loading has been acknowledged as among the most pressing concerns in the marine environment. To what extent has eutrophication altered the systemic integrity of the planet now that the human population has increased eightfold since the first technological upheaval 2.5 centuries ago?

Quantity and Composition of Human Waste Load

Under pristine conditions, nutrient load to recipient waters is driven by biogeochemical reactions from the landscape and by runoff (Smith et al. 2003). Since the Industrial Revolution, the human factor has become increasingly important, if not the dominant driver of total nutrient loading. The use of fertilizer and the disposal of organic waste from various production activities and sewage have considerably changed the composition of materials delivered to the coast.

Dissolved inorganic nutrients are most biologically reactive in that they are readily taken up by plants in photosynthesis. The most recent global estimates of inorganic nutrient loading to coastal waters by Smith et al. (2003) indicate that natural sources contribute annually about 5.6 Tg of dissolved inorganic nitrogen (1 Tg $= 10^{12}$ g) and 13.3 Tg of anthropogenic waste. The latter represents nearly a sixfold increase over a period of 20 years in comparison to Meybeck's (1982) 1970 values. The delivery of dissolved inorganic phosphorus increased from 0.8 Tg yr^{-1} in the 1970s to 2.2 Tg yr^{-1} in the 1990s. The anthropogenic contribution accounted for 50% in the 1970s but increased to 73% twenty years later.

Dissolved organic matter is remineralized into its inorganic and refractory components through the action of microbes and is mostly derived from organic waste generated by human activities. Meybeck (1982) provides estimates of annual riverine delivery of dissolved organic nitrogen (14.8 Tg) and dissolved organic phosphorus (1.2 Tg) for the 1970s. Both values are greater than their inorganic counterparts for the same time period. San Diego-McGlone et al. (2000) indicate that inorganic forms account for about 24–55% of total N and about 45–54% of total P in organic waste. Applying this to the 1970s scenario, fluxes of biologically reactive N and P easily become twice that of the loading rates for the inorganic forms. Thus, assessments need to account for both inorganic and labile fractions of the organic waste load to determine fully the fluxes of biologically reactive nutrients.

Ecosystem Responses to Nutrient Loading

Eutrophication is an ecological and biogeochemical response to heavy nutrient loading (Smith et al. 2003). It is indicated by an accumulation of organic matter that is synthesized by primary producers, and which is further increased by the high delivery of decomposable organic matter, primarily from human activity (Rabalais 2002). The high amount of particulate and dissolved organic matter far exceeds that which can be assimilated into growth, and excess organic matter undergoes oxygen-consuming microbial degradation.

There are secondary consequences of eutrophication that have profound impacts on the functioning of the Earth system. The release of greenhouse gases, which are naturally emitted from wetlands, is hastened by eutrophication and the formation of hypoxic zones. A cascade of changes results from shifts in diversity of primary producers, causing structural and functional changes in sediment biota and the upper trophic levels, including fish. Interactions between eutrophication and other coastal activities further reduce the integrity and limit the biogeochemical functioning of the coast as a filter. Among these are non-nutrient pollution, overfishing, aquaculture, and damming. The ultimate feedbacks of a eutrophied coastal system to human and planetary well-being then become profoundly sobering.

Eutrophication and Greenhouse Gases

Denitrification is one form of anaerobic respiration performed by benthic microbes when they reduce nitrate or nitrite, which have been formed in the sediments to produce N_2O or N_2. Evasion of these into the atmosphere provides a gaseous sink for biologically reactive nitrogen. In estuaries, denitrification rates increase linearly with increasing inorganic nitrogen loading, which results in the removal of 40–50% of inorganic inputs (Seitzinger 1988). N_2O is a greenhouse gas and its production in hypoxic coastal areas has been demonstrated by Naqvi et al. (2000) for the western shelf of India, where moderate upwelling occurs and is overlain by low salinity water from land runoff and local rain. Although high production of organic matter is supported both by horizontal delivery from land and through upwelled water, Naqvi et al. (2000) attributes the decline in oxygen levels over the Indian shelf to increased nutrient loads from human activity. This trend may get more pronounced with global warming, which may potentially be reinforced through positive feedback via increasing coastal N_2O efflux.

Hypoxia, however, does not necessarily result in N_2O release. Suppression of sediment denitrification, due to the process of dissimilatory nitrate reduction to ammonium, seems to be the case in the Gulf of Mexico. Thus, whether in bottom sediments as ammonium or denitrified as N_2O, the nitrogen transformation pathways triggered by eutrophication have adverse ecological impacts.

Eutrophication, Overfishing, and Ecosystem Change

Jackson et al. (2001) provide compelling evidence to suggest that overfishing predates current stressors of coastal ecosystems, including eutrophication. The overharvest of top predators, including suspension feeders in temperate estuaries, has rendered coastal ecosystems most vulnerable to the subsequent impacts of nutrient loading, among other anthropogenic disturbances. They note the "microbialization" of coastal systems with the diminished presence of macrobiota and the resulting simplification of trophic relationships because of the loss in redundant functional relationships given fewer species. Simplified trophic webs are less stable and more vulnerable to biotic invasion and population explosions.

Shifts in dominance from light-limited seagrasses and long-lived seaweeds in clean waters to nutrient-limited phytoplankton bloom species and epiphytic macroalgae at medium levels of eutrophication are well documented (Rabalais 2002). With increasing nutrient load, green algae and thick phytoplankton blooms and cyanobacteria proliferate, until benthic macrophytes disappear altogether under hypertrophic conditions. Experimental manipulation of fish–epiphyte–seagrass interactions to control shading by epiphytes on seagrasses indicated limited and occasional control by present herbivorous fish, because the dominant epiphytes were not the preferred food species of herbivores. This reinforces the point made above: reduction in the diversity of top-controlling predators imposes great limitation in the ability of ecosystems to cope with subsequent perturbations. More importantly, changes in plant composition alter the pathways of carbon storage and mineralization processes.

For sediment microbes, increasing eutrophication results in a shift toward heterotrophic microbial processes. One of the major consequences of net heterotrophy for coastal waters is oxygen depletion to hypoxic levels. Oxygen deficiency is suggested to be the critical factor affecting benthic and pelagic biota in eutrophied systems, because of its severely limiting effects on metabolism and physiology (Gray et al. 2002; Rabalais 2002). When chronically high levels of loading remain unabated, toxic hydrogen sulfide is generated so that even sediment infauna is impaired and unable to aerate and bind sediments, which become easier to suspend, further increasing turbidity. Among the higher trophic levels like fish, nutrient enrichment may be accompanied by an increase in biomass up to medium levels of eutrophication, beyond which the demise of demersal species and their replacement by small pelagics may occur because of hypoxia-induced mortalities.

It remains to be seen how altered biogeochemical cycles triggered by eutrophication and hypoxia will result in terms of global inventories of carbon, nitrogen, phosphorus, and silica. There is consensus that the nearshore systems to date are net heterotrophic and microbe dominated. It is less clear whether these same systems are net denitrifiers today, but expanding hypoxic zones worldwide increase this likelihood. More than elevated rates and increased quantities, eutrophication has changed the biotic pathways through which these elements cycle, greatly reducing natural capacities for fishery biomass production and waste assimilation. Coastal ecosystems are thin ribbons around landmasses that provide life-support systems for humans and highly diverse shallow marine life. With eutrophication and overfishing, a cascade of changes takes place and may irreversibly limit this support function.

SYNTHESIS AND OVERVIEW

We have illustrated how human activities in the terrestrial, atmospheric, and marine environments perturb the Earth system. There are some common threads that run through these examples, which we highlight in this final section.

The first is the phenomenon of "surprises." Humankind seems to be amazingly poor at predicting what the important issues are going to be even in the fairly near future (10–30 years). For example, the report of the 1972 Stockholm Conference on the Human Environment makes no mention of the phenomena that are considered as major threats to the environment today: mass extinction of species, tropical deforestation, desertification, ozone depletion, and climate change.

There are two classes of surprises that are relevant in the present context. The first we could call "Cassandra" surprises. Something is going on in plain view and yet we fail to see its relevance, even though one or a few "seers" may try to draw our attention to the threat they perceive. The classic scientific example is the case of rising atmospheric CO_2, where despite the clear documentation of increasing CO_2 levels and the well-known underlying physics of radiative transfer, it took decades to convince the scientific community of the importance of the looming climate change. What can we do to sharpen our perception for the next emerging issue?

The other class of surprises comes from "discontinuities," the jumps and phase changes that are so typical of our world and that we forever fail to anticipate. The standard example for such a discontinuity is the abrupt and dramatic changes in the properties of water as it cools through its freezing point.

Our daily experience and scientific knowledge are full of examples for abrupt change, and as scientists we know that in a nonlinear system with sharp discontinuities, such as the Earth system, it is scientifically unsound to extrapolate beyond the explored parameter space. At a moment in history when humans are moving out of the known parameter space in almost all respects (e.g., atmospheric composition, land cover, demography, to name but a few examples), we should be on the alert for sudden changes, yet we are progressing from one "surprise" to the next: coral bleaching and reef losses, the appearance of polar ozone holes, the emergence of AIDS and other diseases from tropical forests. Instead of suppressing data that point to sudden change, as was done with the satellite data that first indicated the Antarctic ozone hole, we need to develop skills to detect the warning signals of the next big surprise.

A potential source for unpleasant surprises consists of the very inadequately explored interactions of climate and environmental change, socioeconomic development, and human and animal health (Patz and Reisen 2001). Could, for example, AIDS, Marburg disease, or SARS be the first indication that the human epidemiological environment is near a break point? Humans are spreading into the last corners of the tropical forest, where they are encountering diseases without the necessary evolved resistance. In a contiguous worldwide human population with high connectivities and few barriers to transmission, infectious agents can move about quickly. High population densities in close contact with animal reservoirs of infectious disease make possible the rapid exchange of genetic material. Malnutrition is of epidemic proportions in many developing countries, providing large immune-compromised populations where diseases can spread into very rapidly. Warmer and wetter climate conditions may also facilitate the spread of diseases. The 1918 Spanish Flu pandemic is thought to have cost 20–40 million lives worldwide (CDC 2003). Medical science may be more advanced now than in 1918, but there is little chance to deliver its benefits to 6 billion people fast enough. Signs of an impending pandemic may lead to rapid economic collapse in a world economy dependent on rapid global exchange of goods and services (e.g., the current economic effects of a very few cases of SARS). It does not take much fantasy to develop a scenario much scarier and much more imminent than what comes out of the current stock of gradual climate change.

The obvious conclusion is that we have to intensify our search for discontinuities and surprises. By definition, we cannot predict surprises, but we can learn to be more aware of their importance and more sensitive to their early warning signs. Just as importantly, we need to reduce the chance of bad surprises. The prudent course would be to maintain the Earth system as much as possible within the known parameter space and to pursue a course of sustainability. In all three cases described above, the tremendous advances in our understanding of Earth system functioning over the past two to three decades (Steffen et al. 2001) has provided substantial progress in the knowledge base required for transitions to sustainability. We start to understand what drives change and its consequences, and we can begin to see alternatives that will reduce negative consequences. To solve the problems outlined above, however, requires a new level of integration in the human–environment system. The challenge for the coming decades is to integrate the biophysical knowledge base with fundamental advances in economic, political, and social science as well as engineering approaches, in order to find workable, economically viable alternatives to business-as-usual.

REFERENCES

Achard, F., H.D. Eva, H.J. Stibig et al. 2002. Determination of deforestation rates of the world's humid tropical forests. *Science* **297**:999–1002.

Andreae, M.O., P. Artaxo, C. Brandão et al. 2002. Biogeochemical cycling carbon, water, energy, trace gases, and aerosols in Amazonia: The LBA-EUSTACH experiments. *J. Geophys. Res.* **107**:8066.

Andreae, M.O., and P.J. Crutzen. 1997. Atmospheric aerosols: Biogeochemical sources and role in atmospheric chemistry. *Science* **276**:1052–1056.

Cassman, K.G., and P.L. Pingali. 1995 Intensification of irrigated rice systems: Learning from the past to meet future challenges. *GeoJournal* **35**:299–305.

CDC (Centers for Disease Control). 2003. Pandemic Influenza. http://www.cdc. gov/od/nvpo/pandemics

Chameides, W., P. Kasibhatla, J. Yienger, and H. Levy II. 1994. Growth of continental-scale metro-agro-plexes, regional ozone pollution, and world food production. *Science* **264**:74–77.

Cloern, J.E. 2001. Our evolving conceptual model of the coastal eutrophication problem. *Mar. Ecol. Prog. Ser.* **210**:223–253.

Cox, P.M., R.A. Betts, C.D. Jones, S.A. Spall, and I.J. Totterdell. 2000. Acceleration of global warming due to carbon-cycle feedbacks in a coupled climate model. *Nature* **408**:184–187.

Curran, S., A. Kumar, W. Lutz, and M. Williams. 2002. Interactions between coastal and marine ecosystems and human population systems: Perspectives on how consumption mediates this interaction. *Ambio* **31**:264–268.

Geist, H.J., and E.F. Lambin. 2001. What drives tropical deforestation? LUCC Rept. Series 4. Louvain-la-Neuve: LUCC International Project Office.

Gleick, P. 1993. Water and conflict: Fresh water resources and international security. *Intl. Security* **18**:79–112.

Gray, J.S., R.S. Wu, and Y.Y. Or. 2002. Effects of hypoxia and organic enrichment on the coastal marine environment. *Mar. Ecol. Prog. Ser.* **238**:249–279.

Houghton, R.A. 2003a. Revised estimates of the annual net flux of carbon to the atmosphere from changes in land use and land management 1850–2000. *Tellus* **55B**:378–390.

Houghton, R.A. 2003b. Why are estimates of the terrestrial carbon balance so different? *Global Change Biol.* **9**:500–509.

Howarth, R., G. Billen, D. Swaney et al. 1996. Regional nitrogen budgets and riverine N & P fluxes for the drainages to the North Atlantic Ocean: Natural and human influences. *Biogeochemistry* **35**:75–139.

Jackson, J.B.C., M.X. Kirby, W.H. Berger et al. 2001. Historical overfishing and the recent collapse of coastal ecosystems. *Science* **292**:629–636.

Laurance, W.F., M.A. Cochrane, S. Bergen et al. 2001. The future of the Brazilian Amazon. *Science* **291**:438–439.

Matson, P., K. Lohse, and S. Hall. 2002. The globalization of nitrogen deposition: Consequences for terrestrial ecosystems. *Ambio* **31**:113–119.

Matson, P., R. Naylor, and I. Ortiz-Monasterio. 1998. Integration of environmental, agronomic, and economic aspects of fertilizer management. *Science* **280**:112–114.

Matson, P., W. Parton, A. Power, and M. Swift. 1997. Agricultural intensification and ecosystem properties. *Science* **277**:504–508.

Meher-Homji, V.M. 1991. Probable impact of deforestation on hydrological processes. *Clim. Change* **19**:163–173.

Meybeck, M. 1982. Carbon, nitrogen, and phosphorus transport by world rivers. *Am. J. Sci.* **282**:401–450.

Myers, N. 2002. Exploring the frontiers of environmental science. In: A Better Future for the Planet Earth, vol. 2, pp. 267–278. Tokyo: Asahi Glass Foundation

Myers, N., and A. Knoll. 2001. The biotic crisis and the future of evolution. *Proc. Natl. Acad. Sci. USA* **98**:5389–5392.

Naqvi, S.W.A., D.A. Jayakumar, P.V. Narvekar et al. 2000. Increased marine production of N_2O due to intensifying anoxia on the Indian continental shelf. *Nature* **408**:346–349.

Nepstad, D.C., A. Verissimo, A. Alencar et al. 1999. Large-scale impoverishment of Amazonian forests by logging and fire. *Nature* **398**:505–508.

Nixon, S.W. 1995. Coastal marine eutrophication: A definition, social causes, and future concerns. *Ophelia* **41**:199–220.

Nober, F.J., H.-F. Graf, and D. Rosenfeld. 2002. Sensitivity of the global circulation to the suppression of precipitation by anthropogenic aerosols. *Global & Planet. Change* **37**:57–80.

Patz, J.A., and W.K. Reisen. 2001. Immunology, climate change, and vector-borne diseases. *Trends Immunol.* **22**:171–172.

Postel, S., G. Daily, and P. Ehrlich. 1996. Human appropriation of renewable fresh water. *Science* **271**:785–787.

Prather, M., D. Ehhalt, F. Dentener et al. 2001. Atmospheric chemistry and greenhouse gases. In: Climate Change 2001: The Scientific Basis. Working Group I Contribution, Third Assessment Report of the IPCC, ed. J.T. Houghton, Y. Ding, D.J. Griggs et al., pp. 239–287. Cambridge: Cambridge Univ. Press.

Prentice, I.C., G.D. Farquhar, M.J.R. Fasham et al. 2001. The carbon cycle and atmospheric carbon dioxide. In: Climate Change 2001: The Scientific Basis. Working Group I Contribution, Third Assessment Report of the IPCC, ed. J.T. Houghton, Y. Ding, D.J. Griggs et al., pp. 183–237. Cambridge: Cambridge Univ. Press.

Prueger, J., L. Hipps, and D. Cooper. 1996. Evaporation and the development of the local boundary layer over an irrigated surface in an arid region. *Agric. & Forest Meteorol.* **78**:223–237.

Rabalais, N. 2002. Nitrogen in aquatic ecosystems. *Ambio* **31**:102–112.

Roper, J., and R.W. Roberts. 1999. Deforestation: Tropical Forests in Decline. CIDA Forestry Advisers Network, Canadian International Development Agency (CFAN).

Rosenfeld, D. 1999. TRMM observed first direct evidence of smoke from forest fires inhibiting rainfall. *Geophys. Res. Lett.* **26**:3105–3108.

Ruttan, V. 1999. The transition to agricultural sustainability. *Proc. Natl. Acad. Sci. USA* **96**:5960–5967.

San Diego-McGlone, M.L., S.V. Smith, and V.F. Nicolas. 2000. Stoichiometry interpretations of C:N:P ratios in organic waste materials. *Mar. Poll. Bull.* **40**:325–330.

Seitzinger, S.P. 1988. Denitrification in freshwater and coastal marine ecosystems: Ecological and geochemical significance. *Limnol. & Oceanog.* **33**:702–724.

Silva Dias, M.A., S. Rutledge, P. Kabat et al. 2002. Clouds and rain processes in a biosphere–atmosphere interaction context in the Amazon Region. *J. Geophys. Res.* **107**: 8072.

Smil, V. 1999. Nitrogen in crop production: An account of global flows. *Global Biogeochem. Cycles* **13**:647–662l.

Smil, V. 2000. Phosphorus in the environment: Natural flows and human interferences. *Ann. Rev. Energy Env.* **25**:53–88.

Smith, S.V., D.P. Swaney, L. Talaue-McManus et al. 2003. Humans, hydrology, and the distribution of inorganic nutrient loading to the ocean. *BioScience* **53**:235–245.

Steffen, W., P. Tyson, J. Jager et al. 2001. Earth system science: An integrated approach. *Environment* **43**:21–27.

Tilman, D., K.G. Cassman, P.A. Matson, R. Naylor, and S. Polasky. 2002. Agricultural sustainability and intensive production practices. *Nature* **418**:671–677.

Werth, D., and R. Avissar. 2002. The local and global effects of Amazon deforestation. *J. Geophys. Res.* **107**:8087.

World Resources Institute. 1994. World Resources 1994–1995: A Guide to the Global Environment. Washington, D.C.: World Resources Inst.

14

Atmospheric Chemistry and Climate in the Anthropocene

Where Are We Heading?

P. J. CRUTZEN[1] and V. RAMANATHAN[2]

[1]Max Planck Institute for Chemistry, 55020 Mainz, Germany
[2]Scripps Institution of Oceanography, La Jolla, CA 92037, U.S.A.

ABSTRACT

Humans are changing critical environmental conditions in many ways. Here, the important changes in atmospheric chemistry and climate are discussed. The most dramatic examples of major human impacts are the increase of the "greenhouse" gases, especially carbon dioxide (CO_2), in the atmosphere and the unpredicted breakdown of much of the ozone in the lower stratosphere over Antarctica during the months of September to November, caused by the emissions of chlorofluorocarbons (CFCs). Other, more regional but ubiquitous examples include photochemical smog and acid rain.

Industrial activities are not alone in causing air pollution and in changing the chemical composition of the atmosphere. Biomass burning, which takes place largely in the developing world, also contributes in major ways.

In the future, climate warming due to CO_2 emissions will continue to increase over present levels and pose a major problem for humankind. Current radiative forcing by "greenhouse gases" can, to a substantial degree (up to half), be dampened by increased backscattering of solar radiation, either directly by aerosol particles or indirectly through their influence on cloud albedo, or also by cloud feedbacks independent of anthropogenic aerosols. Cloud and hydrological cycle feedbacks provide major challenges. It is unlikely and undesirable that aerosol emissions will continue to increase, as greater emphasis will be placed on air quality, also in the developing world. However, due to its long atmospheric lifetime and expected growth in global emissions, CO_2 will continue to accumulate, exacerbating climate warming and related problems in the future. Drastic measures are thus needed at the international level to reduce the emissions, in particular, of CO_2 through energy savings, alternative energy sources, and sequestration.

INTRODUCTION

The bulk of the Earth's mass, about 6×10^{27} g, is largely concentrated in the core ($\approx 30\%$) and mantle ($\approx 70\%$). The crust only contributes 0.4% to the total. The mass of the biosphere is less than a millionth of that of the crust. Still, its influence on shaping the surface of the Earth and biogeochemical cycles is profound. With about 10^{14} g of "dry matter," humans constitute only 10^{-5} of the mass of the biosphere. Only a minute fraction of this is human brains, the sites of the enormous collective thinking power of the human race, unfortunately often badly used. Despite mass starvation, epidemics, and wars, the human population has grown by a staggering factor of ten over the past three centuries. Humankind's unique capacity to produce knowledge and technology and transfer it to subsequent generations, its mastery of fire, as well as the invention of agriculture through domestication of plants and animals has increasingly impacted the environment over time, sharply out of proportion to the relatively small brain mass of the human species. Generations of ambitious *Homo sapiens* have played, and will continue to play, a major catalytic role in affecting the basic properties of the atmosphere with impacts on climate, ecosystems, biological diversity, and human health. Humankind's activities accelerated particularly during the past few hundred years and precipitated the entry of a new geological era known as the "Anthropocene" (Crutzen 2002; Crutzen and Steffen 2003; see also Appendix 14.1).

This was already foreseen eighty years ago by the Russian biologist/geologist Vernadsky (1998/1926) who wrote:

> Without life, the face of the Earth would become as motionless as the face of the moon ... the evolution of different forms of life throughout geological time increases the biogenic migration of elements in the biosphere .... In an insignificant time the biogenic migration has been increased by the use of man's skill to a degree far greater than that to be expected from the whole mass of living matter .... The surface of the Earth has been transformed unrecognizably, and no doubt far greater changes will yet come. We are confronted with a new form of biogenic migration resulting from the activity of the human reason.

Vernadsky could only see the beginning of the major changes in land use that followed, but he could not foresee the changes in the chemical composition of the atmosphere that were ahead. He recognized, however, the importance of ozone as a cover against harmful solar radiation. The importance of carbon dioxide as a shield against heat loss to space was discussed near the end of the nineteenth century especially by Svante Arrhenius, among others; however, he too could not possibly imagine that less than a century later, climate warming together with ozone depletion would become issues of great environmental concern requiring an international response at the political level. In Figure 14.1, we depict probably the most cited (here updated) graphs in the history of the effects of human activities on the atmosphere: (a) Keeling's Mauna Loa CO_2 growth curve (Keeling and Whorf 2000) and (b) the "Ozone Hole" graph by Farman et al. (1985). Also shown is a vertical ozone profile in the "ozone hole" by Hofmann et al. (1989).

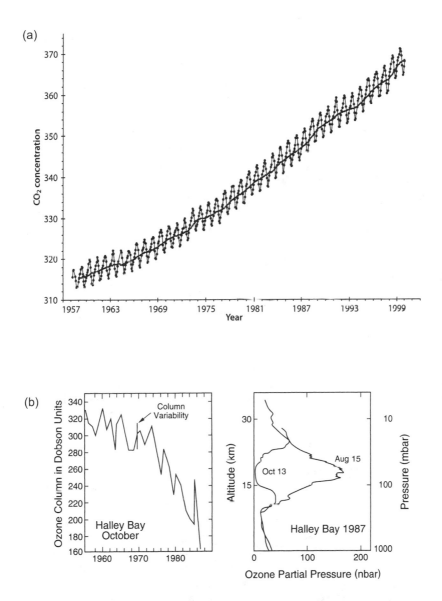

Figure 14.1   (a) The "Keeling curve," which shows the steady increase in atmospheric CO_2 concentration recorded at Mauna Loa in Hawaii, 1958–1999 (adapted from Keeling and Whorf 2000). (b) The Antarctic total ozone as reported by Joe Farman and colleagues in 1985 (lefthand panel) and currently typical vertical ozone profiles (righthand panel) measured by D. Hofmann et al. (1989). Since then, ozone loss has continued (with an unexpected exception during the spring of 2002), reaching levels as low as about 100 DU. Lowest concentrations of ozone occur in the same height range, 14–22 km, in which naturally occurring maximum ozone concentrations are found.

Fossil-fuel and biomass burning, land-use changes, as well as agriculture have caused increases in the atmospheric concentrations of the greenhouse gases: CO_2, CH_4, N_2O, and tropospheric ozone. The same activities have also led to major regional increases in aerosol loadings, in particular in the optically most active submicrometer size range, which also have the longest residence time in the atmosphere. If these particles would only scatter solar radiation, such as is the case for sulfate, they would cause a cooling, directly because of reflection of solar radiation to space and indirectly via cloud brightening, the so-called "indirect aerosol effect" (discussed later in detail). However, if the aerosol contains an absorbing material, in particular black carbon, then the aerosol adds a warming component to the atmosphere as well as a net cooling effect at the Earth's surface. The combination of these factors, which are important parts of the atmospheric chemistry system, has likely been the main driving force behind the observed global average surface warming of 0.6 ± 0.2 K during the twentieth century (IPCC 2001), and is expected to increase during the current century and beyond.

Although important progress has been made in our understanding of atmospheric chemistry and its relation to climate and processes at the land and ocean surface, much remains to be done. In this brief overview, we present some of the main factors that determine the chemical composition of the atmosphere and its impact on climate and biosphere.

PRESSING PROBLEMS RELATED TO THE CHEMICAL COMPOSITION OF THE ATMOSPHERE

The "Greenhouse" Gases CO_2, CH_4, and N_2O

Except for variable amounts of water vapor, which can increase to a few percent in the tropics, more than 99.9% of the atmosphere consists of N_2, O_2, and Ar. The abundance of these gases cannot be affected significantly by human activities. For instance, the current practice of burning, which releases 6 Pg (Pg = Petagram = 10^{15} g) of fossil-fuel carbon per year (i.e., on average about one ton per person, albeit very unevenly distributed around the world), consumes only about a 30 millionth of the atmospheric oxygen reservoir. Nevertheless, these changes can be measured, and because O_2 is much less soluble in seawater than CO_2, changes in atmospheric oxygen have been used to derive important terms in the global carbon budget, the uptake of CO_2 in the oceans and in terrestrial ecosystems (Keeling et al. 1996). A very important question to ask, however, is where the sinks are and to find out whether they are only temporarily active.

The most noticeable influence of humans on the global composition of the atmosphere and on Earth's climate is CO_2, the carbon source for photosynthesis. The volume mixing ratio of CO_2 is currently 370 ppmv (parts per million by volume), and it is growing at an annual rate of about 0.4%. This contrasts values of

280 ppmv during the interglacials and 180 ppmv during glacial periods (IPCC 2001). For the chemistry within the atmosphere, CO_2 does not play a primary role, except for a minor influence on cloud acidity: CO_2 alone would yield a pH of 5.6, but dissolution of other chemicals is normally more important.

In the 1990s, the estimated budget terms for CO_2, in units of Pg C per year, were as follows: emissions from fossil-fuel combustion = 6.3 ± 0.4; growth rate in the atmosphere = $+3.2 \pm 0.1$; uptake by the oceans = 1.7 ± 0.5; uptake by land = 1.4 ± 0.7 (IPCC 2001). The latter favorable condition may, however, not last. By the end of this century, the current terrestrial net carbon sink is expected to turn into a net source of 7 Pg C yr^{-1} — larger than the oceanic net carbon sink which saturates at 5 Pg C yr^{-1} (Jones et al. 2003) — and lead to a further increase in the atmospheric source of CO_2 in addition to the input of CO_2 from fossil-fuel burning. These results do not, however, agree with the findings of Melillo et al. (2002); their soil warming experiments in mid-latitude hardwood forests show only a small, short-lived release of CO_2 to the atmosphere.

Methane is globally the next most abundant greenhouse gas, with a direct "global warming potential" (GWP) of 23 over that of CO_2, integrated over a 100-year horizon (Ramaswamy et al. 2001). It is also chemically active on the global scale. In the troposphere, CH_4 is involved in reactions that determine the concentrations of ozone and hydroxyl radicals, and thus the oxidation power of the atmosphere. In the stratosphere, its oxidation is a significant source of water vapor. There it also serves partially as a sink for highly reactive ozone-destroying Cl and ClO radicals, by converting them to HCl, which does not react with ozone. Produced through the anaerobic decay of organic matter in wetlands, rice fields, in the rumen of cattle and in landfills, with further emissions coming from coal mines and natural gas leaks, the atmospheric methane content is strongly influenced by human activities. Its average atmospheric abundance has more than doubled since preindustrial times, starting from values around 0.7 ppmv (obtained from analysis of air trapped in ice cores) to a current level of about 1.75 ppmv. During glacial periods, its concentration was even lower, 0.4–0.5 ppmv. Recent observations have shown a slowing-down in the growth of atmospheric methane, which may indicate an approaching balance between sources and sinks, the latter largely due to reaction with hydroxyl radicals in the atmosphere (see below). The total source strength of CH_4, which can be derived from its model calculated loss by reaction with OH in troposphere and stratosphere, uptake at the surface, and present atmospheric growth rate, is about 600 million tons per year. If average OH concentrations have not changed much, as most model calculations suggest, then the preindustrial, natural methane source and sink was about 40% of its present value and stemmed primarily from wetlands. Although the total source and sink terms in the methane budget are rather well known, the individual, global emissions from wetlands and human activities, in particular from rice production, biomass burning, and fossil-fuel production, are much less certain. From the decline in the growth rate of CH_4 (Dlugokencky et al. 1998), an equilibrium between sources and sinks of CH_4 may soon become established. In fact, in the future it might even be possible to lower methane concentrations. To reach such a goal, it is important to have better quantitative information about the individual anthropogenic CH_4 sources. It is also imperative to monitor the releases of methane from natural tropical wetlands and high-latitude northern peat- and wetlands, which may occur under the climate warming regime of the Anthropocene, and which causes higher biological activity, a longer growing season, melting of permafrost, increased precipitation, and thus a possible expansion of the anaerobic zone. Similarly, although its current contribution is currently rather small, release of CH_4 from destabilizing methane hydrates may, in the distant future, become another significant source of CH_4.

With a mean residence time of about a century, N_2O is another significant greenhouse gas. It is present in the atmosphere at a mixing ratio of about 315 nmol/mol, increasing annually by 0.2–0.3%. N_2O is an intermediate product in the nitrogen cycle and is, therefore, among other factors, influenced by the application of nitrogen fertilizer in agriculture. It also plays a major role in stratospheric chemistry. Its reaction with $O(^1D)$ atoms produces NO, which, together with NO_2, catalytically destroys ozone. Thereby it controls the natural level of ozone in the stratosphere. Paradoxically, chemical interactions between ClO_x (Cl and ClO) and NO_x (NO and NO_2) radicals substantially reduce otherwise much larger ozone destruction.

As shown in the IPCC 2001 report, and as was also the case for CO_2, the total sinks of CH_4 and N_2O are now rather well known from atmospheric observations and knowledge of their chemistry. However, the contributions by individual sources are not.

A major question is whether the natural sources of CO_2, CH_4, and N_2O will increase as a result of climate change. The biogeochemical cycles of carbon and nitrogen, which are interconnected, have great uncertainties and require continued research.

Ozone, the Cleansing Effect of Hydroxyl in the Troposphere, and Acid Rain

In and downwind of those regions in which fossil-fuel and biomass burning take place, high ozone concentrations are generally observed in the lower troposphere, and this impacts human health and agricultural productivity. Ozone also acts as a greenhouse gas, whose anthropogenic increase is estimated to have contributed a positive radiative forcing of up to 0.5 Wm^{-2} since preindustrial times (IPCC 2001). Ozone is involved in the production of hydroxyl (OH) radicals, the atmospheric oxidizer, also dubbed the "detergent of the atmosphere," which is present in the troposphere at a global average volume mixing ratio of only about 4×10^{-14}. Nevertheless, this ultra-minor constituent is responsible for the removal of almost all gases that are emitted to the atmosphere by natural processes and anthropogenic activity. OH is formed through photolysis of ozone by solar ultraviolet radiation yielding electronically excited $O(^1D)$ atoms, a fraction of which reacts with water vapor to produce OH radicals. Thus, while too much ultraviolet radiation and ozone can harm humans and the biosphere, OH radicals are also indispensable for cleansing the atmosphere. Whether the growth in ozone has led to an overall global increase in hydroxyl, and thus in the oxidizing power of the atmosphere, is not clear, since enhanced OH production has been countered by destruction due to growing concentrations of carbon monoxide and methane, which both react with OH. Although major long-term changes in OH do not appear to have occurred, using methyl chloroform (CH_3CCl_3) as a chemical tracer to derive global average OH concentrations, the two most active research groups in this area (Prinn et al. 2001 and Krol et al. 2003) have reached quite different conclusions about OH trends, especially for the early 1990s. During this period, CO and CH_4, the main gases with which OH reacts in the global troposphere, showed quite anomalous, unexplained behavior. Since 1996, methyl chloroform has no longer been produced under the provisions of the Montreal Protocol and its amendments. Thus, CH_3CCl_3 concentrations are decreasing, and other chemical tracers, such as HFCs or HCFCs (the CFC replacement products), have come into use as markers of change in OH. Accurate data on their release to the atmosphere are critical. Uncertainties in these have caused much of the disagreements between the two research groups, as mentioned above. Thus, Prinn et al. (2001) estimate a global, CH_3CCl_3 weighted increase by $15 \pm 20\%$ between 1979 and 1989, followed by a sharp decline to reach values in the year 2000 of $10 \pm 24\%$ below those in 1979. By contrast, Krol et al. (2003) derive an upward trend of $6.9 \pm 9\%$ from 1978 to 1993. One way to overcome this discrepancy may be to use dedicated tracers, solely fabricated for the purpose of deriving "global average OH" and its trends.

Ozone is a driving force in atmospheric chemistry. It can be produced or destroyed in the troposphere, largely depending on the concentrations of NO. Its concentration distribution is highly variable in time and space. Unfortunately, measurements of ozone are still much too sparse to provide a satisfactory test for photochemical models and to derive the important terms in its budget and those of its precursors, such as hydrocarbons released from forests and from fossil-fuel and biomass burning, as well as NO emitted by soils and lightning discharges. Whereas the level of emissions of NO are quite well known (Galloway et al. 2002) — 25 Tg N yr^{-1} from fossil-fuel burning, including aviation — the emissions from soils, lightning, and biomass burning are very uncertain. Although downward transport from the stratosphere is no longer considered to be the main source of tropospheric ozone, it remains a major contribution to ozone in the upper troposphere, where its effect as a greenhouse gas is maximized.

Industrial emissions of SO_2 and NO, which are oxidized to sulfuric and nitric acid, have led to acidification of precipitation and caused the acidification of lakes and death of fish. This phenomenon was first reported in the Scandinavian countries and in the northeast section of the United States. Regulatory measures, especially against SO_2 release, have somewhat relaxed the situation, but acidity of the rain did not decrease as much as was hoped for, since the emissions of the neutralizing cations Ca, Na, Mg, and K from regional point-sources simultaneously decreased (Hedin et al. 1994). A special issue of the journal *Ambio* (Gunn et al. 2003) was devoted to the biological recovery from lake acidification. "Acid rain" has grown into an environmental problem in several coal-burning regions in Asia.

Biomass Burning and the Consequences of Land-use Change in the Developing World

Air pollution has traditionally been associated with fossil-fuel burning. Over the past 2–3 decades, however, biomass burning, which occurs mainly during the dry season in the poorer nations of the tropics and subtropics, has also been recognized as a major source of air pollution. It is estimated that annually 2–5 Pg of biomass carbon are burned in shifting cultivation, permanent deforestation (a net source for atmospheric CO_2), and savannas, as well as through the combustion of domestic and agricultural wastes and firewood. In the process, mostly recycled CO_2 is produced but so are also many chemically active gaseous air pollutants (e.g., CO, CH_4 and many other pure or partially oxidized hydrocarbons, as well as NO_x), thereby delivering both the "fuel" and the catalysts for photochemical ozone formation. During the dry season, high ozone concentrations are indeed widely observed in rural areas of the tropics and subtropics. Meteorological conditions are important. As shown in Figure 14.2a and 14.2b (Richter and Burrows 2002), for the tropical Southeast Asia/Australia region, interannual variability in atmospheric circulation strongly affects biomass burning and ozone chemistry, and caused concentrations of ozone and its precursors NO_2 and CH_2O that were higher during El Niño/1997 than during La Niña/1996 in September.

Land-use change, especially the removal of forest cover, can have a substantial influence on ozone and particulate matter. In an intact forest, NO_x, which is released from the soil, is largely recycled by uptake within the forest canopy, and ozone production is small. In fact, it is quite possible that reactions with isoprene and other reactive hydrocarbons, emitted by the vegetation, are a sink for OH, thereby suppressing regional photochemical activity. In a disturbed forest or savannas, however, NO_x can more easily escape into the troposphere. There, in the presence of natural hydrocarbons and NO_x as catalysts, ozone can be produced, which in turn can harm the remaining vegetation. Note that the emissions of natural hydrocarbons (e.g., isoprene, C_5H_8) are very large (10^{15} g yr^{-1} or ten times larger than their anthropogenic emissions; Guenther et al. 1995), so that availability of NO_x is the limiting factor for ozone production. This explains why high ozone concentrations are generally not found in the vicinity of tropical

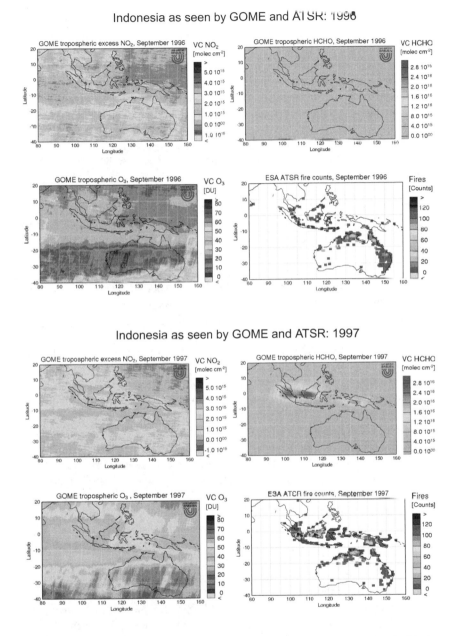

Figure 14.2   The distributions of tropospheric NO_2, O_3, and CH_2O measured with the DOAS (different optical absorption spectrometer) instrument on board of the ENVISAT satellite during two consecutive September months. Also shown are the fire counts (Burrows et al., pers. comm.; figure used with kind permission of J. P. Burrows and his team from the University of Bremen).

forests. In disturbed environments, especially where fires are a source of NO_x during the dry season, high ozone concentrations, approaching 100 nmol/mol, can be reached. This can also have consequences for aerosol production. Enhanced ozonolysis of mono- and sesquiterpenes and other higher terpenoid hydrocarbons can in supra-linear fashion lead to drastically enhanced production of secondary organic aerosols, which affect solar radiation scattering and which can also serve as cloud condensation nuclei (Kanakidou et al. 2000).

Light-absorbing and Light-scattering Particles and Regional Surface Climate Forcing

During the last decade, the role of aerosol particles in the radiation budget of the atmosphere and the hydrological cycle became a topic of great significance. Initially, interest was centered on the radiative cooling properties of nonabsorbing sulfate aerosol (Charlson et al. 1991). Recent international field research programs (BIBLE, TARFOX, ACE-2, INDOEX, ACE-ASIA) and space observations have impressively documented the widespread occurrence of light-scattering and light-absorbing smoke particles over many regions of the globe and their effect on the Earth's radiation budget. Furthermore, in particular, results from the INDOEX measurement campaign of 1999 (Ramanathan et al. 2001a) strongly suggest that much more emphasis than before must be given to the order of magnitude larger radiation energy disturbances at the Earth's surface and in the lower troposphere than at the top of the atmosphere (TOA) in- and downwind of heavily polluted regions (see Figure 14.3). Of particular importance are the atmospheric heating and surface cooling effects that are caused by the absorption of solar radiation by the black carbon in the smoke, which, through thermal stabilization of the boundary layer, reduces the strength of hydrological cycle. To date, model studies suggest that aerosols with single scattering albedo (SSA) greater than 0.95 will lead to surface cooling, whereas SSA values less than 0.85 will lead to surface warming. For SSA values in between, the sign of the net forcing will depend critically on cloud fraction, cloud type, and the reflectivity of the surface. High precision radiation measurements taken from space and from the surface during INDOEX have been used to demonstrate that over the polluted Arabian Sea, radiation forcing at the surface was three to four times greater than at the TOA (Satheesh and Ramanathan 2000). Direct aerosol radiative forcing measurements similar to those obtained from INDOEX have been performed in other parts of the world (e.g., for the Mediterranean, see Markowicz et al. 2002; for other regions of the world, see Kaufman et al. 2002), and these data should significantly help improve the accuracy of purely model-derived forcing estimates used in IPCC-type assessment studies. Worldwide SSA data from the AERONET network were published by Dubovik et al. (2002) and are reproduced in Figure 14.4. A rough compilation of the sources,

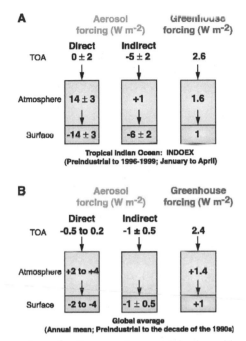

Figure 14.3 Comparison of anthropogenic aerosol forcing with greenhouse forcing. (a) The Indo-Asian region. The greenhouse forcing was estimated from the NCAR community climate model with an uncertainty of ± 20%. (b) Same as above, but for global and annual average conditions. The global average values are a summary of published estimates from Ramanathan et al. (2001b).

lifetimes, distributions, and optical properties of various kinds of aerosol in the atmosphere is given in Table 14.1.

Especially significant may be the possibilities for positive feedbacks. During the dry season, the residence time of aerosols is strongly enhanced by the lack of rain. With the addition of smoke from biomass burning, causing greater dynamic stability of the boundary layer, precipitation is suppressed, thereby enhancing the residence time of the particles, etc. In fact, even when clouds are formed, precipitation efficiency will decrease as a result of overseeding; this will bring about an increase in longevity, global cover, and reflectivity, and lead to a further cooling of the Earth's surface, less rainfall near populated regions, and more in remote areas.

In Asia, air pollution is attributed to a mixture of sources: biomass burning and fossil-fuel burning without, or with inadequate, emission controls. With the expected growth in population and industry, emissions from the Asian continent will become a major factor affecting not only the regional but most likely also global climates. Asia, however, is not the only part of the world with this kind of problem. Biomass burning is also heavily practiced in South America and Africa, causing high aerosol loadings, in particular over the savannas, as shown in

Table 14.1 Global source strength, atmospheric burden, and optical extinction due to the various types of aerosols (for the 1990s).

Source	Flux ($Tg\ yr^{-1}$)	Lifetime (days)	Column Burden ($mg\ m^{-2}$)	Specific Scattering/ Absorption (m^2/g)	Optical Depth ($\times 100$) Scattering/Absorption
Natural					
Primary					
Dust (desert)	900–1500	4	19–33	0.6	1–2
Sea salt	2300	1	3	1.5	2
Biological debris	50	4	1	2	0.2
Secondary					
Sulfates from biogenic gases	70	5	2	8	1.6
Sulfates from volcanic SO_2 (troposphere)	20	10	1	8	0.8
Sulfates from Pinatubo (1991) (stratosphere)	(40)	(400)	(80)	(2)	(16)
Organic matter from biogenic hydrocarbons	20	5	0.6	8	0.5
Total Natural	2400–3000		32–45		6–7 (± 3)
Anthropogenic					
Primary					
Dust (soil + desert)	0–600	4	0–13	0.7	0–0.9
Industrial dust	40	4	0.9	4	0.4
Black carbon (BC)	14	7	0.6	4/10	0.2/0.6
Organic carbon in smoke	54	6	1.8	6	1
Secondary					
Sulfates from SO_2	140	5	3.8	8	3
Organic hydrocarbons	20	7	0.8	8	0.6
Total Anthropogenic	270–870		8–21		5–6 (± 3)/0.6 (± 3)
TOTAL (Natural and Anthropogenic)					12 (± 4)/0.6 (± 3)

Explanation to Table 14.1   This is an update of the compilation by Andreae (1995) with the following changes:

Desert and soil dust. From a range of 1000–3000 Tg yr^{-1}, Andreae chose 1500 Tg and included this under natural sources. However, Tegen and Fung (1995) propose that 30–50% of the source might be derived from lands disturbed by human action. As this view is disputed, we adopt a range 0–40% as anthropogenic. The two values given for dust should not be interpreted as the range due to uncertainty; rather, the lower value (i.e., the 900 Tg yr^{-1} estimate for primary dust) for the natural aerosol is based on the assumption that some of the dust (i.e., the 600 Tg yr^{-1} value) may be anthropogenic.

Nitrate aerosol was not considered explicitly as much HNO$_3$ formed from NO$_2$ oxidation will be deposited on already existing particles. The fraction that forms new particles is highly uncertain. Published estimates of nitrate direct radiative forcing range from near zero to values similar to those for sulfates.

Sea salt. The given source strength is the average of Andreae (1995) and Penner et al. (2001). The scattering coefficient recommended by Andreae (1995) was about 0.4 m^2/g; but the inclusion of fine sea salt particles (radius smaller than 1 μm) has led to a substantial upward revision to a value of about 2.5 m^2/g at a relative humidity of about 80% (Haywood et al. 1999). We have adopted the average of these values.

Volcanic emissions. Ash is not considered because of the very short residence time of these larger particles. The sulfate flux of 20 Tg yr^{-1} (Andreae 1995) to the troposphere derived from SO$_2$ oxidation is included. Also shown, but not included in the sum, is the contribution by the rare event of the Pinatubo eruption of 1991.

Black and organic carbon. We adopt 7 Tg for fossil-fuel burning (Penner et al. 2001) and 7 Tg for biomass burning (Penner et al. 2001; Haywood et al 1999), with a total emission of 14 Tg.

Most aerosol source estimates, their lifetimes, and their optical effects are uncertain by at least a factor of 2, in particular sea salt and soil dust (see also Tegen et al. 2000). Despite the large uncertainties, this table indicates that the global average optical depths from natural and anthropogenic sources may be of similar magnitude with about 10% of the anthropogenic part due to black carbon. Note that, although the natural sources for aerosol are much larger than the anthropogenic inputs, the optical depths are very similar because the former produces fewer, but larger size, particles.

We adopted the following guidelines in arriving at the radiation parameters: The scattering coefficients account for the increase in scattering due to the hydration of the aerosols with a relative humidity of 80% in the first 1 km and 60% above 1 km. For sulfates, organics, and sea salt, we assume a scale height of 1 km for the vertical variation. Because of the relative humidity dependence, the sulfate scattering coefficient decreases from 8.5 m^2/g near the surface to about 7 m^2/g above 1 km. We assume that the uncertainties in aerosol optical depth (AOD) for natural and anthropogenic quantities are uncorrelated. Also note that we do not cite the uncertainties in individual terms but the overall uncertainty in AOD. Although we separate the scattering and absorbing optical depth, the sum of the two yields the so-called extinction optical depth. When we sum the individual optical depths to estimate the total optical depth, we assume that the aerosols are externally mixed, i.e., the particles exist as chemically distinct species. If we make the other extreme assumption that all of the species are internally mixed, the scattering coefficient will be reduced and the absorption coefficient will increase (Jacobson 2001).

278

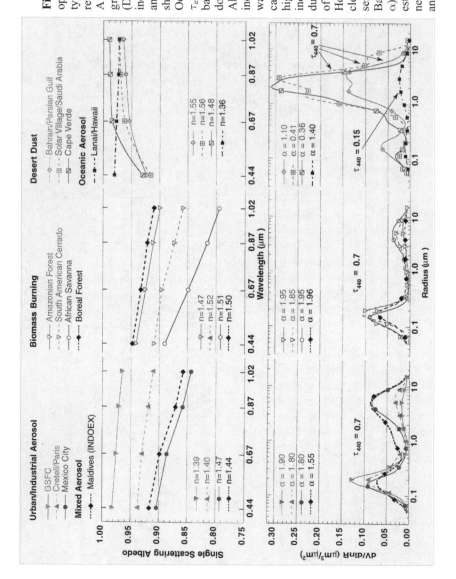

Figure 14.4 The averaged optical properties of different types of tropospheric aerosol retrieved from the worldwide AERONET network of ground-based radiometers (Dubovik et al. 2002). Urban industrial, biomass burning, and desert dust aerosols are shown for $\tau_{ext}(440) = 0.7$. Oceanic aerosol is shown for $\tau_{ext}(440) = 0.15$ since oceanic background aerosol loading does not often exceed 0.15. Also, $\omega_0(\lambda)$ and the refractive index n shown for Bahrain was obtained only for the cases when $\alpha \leq 0.6$ (for higher α, $\omega_0(\lambda)$ and refractive index n were very variable due to a significant presence of urban-industrial aerosol). However, we show the particle size distribution representing all observations in Bahrain (complete range of α). Ångström parameter α is estimated using optical thickness at two wavelengths: 440 and 870 nm.

Figure 14.4. Traditional air pollution is on the decline in the industrialized world and now highest in the developing world. For many decades, developed nations were champions in CO_2 emissions on a per capita basis; now, however, emissions are rapidly growing, most noticeably in the Asian countries, even though traditional air pollution is declining due to health-related regulations. Thus, because of the long atmospheric residence time of CO_2 of more than a century, compared to only about a week for the aerosol particles, in the future the cumulative global climatic warming by CO_2 will outpace any cooling effect that comes from the backscattering of solar radiation into space by the aerosol particles.

POLICY-RELATED ISSUES, LESSONS TO BE LEARNED

The "Ozone Hole"

Human activity has caused major decreases in stratospheric ozone. Most dramatic has been the development of the stratospheric "ozone hole," the complete destruction of ozone, which has been recurring almost annually during September–November, sometimes until midsummer, since the beginning of the 1980s over Antarctica in the 12–22 km height interval, precisely where maximum ozone concentrations had earlier always been observed (see Figure 14.1 and Appendix 14.2). (The weak ozone hole of the spring of 2002 was a surprising exception, most likely caused by unusual meteorological conditions.) Stratospheric ozone loss is caused by emissions of relatively small quantities of CFC gases, accumulating to a little over 3.5 nmol/mol levels of total chlorine, among which are the very reactive ozone-destroying Cl and ClO catalysts. Under the prevailing atmospheric conditions during late winter/early spring, production of these catalysts is strongly enhanced, leading to a chemical instability, especially over Antarctica and to a lesser extent over the Arctic. The development of the ozone hole takes place in a location and at a time of the year that were least expected. Conventional wisdom held that, especially over Antarctica, ozone should be chemically inert. This is indeed true for the natural stratosphere, but it does not hold in a stratosphere containing about six times more chlorine than under natural conditions. For the ozone hole to develop, five criteria must be fulfilled simultaneously (see Appendix 14.2 and accompanying figure), which explains why it was not predicted. In fact, when satellite measurements started to reveal a depletion of ozone, the data were first put aside and attributed to measurement errors. From this experience we should learn several lessons:

1. Do not release chemicals into the environment before their impacts are well studied. After their first release into the atmosphere, it took approximately four decades before Molina and Rowland (1974) could demonstrate that CFCs harm the ozone layer, and another two decades before their production was halted by international regulations. It will take up to

an additional half century before the ozone hole will close again, which attests to the long timescales involved.

2. Although major strides are constantly being made in the development of comprehensive climate models, we should not yet overly rely on models to make predictions. With their discrete space/time coordinate systems, which do not resolve important smaller-scale meteorological features, replacing them by so-called "subgrid-scale parameterizations," models remain approximations of the complex, multidisciplinary feedback systems of the real world. Global climate models have particular difficulties with describing the various elements of the hydrological cycle and the physical–meteorological–chemical processes, which determine the distribution of aerosol and its impact on cloud properties and Earth's radiation budget — a topic to which we will return below. Numerical models are based on incomplete knowledge and have at best been partially tested against observations in the present world, which may not apply in the future, given the great uncertainty of societal responses.

3. Nevertheless, models are indispensable tools to combine available knowledge from several disciplines. Used in the research mode, they serve best when producing results that deviate from observations, thus indicating gaps in knowledge. Models can also be very valuable devices to identify potential low-probability/high-impact features in the complex environmental system. Under certain circumstances, their results should at least be taken as warning signals for impending dangers.

4. Major changes may be triggered by relatively small disturbances. This was dramatically shown in the CFC case. Also, as we discuss below, in comparison with the average influx of 340 Wm^{-2} of solar radiation at the TOA, radiation flux disturbances due to human activities of a few $W\,m^{-2}$ are very small indeed but, nevertheless, very important.

5. If unexpected data (i.e., real data or calculated by models) appear, the so-called "outliers" take them seriously, as they may be the first signals of shifts in environmental conditions (cf., e.g., the presence of "negligibly" few automobiles on postcards from the beginning of the past century against pictures of today's ubiquitous traffic jams).

6. Do not assume that scientists always exaggerate; the ozone loss over Antarctica was much worse than originally thought.

How Well Do We Know the Science of Climate Change? CFCs Compared to CO_2

The Mauna Loa CO_2 records of C.D. Keeling of the Scripps Institution of Oceanography and the ozone hole graph by Joe Farman and colleagues of the British Antarctic Survey constitute the most striking examples of human impacts on the global atmosphere (see Figure 14.1). Following intensive research efforts, which clearly showed that the ozone loss was caused by catalytic reactions involving chlorine radicals, international agreements were set in force in 1996 to stop the production of CFCs as well as several other chlorine- and bromine-containing gases in the developed world. Due to the longevity of these products, however, it will take about half a century before the "ozone hole" will largely close again. In the case of CO_2, regulations are much more difficult to achieve; it is estimated that it will take many more years before agreements on emissions can be reached to limit fossil-fuel use sufficiently. To demonstrate the magnitude of the task, consider the following: to prevent a further increase in levels of CO_2, the present level of worldwide CO_2 emissions from fossil fuels would need to be reduced by as much as 60%, a practically unattainable task, since without abundant renewable energy sources, much of the developing world can be expected to increase their fossil-fuel use. All efforts should be made to keep CO_2 growth at a climatically acceptable level. Energy savings and improved technology will have to make major contributions; however, technological fixes, such as carbon sequestrations, may become inevitable (Lackner 2003). Because of the long lifetime of atmospheric CO_2, it may take centuries before "global warming" and its effects will peak; in terms of sea-level rise, this may take a millennium. One often cited reason for a delay in international regulations is that the science is so uncertain. Here, however, we must call attention to a potentially disturbing signal. The Earth's current radiation budget at the TOA for the period of 1955 to 1996 was out of balance by 0.32 ± 0.15 Wm^{-2}, corresponding to the heat uptake in the oceans, plus lesser contributions from ice melting and heating of the atmosphere (Levitus et al. 2001). In comparison to the greenhouse gas climate forcing of 2.5–3.5 Wm^{-2} (see Figure 14.5; IPCC 2001), this term may seem small, but it corresponds to the net heating of the Earth system, which accumulates from year to year. Note that this net global warming of 0.32 ± 0.15 Wm^{-2} cannot be confirmed by measurements at the TOA because of the lack of shortwave and infrared measurement capabilities with sufficient accuracy and precision. It is, however, close to the heat uptake in the oceans of 0.32 ± 0.15 Wm^{-2}, derived by Levitus et al. (2001). It also agrees rather well with the results from model calculations, which take into account the growth of "greenhouse gases" and the direct cooling effect of anthropogenic sulfate particles during the same period (Barnett et al. 2001).

We next present in broad terms the changes in the Earth's radiative forcing between 1860 and the early 1990s. The increase in the infrared downward radiation flux forcing by greenhouse gases at the TOA is 2.3 ± 0.5 Wm^{-2}. From this flux we must subtract an upward radiation flux of 1 ± 0.3 Wm^{-2} due to the surface temperature rise of 0.6 ± 0.2 K, calculated for the case of a water vapor/surface temperature feedback (Ramanathan 1981). Because the atmosphere loses 0.3 Wm^{-2} as a result of heat uptake in the ocean (Levitus et al. 2001), the atmosphere would gain 1.0 ± 0.6 Wm^{-2} of energy, but it is unable to do so because of its small heat capacity. Consequently, this energy is backscattered as shortwave

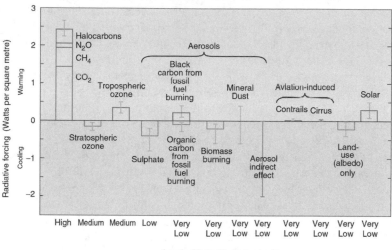

Figure 14.5   Over the past decade, the atmospheric chemistry research community has identified and quantified the distributions of a number of radiatively active substances. However, especially regarding the cooling (parasol) effect by aerosol in the hydrological cycle, as shown in the blue field, the level of scientific understanding in the calculated radiative forcing of these various substances is still very low (IPCC 2001).

radiation, corresponding to $43 \pm 25\%$ of the greenhouse gas warming. We call this "the parasol effect." This energy loss must have occurred by enhanced reflection of solar radiation by aerosols and clouds. (We assume no change in surface albedo and neglect the variations in solar radiation by 0.3 ± 0.2 W m^{-2}). Remarkably, the range is also significantly caused by uncertainty in the trends in global average temperature.

In line with this analysis, there has been a 2% increase in cloud cover over the mid- to high-latitude land areas during the twentieth century (IPCC 2001). Enhanced albedo is also supported by the pan evaporation measurements of Roderick and Farquhar (2002) and cloud darkening observations in eastern Europe after the industrial cleanup as a result of the collapse of the communist regimes in Europe (Krüger and Graßl 2002). In addition, satellite-derived cloud and surface properties have shown that the Arctic has warmed, accompanied by increased cloudiness in all seasons, except spring (Wang and Key 2003), which has prevented even greater Arctic warming.

Without the water vapor/surface temperature feedback, 2 Wm^{-2} instead of 1 Wm^{-2} must be subtracted from the greenhouse forcing term (Ramanathan 1981). This no-feedback case is unlikely: Higher temperatures should allow for more evaporation and a greater propensity of the atmosphere to hold water vapor, agreeing with what is shown by climate models and confirmed by obser vations, which show an increase by several per cent in water vapor concentrations per decade over many regions of the Northern Hemisphere (IPCC 2001). Data are not available for the Southern Hemisphere.

Despite compensation of greenhouse warming by aerosols and clouds, due to the much shorter lifetimes of the aerosol, cumulative forcing by the greenhouse gases, especially CO_2, will in the future outpace the compensation provided by aerosols, implying accelerated climate warming. This was also pointed out by Anderson et al. (2003). The power of CO_2 and CH_4 as greenhouse gases has been highlighted in several recent studies, showing their importance for climate change during the glacials and interglacials and other geological periods (e.g., Hinrichs et al. 2003; Barrett 2003; DeConto and Pollard 2003).

With an inevitable, continued growth in CO_2 concentrations and model predictions of an average global warming in the range of 1.4–5.8 K, there is great risk that humankind is rapidly moving into an uncertain and uncharted climate future with great risks of major climate warming and related problems, such as higher frequency of extreme events, sea-level rise, and regional lack of freshwater availability. The fourfold range in the estimated warming is due to a combination of uncertainties in scenarios, climate model feedbacks, and adopted future scenarios in greenhouse gas emissions and aerosol loadings. In all likelihood, climate change will represent — even more than now — a major issue that humankind must face in this and, probably also, the next century.

Impacts of Particles on Human Health and Climate: A Dilemma?

Health problems associated with air pollution are substantial. The World Health Organization estimates that worldwide each year about 600,000 people fall victim to outdoor (Stone 2002) and 1.6 million to indoor air pollution (Smith 2002). Biomass burning in the households of the developing world for cooking and heating produces as much as 7–20% of the total emissions of CO, NO, and hydrocarbons (Ludwig et al. 2003). Thus, global, regional, local, and even domestic air pollution effects intertwine. It is clear that for health reasons, priority should and will be given to the abatement of the heavy air pollution, which affects large tracts of land and people, especially in the developing world. As stated above, this can, however, have important consequences for regional and also global climate. Regionally, it will be advantageous as more solar radiation will again reach the surface with a positive effect on precipitation. In addition, cloudiness may decrease via a lower second indirect aerosol effect and lead to a further increase in solar radiation. Black carbon is a special case. Its removal will cool the planet, but its budget is uncertain. Globally, however, if the cooling effect of the aerosol is greater than the heating effect by black carbon particles, the cleanup will add to global warming.

In summary, greenhouse gas emissions will increase surface temperature and rainfall. Aerosols counteract through surface cooling and drying. Since aerosol effects are concentrated regionally, global warming and regional cooling will happen simultaneously. Furthermore, as shown by Novakov et al. (2003), the aerosol effect itself fluctuates between warming and cooling, depending on how the various countries shift from inefficient technology (which produces more absorbing aerosols) to more efficient technology (which produces cooling aerosols). The opposite effects of aerosols and greenhouse gases on precipitation, on timescales of decades, may regionally lead intermittently to more frequent droughts and excess rainfall conditions.

FINAL REMARK

Humankind is carrying out a grand experiment on its own planet with uncertain, but, most likely, major environmental consequences. This has been emphasized here, in part with some new arguments (as far as we know) on the climate side. Thus far, the political/economical actors have not been sufficiently impressed with the warnings by the climate research community (IPCC 2001), pointing to a major failure in communications.

ACKNOWLEDGMENTS

Thanks go to Andi Andreae, Bert Bolin, Peter Cox, Ulrich Cubasch, Roland von Glasow, Jos Lelieveld, and Will Steffen for productive discussions and advice.

APPENDIX 14.1   THE "ANTHROPOCENE"

Supported by great technological and medical advancements and access to plentiful natural resources, the expansion of humankind, both in numbers and per capita exploitation of Earth's resources, has been astounding (Turner et al. 1990). Several examples of the growth of human activities and economic factors impacting on the environment during the twentieth century are given in Table A-1.

During the past three centuries, human population increased tenfold to 6000 million, growing by a factor of four during the twentieth century alone (McNeill 2000). This growth in human population was accompanied, e.g., by a growth in the methane-producing cattle population to 1400 million (about one cow per average-size family).

In a few generations humankind will have exhausted the fossil fuels that were generated over several hundred million years, resulting in large emissions of air pollutants. The release of SO_2 (globally about 160 Tg yr^{-1}) to the atmosphere by coal and oil burning is at least two times larger than the sum of all natural emissions, occurring mainly as marine dimethyl-sulfide from the oceans (Houghton et al. 1990, 1996, 2001). Oxidation of SO_2 to sulfuric and of NO_x to nitric acid has led to acidification of precipitation and lakes, causing forest damage and fish death in biologically sensitive lakes in regions such as Scandinavia and northeastern North America. Due to substantial reduction in SO_2 emissions, the situation has improved. However, the problem is now getting worse in Asia.

From Vitousek et al. (1997) we learn that 30–50% of the world's land surface has been transformed by human action; the land under cropping has doubled during the past

Table A-1 Partial record of the growths and impacts of human activities during the twentieth century (McNeill 2000).

Item	Increase Factor, 1890s–1990s
World population	4
Total world urban population	13
World economy	14
Industrial output	40
Energy use	16
Coal production	7
Carbon dioxide emissions	17
Sulfur dioxide emissions	13
Lead emissions	≈8
Water use	9
Marine fish catch	35
Cattle population	4
Pig population	9
Irrigated area	5
Cropland	2
Forest area	20% decrease
Blue whale population (Southern Ocean)	99.75% decrease
Fin whale population	97% decrease
Bird and mammal species	1% decrease

century at the expense of forests, which declined by 20% (McNeill 2000) over the same period. More nitrogen is now fixed synthetically and applied as fertilizers in agriculture than fixed naturally in all terrestrial ecosystems ($120 \, Tg \, yr^{-1}$ vs. $90 \, Tg \, yr^{-1}$) (Galloway et al. 2002). The Haber–Bosch industrial process to fix N from N_2 in the air made human population explosion possible. It is remarkable to note the importance of a single invention for the evolution on our planet. Only $20 \, Tg \, N \, yr^{-1}$ is contained in the food for human consumption. Wasteful application of nitrogen fertilizers in agriculture and especially its concentration in domestic animal manure have led to eutrophication of surface waters and even groundwater in many locations around the world. Fossil-fuel burning adds another $25 \, Tg \, N \, yr^{-1}$ highly reactive NO_x to the atmosphere, causing photochemical ozone formation in extensive regions around the globe. Additional input of altogether $165 \, Tg \, N \, yr^{-1}$ is almost double as large as natural biological fixation. Disturbance of the N cycle also leads to the microbiological production of N_2O, a greenhouse gas and a source of NO in the stratosphere where it is strongly involved in stratospheric ozone chemistry. Human disturbance of the N cycle has recently been treated in a special publication of *Ambio* (vol. 31, March 2002). As a result of increasing fossil-fuel burning, agricultural activities, deforestation, and intensive animal husbandry, especially cattle holding, several climatically important "greenhouse" gases have substantially increased in the atmosphere over the past two centuries: CO_2 by more than 30% and CH_4 by more than 100%, contributing substantially to the observed global average temperature increase by about 0.6°C that has been observed during the past century. In 1995, IPCC stated: "The balance of evidence suggests a discernable human influence on global climate." In 2001: "There is new and stronger evidence that most of the warming observed over the last 50 years is attributable to human activities" (Houghton et al. 1990, 1996, 2001). Depending on the scenarios of future energy use and model uncertainties, increasing emissions and the resulting growth in atmospheric concentrations of CO_2 are estimated to cause a rise in global average temperature by 1.4°–5.8°C during the present century, accompanied by sea-level rise of 9–88 cm (and 0.5–10 m until the end of the current millennium). The largest anthropogenic climate changes are thus awaiting future generations.

Furthermore, humankind also releases many toxic substances in the environment and some, the chlorofluorocarbon gases ($CFCl_3$ and CF_2Cl_2), which are not toxic at all, have led to the Antarctic springtime "ozone hole" and would have destroyed much more of the ozone layer if international regulatory measures to end their production by 1996 had not been taken. Nevertheless, due to the long residence times of CFCs, it will take at least until the middle of this century before the ozone layer will have largely recovered and the ozone hole will have disappeared.

Considering these and many other major and still growing impacts of human activities on Earth and atmosphere, at all scales, it is more than appropriate to emphasize the central role of humankind in geology and ecology by using the term "Anthropocene" for the current geological epoch. The impact of current human activities is projected to last for very long periods. According to Loutre and Berger (2000), due to past and future anthropogenic emissions of CO_2, climate may depart significantly from natural behavior even over the next 50,000 years.

To assign a more specific date to the onset of the "Anthropocene" is somewhat arbitrary, but we propose the latter part of the eighteenth century, although we are aware that alternative proposals can be made. However, we choose this date because, during the past two centuries, the global effects of human activities have become clearly noticeable. This is the period when data retrieved from glacial ice cores show the beginning of a growth in the atmospheric concentrations of several "greenhouse gases," in particular CO_2, CH_4, and N_2O (Houghton et al. 1990, 1996, 2001). Such a starting date coincides with James Watt's invention of the steam engine in 1784.

Without major catastrophes (e.g., enormous volcanic eruptions, an unexpected epidemic, a large-scale nuclear (or biological) war, an asteroid impact, a new ice age, or continued plundering of Earth's resources by wasteful technology), humankind will remain a major geological force for many millennia, maybe millions of years, to come. To develop a worldwide accepted strategy leading to sustainability of ecosystems against human-induced stresses will be one of the great tasks of human societies, requiring intensive research efforts and wise application of the knowledge thus acquired.

Exciting, but also difficult and daunting tasks lie ahead of the global research and engineering community to guide humankind toward global, sustainable, management of the ecosphere in the Anthropocene (Schellnhuber 1999).

APPENDIX 14.2   THE OZONE HOLE

Stratospheric ozone is formed through the photolysis of O_2 and recombination of the two resulting O atoms with O_2: $3O_2 \rightarrow 2O_3$. These reactions are clearly beyond human control. Reactions are also needed to reproduce O_2, otherwise within 10,000 years all oxygen would be converted to ozone. Because laboratory simulations of rate coefficients in the late 1960s had shown that the originally proposed reactions by Chapman in 1930 were too slow to balance ozone production, additional reactions involving several reactive radical species were postulated. The additional ozone destroying reaction chains can be written as:

$$X + O_3 \rightarrow XO + O_2$$
$$O_3 + h\nu \rightarrow O + O_2 \ (\lambda < 1140 \ nm)$$
$$O + XO \rightarrow X + O_2$$
$$\text{net:} \quad 2O_3 \rightarrow 3O_2,$$

where X stands for OH, NO, Cl, or Br, and XO correspondingly for HO_2, NO_2, ClO, and BrO. These catalysts are influenced by human activities, especially the emissions of industrial chlorine compounds which are transferred to the stratosphere, such as CCl_4, CH_3CCl_3, and most importantly, the chlorofluorocarbon ($CFCl_3$ and CF_2Cl_2) gases. The current content of chlorine in the stratosphere, about 3 nmol/mol, is about six times higher than what is naturally supplied by CH_3Cl.

For a long time it was believed that chemical loss of ozone by reactive chlorine would mostly take place in the 25–50 km height region and that at lower altitudes in the stratosphere, which contains most ozone, only relatively little loss would take place. The reason is that the NO_x and the ClO_x radicals, like two "Mafia" families, kill each other by forming $ClONO_2$ and HCl:

$$ClO + NO_2 + M \rightarrow ClONO_2 + M, \text{ and}$$
$$ClO + NO \rightarrow Cl + NO_2$$
$$Cl + CH_4 \rightarrow HCl + CH_3.$$

Most inorganic chlorine is normally present as HCl and $ClONO_2$, which react neither with each other nor with ozone in the gas phase, thus protecting ozone from otherwise much larger destruction.

This preferred situation does not always exist. In 1985, scientists from the British Antarctic Survey presented their observations showing total ozone depletions over the Antarctic by more then 50% during the late winter/springtime months (September to November), with ozone depletions taking place in the 14–22 km height region where normally maximum ozone concentrations are found; within a few weeks after polar sunrise, almost all ozone is destroyed, creating the "ozone hole." How was this possible? Nobody anticipated this; in fact, it was believed that at high latitudes, ozone in the lower stratosphere was largely chemically inert.

It only took some two years of research to identify the main processes that led to these large ozone depletions and to show that the CFCs were the culprits. The explanation involves each of five necessary conditions:

1. Low temperatures, below about –80°C, are needed to produce ice particles consisting of nitric acid and water (nitric acid trihydrate) or water molecules. In this process, NO_x catalysts are also removed from the stratosphere through the reactions:

$$NO + O_3 \rightarrow NO_2 + O_2$$
$$NO_2 + NO_3 + M \rightarrow N_2O_5 + M$$
$$N_2O_5 + H_2O \rightarrow 2 \ HNO_3$$

thereby producing HNO_3 which is incorporated in the particles.

2. On the surface of the ice particles, HCl and $ClONO_2$ react with each other to produce Cl_2 and HNO_3:

$$HCl + ClONO_2 \quad \rightarrow \quad Cl_2 + HNO_3.$$

The latter is immediately incorporated in the particles.

3. After the return of daylight following the polar night, Cl_2 is photolyzed to produce 2 Cl atoms:

$$Cl_2 + h\nu \quad \rightarrow \quad 2\ Cl.$$

4. The chlorine atoms start a catalytic chain of reactions, leading to the destruction of ozone:

$$
\begin{aligned}
Cl + O_3 &\rightarrow ClO + O_2 \\
Cl + O_3 &\rightarrow ClO + O_2 \\
ClO + ClO + M &\rightarrow Cl_2O_2 + M \\
Cl_2O_2 + h\nu &\rightarrow Cl + ClO_2 \rightarrow 2Cl + O_2 \\
\text{net:} \quad 2\ O_3 &\rightarrow 3\ O_2.
\end{aligned}
$$

Note that the breakdown of ozone is proportional to the square of the ClO concentrations. As these grew for a long time by more than 4% per year, ozone loss increased by 8% from one year to the next. Also, because there is now about six times more chlorine, about 3 nmol/mol, in the stratosphere, compared to natural conditions when chlorine was solely provided by CH_3Cl, the ozone depletion is now 36 times more powerful than prior to the 1930s when CFC production started. Earlier, under natural conditions, chlorine-catalyzed ozone destruction was unimportant and it will be so again in 1–2 centuries.

5. Enhanced inorganic chlorine (Cl, ClO, HCl, $ClONO_2$, Cl_2O_2) concentrations, produced by CFC photolysis above 25–30 km are brought down during winter into the lower stratosphere by downwind transport from the middle and upper stratosphere within a meteorologically stable vortex with the pole more or less at the center. This is important because at the higher altitudes more organic chlorine is converted to much more reactive inorganic chlorine gases, including the ozone-destroying catalysts Cl, ClO, and Cl_2O_2.

All five factors have to come together to cause the ozone hole (see Figure A-2). It is thus not surprising that the ozone hole was not predicted. This experience shows the critical importance of measurements. What other surprises lie ahead involving instabilities in other parts of the complex Earth system?

REFERENCES

Anderson, J.G., W.H. Brune, and M.H. Proffitt. 1989. Ozone destruction by chlorine radicals within the Antarctic vortex: The spatial and temporal evolution of ClO–O_3 anticorrelation based on *in situ* ER-2 data. *J. Geophys. Res.* **94**:11,465–11,479.
Anderson, T.L., R.J. Charlson, S.E. Schwartz et al. 2003. Climate forcing by aerosols — A hazy picture. *Science* **300**:1103–1104.

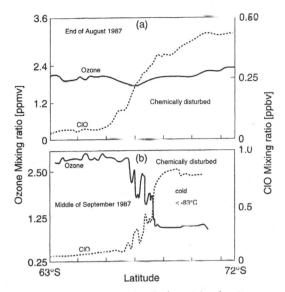

Figure A-2 High concentrations of ClO radicals and the simultaneous rapid ozone destruction occur in winter when the temperature becomes very low (< −80°C). Measurements by J.G. Anderson et al. (1989) show ozone and ClO mixing ratios near 20 km during two ER-2 aircraft flights. ClO radicals, together with Cl the catalysts that destroy ozone, show sharp increases in the low-temperature section south of about 68°S, resulting in strong depletions during the first half of September.

Andreae, M.O. 1995. Climatic effects of changing atmospheric aerosol levels. In: Future Climates of the World: World Survey of Climatology, vol. 16, ed. A. Henderson-Sellers, pp. 341–392. Amsterdam: Elsevier.

Barnett, T.P., D.W. Pierce, and R. Schnur. 2001. Detection of anthropogenic climate change in the world's oceans. *Science* 292:270–274.

Barrett, P. 2003. Cooling a continent. *Nature* 421:221–223.

Charlson, R.J., J. Langner, H. Rohde, C.B. Leovy, and S.G. Warren. 1991. Perturbation of the Northern Hemisphere radiative balance by backscattering from anthropogenic sulfate aerosols. *Tellus* 43 AB:152–163.

Crutzen, P.J. 2002. Geology of Mankind: The Anthropocene. *Nature* 415: 23.

Crutzen, P.J., and W. Steffen. 2003. How long have we been in the Anthropocene: An editorial comment. *Clim. Change* 61:251–257.

DeConto, R.M., and D. Pollard. 2003. Rapid Cenozoic glaciation of Antarctica induced by declining atmospheric CO_2. *Nature* 421:245–249.

Dlugokencky, E.J., K.A. Maserie, P.M. Lang, and P.P. Tans. 1998. Continuing decline in the growth rate of the atmospheric methane burden. *Nature* 393:447–450.

Dubovik, O., B.N. Holben, T.F. Eck et al. 2002. Variability of absorption and optical properties of key aerosol types observed in worldwide locations. *J. Atmos. Sci.* 59:590–608.

Farman, J.C., B.G. Gardiner, and J.D. Shanklin. 1985. Large losses of total ozone in Antarctica reveal seasonal CLO_x/NO_x interaction. *Nature* 315:207–210.

Galloway, J.N., E.B. Cowling, S. Seitzinger, and R.H. Socolow. 2002. Reactive nitrogen: Too much of a good thing? *Ambio* **31**:60–63.

Guenther, A., C.N. Hewitt, D. Erickson et al. 1995. A global model of natural volatile organic compound emissions. *J. Geophys. Res.* **100**:8873–8892.

Gunn, J., S. Sandoy, B. Keller et al. 2003. Biological recovery from acidification: Northern lakes recovery study. *Ambio* **32(3)**:161–248.

Haywood, J.M., V. Ramaswamy, and B. Soden. 1999. Tropospheric aerosol climate forcing in clear-sky satellite observations over the oceans. *Science* **283**:1299–1303.

Hedin, L.O., L. Granat, G.E. Likens et al. 1994. Steep declines in atmospheric base cations in regions of Europe and North America. *Nature* **367**:351–354.

Hinrichs, K.U., L.R. Hmelo, and S.P. Sylva. 2003. Molecular fossil record of elevated methane levels in Late Pleistocene coastal waters. *Science* **299**:1214–1217.

Hofmann, D.J., J.W. Harder, J.M. Rosen, J. Hereford, and J.R. Carpenter. 1989. Ozone profile measurements at McMurdo Station, Antarctica, during the spring of 1987. *J. Geophys. Res.* **94**:16,527–16,536.

Houghton, J.T., G.J. Jenkins, and J.J. Ephraums, eds. 1990. Climate Change: The IPCC Scientific Assessment. Cambridge: Cambridge Univ. Press.

Houghton, J.T., L.G. Meiro Filho, B.A. Callander et al. 1996. Climate Change 1995. Cambridge: Cambridge Univ. Press.

Houghton, J.T., Y. Ding, D.J. Griggs et al. eds. 2001. Climate Change 2001: The Scientific Basis. Cambridge: Cambridge Univ. Press.

IPCC (Intergovernmental Panel on Climate Change). 2001. Climate Change 2001. The Scientific Basis. Working Group I Contribution, Third Assessment Report of the IPCC, ed. J.T. Houghton, Y. Ding, D.J. Griggs et al. Cambridge: Cambridge Univ. Press.

Jacobson, M.Z. 2001. Global direct radiative forcing due to multicomponent anthropogenic and natural aerosols. *J. Geophys Res.* **106**:1551–1568.

Jones, C.D., P.M. Cox, R.L.H. Essery, D.L. Roberts, and M.J. Woodage. 2003. Strong carbon cycle feedbacks in a climate model with interactive CO_2 and sulphate aerosols. *Geophys. Res. Lett.* **30(9)**:1479.

Kanakidou, M., K. Tsigaridis, F.J. Dentener, and P.J. Crutzen. 2000. Human activity-enhanced formation of organic aerosols by biogenic hydrocarbon oxidation. *J. Geophys. Res.* **105**:9243–9254.

Kaufman, Y.J., D. Tanré, and O. Boucher. 2002. A satellite view of aerosols in the climate system. *Nature* **419**:215–223.

Keeling, C.D., and T.P. Whorf. 2000. Atmospheric CO_2 records from sites in the SIO sampling network. In: Trends: A Compendium of Data on Global Change. Oak Ridge, TN: Carbon Dioxide Information Analysis Center, Oak Ridge Natl. Lab., U.S. Dept. of Energy.

Keeling, R.F., S.C. Piper, and M. Heimann. 1996. Global and hemispheric CO_2 sinks deduced from changes in atmospheric O_2 concentration. *Nature* **381**:218–221.

Krol, M.C., J. Lelieveld, D.E. Oram et al. 2003. Continuing emissions of methyl chloroform from Europe. *Nature* **421**:131–135.

Krüger, O., and H. Graßl. 2002. The indirect aerosol effect over Europe. *Geophys. Res. Lett.* **29(19)**:1925.

Lackner, K.S. 2003. A guide to CO_2 sequestration. *Science* **300**:1677–1678.

Levitus, S., J.I. Antonov, J. Wang et al. 2001. Anthropogenic warming of Earth's climate system. *Science* **292**: 267–270.

Loutre, M.F., and A. Berger. 2000. Future climatic changes: Are we entering an exceptionally long interglacial? *Clim. Change* **46**:61–90.

Ludwig, J., L.T. Marufu, B. Huber, M.O. Andreae, and G. Hebś. 2003. Domestic combustion of biomass fuels in developing countries: A major source of atmospheric pollutants. *J. Atmos. Chem.* **44**: 23–37.

Markowicz, K.M., P.J. Flatau, M.V. Ramana, P.J. Crutzen, and V. Ramanathan. 2002. Absorbing Mediterranean aerosols lead to a large reduction in the solar radiation at the surface. *Geophys. Res. Lett.* **29(20)**:1967.

McNeill, J.R. 2000. Something New Under the Sun. New York: Norton.

Melillo, J.M., P.A. Steudler, J.D. Aber et al. 2002. Soil warming and carbon-cycle feedbacks to the climate system. *Science* **298**:2173–2176.

Molina, M., and F.S. Rowland. 1974. Stratospheric sink for chlorofluoromethanes: Chlorine atom catalyzed destruction of ozone. *Nature* **249**:810–812.

Novakov, T., V. Ramanathan, J.E. Hansen et al. 2003. Large historical changes of fossil-fuel black carbon aerosol. *Geophys. Res. Lett.* **30(6)**:1324.

Penner, J.E., M.O. Andreae, H. Annegarn et al. 2001. Aerosols, their direct and indirect effects. In: Climate Change 2001. The Scientific Basis. Working Group I Contribution, Third Assessment Report of the IPCC, ed. J.T. Houghton, Y. Ding, D.J. Griggs et al., pp. 289–348. Cambridge: Cambridge Univ. Press.

Prinn, R.G., J. Huang, R. Weiss et al. 2001. Evidence for substantial variations of atmospheric hydroxyl radicals in the past two decades. *Science* **292**:1882–1888.

Ramanathan, V. 1981. The role of ocean–atmosphere interactions in the CO_2 climate problem, *J. Atmos. Sci.* **38**:918–930.

Ramanathan, V., P.J. Crutzen, J. Lelieveld et al. 2001a. Indian Ocean Experiment: An integrated analysis of the climate forcing and effects of the great Indo-Asian haze, *J. Geophys. Res.* **106**:28,371–28,398.

Ramanathan, V., P.J. Crutzen, J.T. Kiehl, and D. Rosenfeld. 2001b. Aerosols, climate, and the hydrological cycle. *Science* **294**:2119–2124.

Ramaswamy, V., O. Boucher, J. Haigh et al. 2001. Radiative forcing of climate change. In: Climate Change 2001. The Scientific Basis. Working Group I Contribution, Third Assessment Report of the IPCC, ed. J.T. Houghton, Y. Ding, D.J. Griggs et al., pp. 349–416. Cambridge: Cambridge Univ. Press.

Richter, A., and J.P. Burrows. 2002. Retrieval of tropospheric NO_2 from GOME measurements. *Adv. Space Res.* **29**:1673–1683.

Roderick, M.L., and G.D. Farquhar. 2002. The cause of decreased pan evaporation over the past 50 years. *Science* **298**:1410–1411.

Satheesh, S.K., and V. Ramanathan. 2000. Large differences in tropical aerosol forcing at the top of the atmosphere and Earth's surface. *Nature* **405**:60–63.

Schellnhuber, H.J. 1999. "Earth system" analysis and the second Copernican revolution. *Nature* **402**:C19-C23.

Smith, K.R. 2002. In praise of petroleum? *Science* **298**:1847.

Stone, R. 2002. Counting the cost of London's killer smog. *Science* **298**:2106–2107.

Tegen, I.D., and I. Fung. 1995. Contribution to the atmospheric mineral aerosol load from land-surface modification. *J. Geophys Res.* **100(D9)**:18,707–18,726.

Tegen, I.D., A. Koch, A. Lacis, and M. Sato. 2000. Trends in tropospheric aerosol loads and corresponding impact on direct radiative forcing between 1950 and 1990: A model study. *J. Geophys. Res.* **105(D22)**:26,971–26,989.

Turner, B.L., II, W.C. Clark, R.W. Kates et al., eds. 1990. The Earth as Transformed by Human Action. Cambridge: Cambridge Univ. Press.

Vernadsky, V.I. 1998/1926. The Biosphere (translated and annotated version from the original of 1926). New York: Springer.

Vitousek, P.M., H.A. Mooney, J. Lubchenco, and J.M. Melillo. 1997. Human domination of Earth's ecosystems. *Science* **277**:494–499.

Wang, X., and J.R. Key. 2003. Recent trends in Arctic surface, cloud, and radiation properties from space. *Science* **299**:1725–1728.

15

Assessing and Simulating the Altered Functioning of the Earth System in the Anthropocene

P. M. COX[1] and N. NAKICENOVIC[2]

[1]Hadley Centre for Climate Prediction and Research, Met Office,
Exeter, Devon EX1 3PB, U.K.
[2]International Institute for Applied Systems Analysis (IIASA) and
Vienna University of Technology, Vienna, Austria

ABSTRACT

The current human-dominated era or "Anthropocene" has been a time of unprecedented rates of change within the Earth system. As such the Anthropocene poses a no-analogue situation in which evidence of past Earth system changes cannot, on their own, tell us of how the system will respond in the future. Instead, we need to develop mechanistic and phenomenological models based on robust underlying principles to extrapolate our knowledge of the past Earth system into projections of the future. This requires an entire spectrum of Earth system models that can be related to each other in a more rigorous way than has been achieved to date. It also requires new integrated models of the two-way coupling between the biophysical and human aspects of the Earth system. Only then will we be able to approach all of the paleodata constraints on Earth system dynamics (which requires long simulations), while also providing regionally specific Earth system projections and spatially specific human development paths for policy makers (which requires, in both cases, high-resolution comprehensive models as well as models that can bridge spatial and temporal scales). Such a spectrum of models offers a solution to the tension between "simulation," which inevitably leads to increasing model complexity, and "understanding," which often involves reducing the system to its bare necessities.

INTRODUCTION

As a working definition, we consider the Earth system to be the sum of the components, interactions, and feedbacks that determine the physical, chemical, and biological environment of the planet, as well as its variations in space and time.

The current Anthropocene era is a period of unprecedented rate of change in the Earth system brought about by human activities. Atmospheric CO_2 levels are now higher than they have been for at least 2 million years and probably much longer (Prentice et al. 2001). Methane concentrations are double what they were in the preindustrial era. Anthropogenic inputs have also almost doubled the amount of available N coming into the global terrestrial ecosystems each year (Vitousek et al. 1997). Throughout human history, land-use change has reduced the global fraction of forests from 45% to 27% (Andreae et al., this volume), and the rate of deforestation has generally accelerated with 16 million ha of tropical forest lost per year through the 1990s (Houghton 2003). An even more significant human perturbation of the Earth system is projected for the next 100 years.

The absence of a past analogue for the Anthropocene implies that we need to extrapolate our knowledge of the functioning of the Earth system into new areas of its phase space. Although purely empirical relationships between variables (e.g., global temperature and CO_2 or energy services and CO_2) can be used to interpolate data points, they cannot be reliably utilized outside the region for which they were fitted. Instead, mechanistic and phenomenological models are required based on principles that apply to both the past and future Earth system states (in this way, mechanistic and phenomenological models extrapolate mechanisms and the underlying systems dynamics rather than the empirical correlations between variables).

An Earth system model is a numerical representation of the Earth system. The precise definition of such a model depends on the timescale of interest. Much slower processes can be considered as constant boundary conditions (e.g., locations of continents in century-scale projections). Much faster processes can be considered as stochastic forcing of the slower processes (e.g., the impact of weather systems on the deep-ocean circulation). Components of the natural Earth system, along with their characteristic timescales for change, are shown in Table 15.1.

Processes with similar timescales need to be explicitly modeled if they are a sufficiently strongly interacting part of the overall system. Weakly coupled subsystems may, however, be considered as time-varying boundary conditions if they respond little to changes in the system of interest (e.g., variations in solar input).

Earth system models span a massive range of complexities, from conceptual models designed to enhance understanding, to general circulation models (GCMs) developed to predict future climate change. This spectrum of Earth system model complexities reflects a natural tension between the scientist's desire to understand the system and the policy makers' desire for useful predictions. To develop an understanding of the workings of a system, a scientist will tend to simplify and idealize, taking the model of the system away from reality but generating more robust and reproducible findings. The key motivation is often to isolate the dominant factors in the system's behavior rather than to predict its evolution in detail. This driver leads to simpler conceptual models. By contrast, to produce predictions of a system's behavior (e.g., as required by policy makers) requires complex simulation models (e.g., GCMs). These models need to be

Table 15.1   Components of the Earth system and their characteristic timescales

Earth System Component	Subcomponent	Timescale of Change
Atmosphere	Boundary layer	Day
	Free atmosphere	Days to months
Ocean	Mixed layer	Months to years
	Deep ocean	100–1000 years
	Sea ice	Days to 100 years
Cryosphere	Snow	Days to years
	Mountain glaciers	100 years
	Ice sheets	10,000 years
Biosphere	Vegetation	10 days to 1000 years
	Soil carbon	1 to 1000s years
Geosphere	Rock weathering	$> 10^5$ years
	Plate tectonics	$> 10^7$ years
Anthroposphere	Technological change	20–100 years
	Life expectancy	35–75 years
	Doubling of CO_2 emissions	30–40 years (at current rates)
	Doubling of human population	80 years (at current rates)

mechanistic, where they are used to extrapolate into no-analogue situations (e.g., the Anthropocene), and complete so that all plausibly important processes are included.

Table 15.2 shows three classes of models that make up the spectrum of Earth system models. Conceptual models include "box models" in which a substance or quantity (e.g., carbon in vegetation) is often represented by a single compartment (or "box"). Aspects of the coupled Earth system are studied by coupling a number of boxes together through parameterized exchanges. Box models are very economical and thus can be used to study Earth system dynamics on long timescales (billions of years in some cases). Their relative simplicity makes it possible to understand their outputs in terms of the underlying model assumptions; however, they tend to have a high ratio of tunable parameters to internal degrees of freedom.

Comprehensive models attempt to model the key Earth system processes explicitly. For example, GCMs were born of weather forecasting models, which aim to simulate the movement and development of weather systems. Unfortunately, it is not feasible to run very high-resolution models for periods of interest with regard to climate change (10–100 years), and therefore lower-resolution variants of atmospheric and oceanic models are employed in GCMs. Unresolved fluid transports, such as convection, need to be parameterized (i.e., represented in terms of the resolved large-scale variables) because these occur on scales much less than the GCM gridscale. In addition, many nonfluid diabatic

Table 15.2   The spectrum of Earth system models.

	Conceptual	Intermediate Complexity	Comprehensive
Purpose	To develop understanding of emergent properties	Used to simulate past climate change	Used to predict future climate change
Spatial dimensions	0D–1D	2D–3D	3D
Typical timescales	10^3–10^9 years	10–10^4 years	0.1–10^3 years
CPU usage	~10^3 years/min on PC	~10^3 years/day on workstation	~10^3 years/year on supercomputer
Examples	Box models Daisyworld	CLIMBER Statistical Dynamical	GCMs

processes need to be represented in GCMs (e.g., land–atmosphere exchange, cloud and condensation processes, atmospheric radiative transfer). Uncertainty in these parameterizations leads to most of the spread seen among GCM projections of climate change (IPCC 2001). Nevertheless, GCMs have advanced significantly over the last few decades, such that they offer our best guidance on the consequences of continuing anthropogenic emissions of greenhouse gases and aerosol precursors.

Between these two extremes of complexity lie "Earth system models of intermediate complexity" (EMICs), which are designed to address longer timescales than currently feasible with a GCM. EMICs are spatially explicit (i.e., they resolve differences between different parts of the globe) but generally use economical parameterizations (e.g., of atmospheric circulation) rather than explicitly modeling all of the flows. Since they can be run for long periods, EMICs have been invaluable for the interpretation of paleoclimate data in terms of Earth system processes.

In socioeconomic modeling, the current state-of-the-art includes quite complex models used to assess the development of different components of the anthropogenic sphere. Demographic models are usually simulation frameworks that describe future population development as a function of regional or national fertility and mortality trends, as well as, to a lesser degree, a function of migration. Economic models usually maximize utility of consumption as a function of resource availabilities, labor, and capital within an equilibrium framework, either for various world regions or for individual countries. Some of the economic models include energy system submodules so as to determine the emissions. Energy models themselves are often also optimization frameworks that minimize overall energy systems costs or resulting emissions under the resource, demand,

and other constraints. Impact models range from simulation frameworks to economic models of agricultural and other sectors. Recently, such models have been coupled into "integrated assessment" (IA) frameworks to provide a more holistic perspective of future developments and trends that may characterize the Anthropocene and the interactions with Earth system. Another approach has been the development of "reduced form" IA models, which capture essential developments based only on the essential driving forces and variable interactions. These models often have very coarse resolution in time and space. They are used to develop scenarios of future developments.

THE EMERGENCE OF THE ANTHROPOCENE

The main driving forces of the Anthropocene started unfolding with the onset of the Industrial Revolution some two centuries ago. However, it is only during the last decade or two that it has become evident that human activities have resulted in salient intervention and changes of planetary processes which have irreversibly changed the Earth system. Anthropogenic climate change is the most widely known example; however, humans are also impacting many other aspects of Earth system function.

In 1750, after 10,000 years of slow but steady increase, the world's population was approximately 750 million people — the result of the spread of ever better agricultural practices throughout the world. It is estimated that during the last 1000 years, the population has been growing at a rate of 0.1% per year, or doubling on the order of 500 years. The characteristic size of the global population was less than 50 million throughout the age of agriculture (Zabel 2000).

With the emergence of the Industrial Revolution, conditions changed radically. In 1800, the global human population was about 1 billion; today it is ca. 6 billion. This sixfold increase corresponds to an annual growth rate of 0.9% per year. At this rate, the global population doubles every 80 years. A drastic decrease in mortality, improved water quality, diet, sanitary conditions, and medicine have all contributed to this explosive rate of growth. Today, life expectancy in industrialized regions is twice as high as in preindustrial areas of the world (75 years vs. 35 years).

As a consequence of industrial development, productivity has increased enormously in terms of physical and monetary output per hour of work. Thus, those who were fortunate to benefit from formal employment worked roughly half as many hours as workers and farmers at the beginning of the Industrial Revolution. The annual work effort is somewhere between 1500 and 2000 hours per annum in most of the developed countries. The reduction in working time in conjunction with long life expectancy means that those engaged in formal work today have about four times as much "leisure time" over their life time, compared to 200 years ago. Free or leisure time is often expended in a consumptive mode, such as increased mobility. Mobility has increased by three orders of magnitude during the last three centuries. For example, in 1800 an average person in France traveled some 20 m (meters) per day, mostly on horseback and to a much lesser extent by boat; today, average mobility is some 50 km per person per day and is mostly accomplished by car or bus, with about 1 km per person per day by rail and air (Grübler 1998b).

Industrialization has changed the very fabric of human society and existence. An aggregate characterization of the multitude of changes is the gross world (economic) product, which measures the value of all goods and services associated with formal economies. This has grown seventyfold over the last 200 years, corresponding to an annual increase of 2% per year and a doubling in 35 years.

An essential prerequisite for this continuous development process has been the ever-increasing quantity and quality of energy services. In 1800 the world depended on biomass (mostly fuelwood and agricultural waste) as the main energy source for cooking, heating, and manufacturing. Human physical labor and work of animals were the main sources of mechanical energy, with some much more humble contributions of wind and hydraulic power. Figure 15.1 shows how drastically the nature of energy services has been transformed through replacement of traditional (noncommercial) energy sources by coal. In 1850, coal provided some 20% of global primary energy needs and peaked in the 1920s at almost 70% (Nakicenovic et al. 1996). This can be characterized as the first energy transition. The coal age brought railways, steam power, steel, manufacturing, and the telegraph to mention some of the technologies that constituted the "coal cluster."

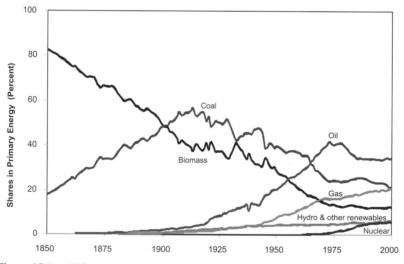

Figure 15.1   Global primary energy requirements since 1850 shown as fractional market shares of the six most important energy sources. Hydropower includes all other renewable energy sources except biomass.

Around 1900, motor vehicles were introduced along with petrochemicals, electricity, and many other technologies that constitute the "oil cluster." It took another 70 years for oil to replace coal as the dominant source of energy in the world. Today, the global energy system is much more complex, with many competing sources of energy and many high-quality and convenient energy carriers ranging from grid-oriented forms, such as natural gas and electricity, to liquids that are mostly used in transportation to solids (coal and biomass) still used in the developing parts of the world (where a third of global population lives that still does not have access to modern energy services). Taken together, fossil energy sources provide some percentage of global energy needs, whereas fuelwood, hydropower, and nuclear energy provide the rest.

Affordable and convenient energy services were one of the primary drivers for the emergence of the Anthropocene. They helped increase human productivity in all sectors, from agriculture to industrial production and provision of services, and replaced human and animal work by abundant sources of power. Consequently, the global primary energy requirement has increased 35-fold during the last 200 years. The energy intensity of economic activities has in fact declined twofold but the 70-fold increase in economic activities required evermore energy. As the share of fossil energy sources, taken together, increased from 20–80% between 1850 and the present, so did the emissions of CO_2 (as an unavoidable byproduct of combustion). Consequently, energy-related emissions of CO_2 increased 22-fold to about 6 billion tons of carbon (6 GtC) today.

Emissions of other radiatively active gases in the atmosphere have accompanied the increase in CO_2. Carbon dioxide concentrations in the atmosphere have increased from about 280 parts per million (ppm) volume in 1750 to over 370 ppm today. Methane concentrations have doubled over the same period. Chlorofluorocarbons (CFCs) are a fundamentally new, anthropogenic constituent of the atmosphere that did not exist in nature. Another indication of the involved complexities is that emissions of sulfur aerosols and particulate matter increase along with energy consumption and emissions of greenhouse gases. Aerosol emissions are now regulated in most of industrialized countries and are declining, but they have resulted in regional cooling, which has offset some of the climate warming caused by increasing concentrations of greenhouse gases.

ENERGY USE, EMISSIONS, AND CLIMATE CHANGE SCENARIOS FOR THE TWENTY-FIRST CENTURY

Scenarios offer a technique for assessing (alternative) evolutionary paths of the Anthropocene and possible response measures and strategies in the face of deep uncontrollable and irreducible uncertainties. Scenario planning is a systemic method for thinking creatively about possible complex and uncertain futures. The central idea of scenario development is to consider a variety of possible futures (Peterson et al. 2003) that include many of the important uncertainties in the human systems rather than to focus on the accurate prediction of a single out-come (which is not possible). Scenarios are images of the future, or alternative futures. They are neither predictions nor forecasts. Rather, each scenario is one alternate image of how the future might unfold. A set of scenarios assists in the understanding of possible future developments of complex coupled systems. Scenarios can be viewed as a linking tool that integrates qualitative narratives or stories about the future and quantitative formulations based on formal modeling. They enhance our understanding of how systems work, behave, and evolve. Scenarios are useful tools for scientific assessments, for learning about complex systems behavior, and for policy making (Jefferson 1983; Davis 1999). In scientific assessments, scenarios are usually based on an internally consistent and reproducible set of assumptions or theories about the key relationships and driving forces of change, which are derived from our understanding of both history and the current situation. Often, scenarios are formulated with the help of numerical or analytic formal models.

The orthodox process of producing climate change projections for the twenty-first century is multistage and multidisciplinary. The drivers of climate change (anthropogenic emissions of greenhouse gases and aerosols as well as land-use change) are derived using socioeconomic and IA models, based on a range of "storylines" regarding population growth, economic development, and technological change. The emissions are then fed into atmospheric chemistry models, which produce corresponding scenarios of changes in the concentrations of greenhouse gases and aerosols for use within GCM climate models (see Steffen et al., this volume). Finally, climate change projections and socioeconomic scenarios are used to assess impacts, adaptation, and mitigation measures and strategies, which are of interest to the policy makers.

Figure 15.2 compares projected future energy requirements across the range of the IPCC SRES scenarios (Nakicenovic et al. 2000), each consistent with a given possible storyline for global economic development. These future IPCC scenarios indicate a sevenfold increase in primary energy requirements by 2100, at the high end of the scale, and almost a twofold increase at the low end. It is interesting to note that the scenarios in the lower range represent sustainable futures with a transition to very efficient energy use and high degrees of conservation. Generally, these are also the scenarios in which energy sources with low carbon intensity play an important role. These are also the scenarios where land-use patterns allow the return of some land to "nature," both because of more sustainable agricultural practices and because of changing dietary practices toward less meat. Scenarios with high shares of renewable energy sources, biomass in particular, would however need substantial shares of nonagricultural land for energy farming and plantations. This is again linked to additional interference in the Earth system. Currently, such complex feedbacks cannot be included in the energy and land-use models, or in IA models, except at the level of explicit assumptions.

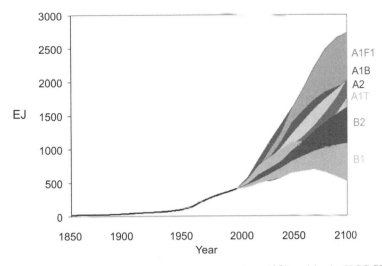

Figure 15.2   Global primary energy requirements since 1850 and in the IPCC SRES scenarios to 2100 (EJ per year). Based on IPCC SRES (Nakicenovic et al. 2000).

Figure 15.3 illustrates how world primary energy structure evolves for each scenario. It shows the contributions of individual primary energy sources — the percentage supplied by coal, oil, and gas, as well as by all nonfossil sources taken together. Each corner of the triangle corresponds to a hypothetical situation in which all primary energy is supplied by a single source — oil and gas at the top, coal to the left, and nonfossil sources (renewables and nuclear) to the right. Historically, the primary energy structure has evolved clockwise according to the two "grand transitions," which are shown by the two segments of the "thick black" curve (see also Figure 15.1). From 1850 to 1920 the first transition can be characterized as the substitution of traditional (nonfossil) energy sources by coal. The share of coal increased from 20% to about 70%, during which time the share of nonfossils declined from 80% to about 20%. The second transition, from 1920 to 1990, can be characterized as the replacement of coal by oil and gas (while the share of nonfossils remained essentially constant). The share of oil and gas increased to about 50% and the share of coal declined to about 30%.

Looking into the future, different possibilities unfold across the scenarios. Figure 15.3 shows the divergent evolution of global primary energy structures across scenarios between 1990 and 2100, regrouped into their respective scenario families and four A1 scenarios groups that explore different technological developments in the energy systems. In a clockwise direction, A1 and B1 scenario groups map the structural transitions toward higher shares of nonfossil energy in the future, which almost closes the historical "loop" that started in 1850. The B2 scenarios indicate a more "moderate" direction of change with about half of the energy coming from nonfossil sources and the other half shared by coal on one side and oil and gas on the other. Finally, the A2 scenario group

marks a return back to coal. This is especially important for those regions of the
world that have ample coal resources (e.g., India and China). Shares of oil and
gas decline while nonfossils increase moderately. What is perhaps more significant than the diverging developments in the four marker scenarios is that the
whole set of 40 scenarios covers virtually all possible directions of change, from
high shares of oil and gas to high shares of coal and nonfossils. In particular, the
A1 scenario family covers basically the same range of structural change as all
other scenarios together. In contrast, the IPCC IS92 scenarios cluster into two
groups: (a) IS92c and IS92d and (b) the four others. In all of these, the share of

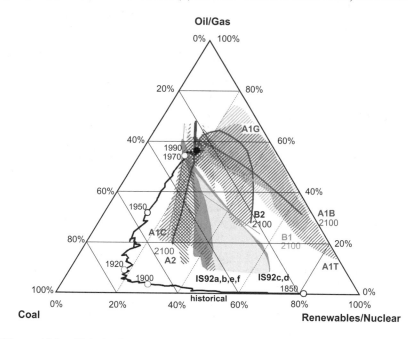

Figure 15.3 Global primary energy structure, shares (%) of oil and gas, coal, and
nonfossil (zero carbon) energy sources: Historical development from 1850 to 1990 and
in SRES scenarios. Each corner of the triangle corresponds to a hypothetical situation in
which all primary energy is supplied by a single source: oil and gas on the top; coal to the
left; nonfossil sources (renewables and nuclear) to the right. Constant market shares of
these energies are denoted by their respective isoshare lines. Historical data from 1850 to
1990 are based on Nakicenovic et al. (1998). For 1990 to 2100, alternative trajectories
show the changes in the energy systems structures across SRES scenarios. They are
grouped by shaded areas for the scenario families A1, A2, B1, and B2 with respective
markers shown as lines. In addition, the four scenario groups within the A1 family (A1,
A1C, A1G, and A1T) explore different technological developments in the energy systems and are shaded individually. For comparison, the IPCC IS92 (Leggett et al. 1992)
scenario series are also shown, clustering along two trajectories (IS92c, d, and IS92a, b,
e, f). For model results that do not include noncommercial energies, the corresponding
estimates from the emulations of the various marker scenarios by the MESSAGE model
were added to the original model outputs. Source: Nakicenovic et al. (2000).

oil and gas declines, and the main structural change occurs between coal and nonfossils. This divergent nature in the structural change of the energy system and in the underlying technological base of the SRES results in a wide span of future greenhouse gas and sulfur emissions.

Here we summarize the related scenarios of CO_2 and SO_2 emissions, although it should be noted that similar projections have been made for other radiatively active species. Figure 15.4 illustrates the range of CO_2 emissions from energy for each of the SRES scenarios, normalized to unity at the current day (Morita et al. 2000). The range of future emissions is very large so that the highest scenarios envisage more than a sevenfold increase of global emissions by 2100, whereas the lowest have emissions that are less than today. Together, the SRES scenarios span jointly from the 95th percentile to just above the 5th percentile of the distribution of energy emissions scenarios from the literature. There is a substantial overlap in the emissions ranges across the 40 SRES

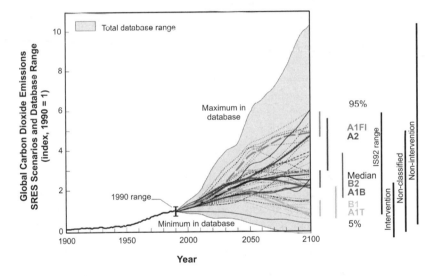

Figure 15.4   Global CO_2 emissions from energy and industry: Historical development from 1900 to 1990 and in 40 SRES scenarios from 1990 to 2100, shown as an index (1990 = 1). The range is large in the base year 1990, as indicated by an "error" bar. The dashed time paths depict individual SRES scenarios and the shaded area the range of scenarios from the literature (see also Figure 15.1). The 5th, 50th (median), and 95th percentiles of the frequency distribution are shown. Jointly, the scenarios span most of the range of the scenarios in the literature. The emissions profiles are dynamic, ranging from continuous increases to those that curve through a maximum and then decline. Colored vertical bars indicate the range of the four SRES scenario families in 2100. Also shown as vertical bars on the right are the ranges of emissions in 2100 of IPCC IS92 scenarios and of scenarios from the literature that include additional climate initiatives (designated as "intervention" scenarios emissions range), those that do not ("non-intervention"), and those that cannot be assigned to either of these two categories ("non-classified"). Source: Nakicenovic et al. (2000).

scenarios. Many share common driving forces, such as population and economic growth, yet lead to a wide range of future emissions. This result is of fundamental importance for the assessments of climate change impacts and possible mitigation and adaptation strategies. It is also important in the broader context of human interference with the Earth system, especially during the future unfolding of the Anthropocene.

Of the greenhouse gases, CO_2 is the main contributor to anthropogenic radiative forcing because of changes in concentrations from preindustrial times. According to Houghton et al. (1996), well-mixed greenhouse gases (CO_2, CH_4, N_2O, and the halocarbons) induced additional radiative forcing of around 2.5 W m^{-2} on a global and annually averaged basis. Carbon dioxide accounted for 60% of the total, which indicates that the other greenhouse gases are significant as well. Whereas CO_2 emissions are attributable primarily to two major sources, energy consumption and land-use change, other emissions arise from many different sources and a large number of sectors and applications.

The SRES emissions scenarios also have different emissions for other greenhouse gases and chemically active species, such as carbon monoxide, nitrogen oxides, and non-methane volatile organic compounds. The uncertainties that surround the emissions sources of the other greenhouse gases, and the more complex set of driving forces behind them, are considerable and unresolved. Therefore, the models and approaches employed for the SRES analyses cannot produce unambiguous and generally approved estimates for different sources and world regions over a century. Emissions of other gases are not shown here; however, they follow dynamic patterns much like those shown in Figure 15.4 for CO_2 emissions (Nakicenovic et al. 2000).

Emissions of sulfur portray even more dynamic patterns in time and space than the CO_2 emissions shown in Figure 15.4. Factors other than climate change (namely regional and local air quality, and transformations in the structure of the energy system and use) intervene to limit future emissions. Figure 15.5 shows the range of global sulfur emissions for all SRES scenarios compared to the emissions range of the IS92 scenarios, 81 scenarios from the literature, and the historical development.

A detailed review of long-term global and regional sulfur emission scenarios is given in Grübler (1998a) and summarized in Nakicenovic et al. (2000). The most important new finding from the scenario literature is recognition of the significant adverse impacts of sulfur emissions on human health, food production, and ecosystems. As a result, scenarios published since 1995 generally assume various degrees of sulfur controls to be implemented in the future and thus have projections substantially lower than previous ones, including the IPCC IS92 scenario series. A related reason for lower sulfur emission projections is the recent tightening of sulfur control policies in the Organization for Economic Cooperation and Development (OECD) countries, such as the Amendments of the Clean Air Act in the U.S.A. and the implementation of the Second European

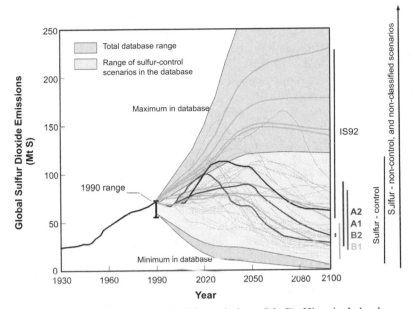

Figure 15.5 Global anthropogenic SO₂ emissions (Mt S): Historical development from 1930 to 1990 and in the SRES scenarios. The dashed colored time paths depict individual SRES scenarios, the solid colored lines the four marker scenarios and two dashed curves the illustrative scenarios, the solid thin curves the six IS92 scenarios, the shaded areas the range of 81 scenarios from the literature, the gray-shaded area the sulfur control and the blue shaded area the range of sulfur noncontrol scenarios or "non-classified" scenarios from the literature that exceeds the range of sulfur control scenarios. The colored vertical bars indicate the range of the SRES scenario families in 2100. Source: Grübler (1998a) and Nakicenovic et al. (2000).

Sulphur Protocol. Such legislative changes were not reflected in previous long-term emission scenarios, as noted in Alcamo et al. (1995) and Houghton et al. (1995). Similar sulfur control initiatives due to local air quality concerns are beginning to impact sulfur emissions also in a number of developing countries in Asia and Latin America (see IEA 1999; La Rovere and Americano 1998; Streets and Waldhoff 2000; Streets et al. 2000). The scenarios with the lowest range project stringent sulfur control levels that lead to a substantial decline in long-term emissions and a return to emission levels that prevailed at the beginning of the twentieth century. The SRES scenario set brackets global anthropogenic sulfur emissions between 27 to 169 Mt S by 2050 and between 11 and 93 Mt S by 2100.

Reflecting recent developments and the literature, it is assumed that sulfur emissions in the SRES scenarios will also be controlled increasingly outside the OECD. As a result, both long-term trends and regional patterns of sulfur emissions evolve differently from carbon emissions in the SRES scenarios. Global sulfur emissions rise initially to decline even in absolute terms during the second half of the twenty-first century, as indicated by the median of all scenarios in Figure 15.5. The spatial distribution of emissions changes markedly. Emissions in the OECD countries continue their recent declining trend (reflecting the tightening of control measures). Emissions outside the OECD rise initially, most notably in Asia, which compensates for the declining OECD emissions. Over the long term, sulfur emissions decline throughout the world; however, the timing and magnitude vary across the scenarios. An important implication of this varying pattern of sulfur emissions is that the historically important, but uncertain negative radiative forcing of sulfate aerosols may decline in the long term. This view is also confirmed by the model calculations reported in Subak et al. (1997), Nakicenovic et al. (1998), and Smith et al. (2000) based on recent long-term greenhouse gas and sulfur emission scenarios.

The alternative developments of the energy systems structure and land-use patterns in the future imply the need to develop a whole host of new energy technologies, and have different implications for climate change. Figure 15.6 is from the IPCC Third Assessment Report (IPCC 2001). It shows projections of changes in the mean global surface temperature for the twenty-first century, based on the various SRES storylines and a range of GCM climate sensitivities. The full range of projections (as indicated by the light grey shading) is approximately half due to divergences among the SRES scenarios, and half due to uncertainties in GCM responses. Neglected feedbacks, such as those between climate and the carbon cycle (Cox et al. 2000; Jones et al. 2003), are likely to extend the range of uncertainty still further.

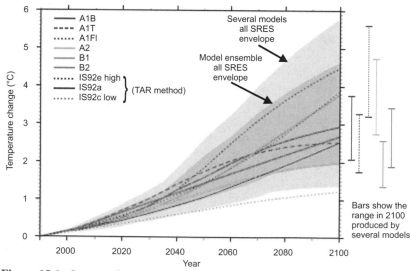

Figure 15.6 Impact of uncertainties in GCMs and emissions scenarios on projected twenty-first century temperature rise (IPCC Third Assessment Report). About half of the uncertainty in the range of future temperature change is due to the range of emissions scenarios while the other is due to climate models.

NEW TOOLS REQUIRED TO GUIDE POLICY
IN THE ANTHROPOCENE

The methodology outlined above has been remarkably successful in coordinating different research disciplines to produce projections of anthropogenic climate change. This approach, however, is not capable of answering some of the most critical questions posed by scientists and policy makers.

Policy makers require information on national and regional levels, which will continue to drive comprehensive GCMs to higher and higher spatial resolution. This branch of model development is typified by the Japanese Earth Simulator Centre, which contains the largest supercomputer in the world, built largely to run century-scale climate simulations at a spatial resolution of 10 km (equivalent to the resolution of a mesoscale weather forecast model, and to be compared with the 250 km resolution of typical GCMs).

The scientific community is also being asked questions that cannot be addressed purely through increases in resolution or integration across approaches:

- *What are the uncertainties in the projections of climate change?* This question has typically been addressed qualitatively simply by comparing the scatter of results from a range of GCMs; however, this does not supply the necessary information for risk analysis, etc. Attempts are now underway to produce a "probability distribution function (PDF)" of future climate using a given GCM with a whole range of plausible internal parameters. The future projections of each model variant are then weighted according to their ability to reproduce key features of the current climate. Such "physics ensembles" require in excess of 50 members, so this approach requires that further advances in CPU are at least partly devoted to assessing uncertainty in models of the existing resolution, rather than committed to further increases in resolution with a "single shot" model. As future climate change is co-determined by human actions (e.g., as described by the ranges of emissions scenarios), treatment of climate uncertainties needs to include future developments in complex coupled systems of socioeconomic and Earth system that are unpredictable (among other reasons due to as yet unknown nature of human response strategies) in addition to being compounded by deep and irreducible scientific uncertainties.
- *Where are the critical points that define "dangerous climate" change?* This requires the ability to explore model phase spaces much more completely than before (since thresholds by definition occupy small parts of phase space, we have a "needle in a haystack" problem). To utilize the paleodata constraints, the Earth system model also needs to be able to run over long periods of the past. What is "dangerous" will also depend on human response capacities to changes in the Earth system and is thus also

dependent on socioeconomic conditions and progress in science and technology.

- *How will feedbacks between the biophysical and human dimensions of the Earth system influence global change?* Recently, a wide range of "integrated assessment" models have been developed that either link state-of-the-art submodels described in the previous paragraph or take a new more simplified approach of capturing only the most essential relationships. Considerable advances have been made; however, there is still a significant gap between the capacity to assess response strategies directed at reducing emissions and to assess possible impacts and adaptation to changes in Earth system and climate change in particular.

To address these new questions new tools are required:

- *A traceable spectrum of Earth system models*: Impact assessments demand higher-resolution models, whereas estimates of uncertainty and exploration of critical thresholds require faster models (to explore model parameter spaces more completely). The continuing development of the entire spectrum of Earth system models — from conceptual models (which can be used to develop qualitative understanding and to guide the development of more complex models), to intermediate complexity models (for approaching past climate data constraints), to comprehensive GCMs (for detailed projections) — is therefore critical (see Figure 15.7). Although models of all these types already exist, there is now an urgent need to relate them to one another in a more rigorous manner. Only then can we have a traceable spectrum in which findings in a model at one complexity can be used to inform other parts of the modeling spectrum (e.g., where a lower-resolution model can be used to estimate error bars on a high-resolution projection).

- *Earth system models with interactive socioeconomics*: It is clear that socioeconomics is driving change in the Anthropocene, but it is unclear to what extent the human dimension needs to be treated as a fully interactive component of Earth system models. Environmental change can certainly feed back on regional socioeconomic factors (e.g., local population, migration) through impacts on humans (e.g., changes in water availability), but to what extent can such regional feedbacks affect the global socioeconomic scenarios (e.g., GHG emissions) and thereby feedback on the global Earth system? To address this will require the development of Earth system models with fully interactive socioeconomics (see Figure 15.8). This, in turn, necessitates IA models with much higher scales of spatial and temporal resolution in driving forces and, even more critically, in human response measures and strategies. The challenges include the treatment of deep uncertainties that are likely to increase with higher model resolution and emergent properties of coupled complex systems. For example, such IA models would need to treat endogenously learning processes and

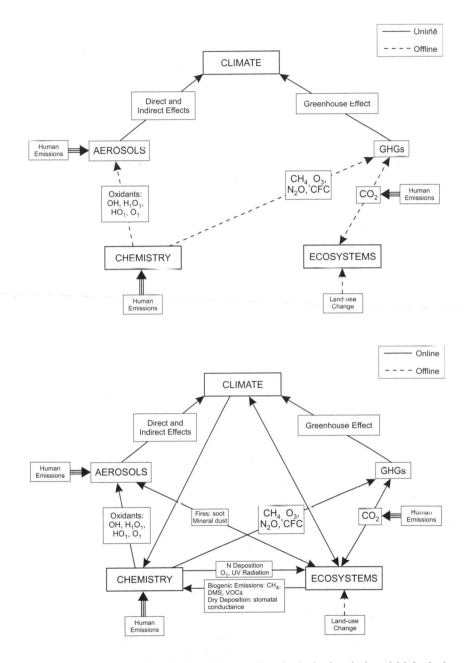

Figure 15.7 Modeled interactions between the physical, chemical, and biological components of the Earth system as modeled in GCMs previously (top) and in the near future (bottom). DMS: dimethyl sulfide; VOCs: volatile organic compounds; GHGs: greenhouse gases; CFC: chlorofluorocarbon.

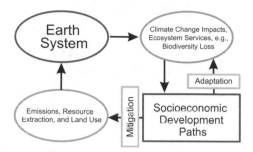

Figure 15.8 Schematic illustration of an integrated modeling framework that includes the future evolution of the Anthropocene characterized by alternative socioeconomic development paths coupled to the changes in the Earth system through mitigation and adaptation measures and strategies in response to impacts such as climate change and biodiversity loss. Adapted from IPCC (2001).

human capacity building in addition to evolution of physical dimensions of the Anthropocene such as the embodied capital and infrastructures.

- *Earth system monitoring*: Data-model fusion, utilizing both ground-based and remotely sensed data, has the potential to produce a more complete diagnosis of the Earth system than has thus far been achieved. More comprehensive data gathering and monitoring of human dimensions of the Anthropocene at higher levels of resolution would help improve IA and socioeconomic models. Routine monitoring of the vital signs of the Earth system will help to constrain projections of the future (because the current Earth system state is better known and because its "observed" evolution can be used to test and develop models) in a way that model development alone is unlikely to achieve.

REFERENCES

Alcamo, J., A. Bouwman, J. Edmonds et al. 1995. An evaluation of the IPCC IS92 emission scenarios. In: Climate Change 1994, Radiative Forcing of Climate Change and An Evaluation of the IPCC IS92 Emission Scenarios, pp. 233–304. Cambridge: Cambridge Univ. Press.

Cox, P.M., R.A. Betts, C.D. Jones, S.A. Spall, and I.J. Totterdell. 2000. Acceleration of global warming due to carbon-cycle feedbacks in a coupled climate model. *Nature* **408**:184–187.

Davis, G.R. 1999. Foreseeing a refracted future. *Scenario & Strategy Planning* **1**:13–15.

Grübler, A. 1998a. A review of global and regional sulfur emission scenarios. *Mitigation & Adaptation Strategies for Global Change* **3**:383–418.

Grübler, A. 1998b. Technology and Global Change. Cambridge: Cambridge Univ. Press.

Houghton, J.T., L.G. Meira Filho, B.A. Callander et al., eds. 1996. Climate Change 1995. The Science of Climate Change. Contribution of Working Group I to the Second Assessment Report of the IPCC. Cambridge: Cambridge Univ. Press.

Houghton, R.A. 2003. Revised estimates of the annual net flux of carbon to the atmosphere from changes in land use and land management 1850–2000. *Tellus* **55B**:378–390.

IEA (International Energy Agency). 1999. Non-OECD coal-fired power generation; Trends in the 1990s. London: IEA Coal Research.

IPCC (Intergovernmental Panel for Climate Change). 2001. Climate Change 2001: The Scientific Basis. Working Group I Contribution, Third Assessment Report of the IPCC, ed. J.T. Houghton, Y. Ding, D.J. Griggs et al. Cambridge: Cambridge Univ. Press.

Jefferson, M. 1983. Economic uncertainty and business decision-making. In: Beyond Positive Economics, ed. J. Wiseman, pp. 122–159. London: Macmillan.

Jones, C.D., P.M. Cox, R.L.H. Essery, D.L. Roberts, and M.J. Woodage. 2003. Strong carbon cycle feedbacks in a climate model with interactive CO_2 and sulphate aerosols. *Geophys. Res. Lett.* **30**:1479.

La Rovere, E.L., and B. Americano. 1998. Environmental Impacts of Privatizing the Brazilian Power Sector. Proc. Intl. Assoc. of Impact Assessment Annual Meeting, Christchurch, New Zealand, April 1998.

Leggett, J., W.J. Pepper, and R.J. Swart. 1992. Emissions scenarios for IPCC: An update. In: Climate Change 1992. The Supplementary Report to the IPCC Scientific Assessment, ed. J.T. Houghton, B.A. Callander, and S.K. Varney, pp. 69–95. Cambridge: Cambridge Univ. Press.

Morita, T., N. Nakicenovic, and J. Robinson. 2000. Overview of mitigation scenarios for global climate stabilization based on new IPCC emissions scenarios (SRES). *Env. Econ. & Policy Stud.* **3**:65–88.

Nakicenovic, N., A. Grubler, H. Ishitani et al. 1996. Energy primer. In: Climate Change 1995. Impacts, Adaptations and Mitigation of Climate Change: Scientific Analyses, ed. R. Watson, M.C. Zinyowera, and R. Moss, pp. 75–92. Cambridge: Cambridge Univ. Press.

Nakicenovic, N., A. Grübler, and A. McDonald, eds. 1998. Global Energy Perspectives. Cambridge: Cambridge Univ. Press.

Nakicenovic, N., J. Alcamo, G. Davis et al. 2000. Special Report on Emissions Scenarios of Working Group III, IPCC. Cambridge: Cambridge Univ. Press.

Peterson, G.D., G.S. Cumming, and S. Carpenter. 2003. Scenario planning: A tool for conservation in an uncertain world. *Conserv. Biol.* **17**:358–366.

Prentice, I.C., G.D. Farquhar, M.J.R. Fasham et al. 2001. The carbon cycle and atmospheric carbon dioxide. In: Climate Change 2001: The Scientific Basis. Working Group 1 Contribution, Third Assessment Report of the IPCC, ed. J.T. Houghton, Y. Ding, D.J. Griggs et al., pp. 183–237. Cambridge: Cambridge Univ. Press.

Smith, S.J., T.M.L. Wigley, N. Nakicenovic, and S.C.B. Raper. 2000. Climate implications of greenhouse gas emissions scenarios. *Technol. Forecasting & Soc. Change* **65**: 195–204.

Streets, D.G, N. Tsai, S. Waldhoff, H. Akimoto, and K. Oka. 2000. Sulfur Dioxide Emission Trends for Asian Countries 1985–1995. Proc. of the Workshop on the Transport of Air Pollutants in Asia, July 22–23, 1999, IIASA, Laxenburg, Austria. Revised sulfur dioxide emissions in Asia in the period 1985–1997. *Atmosph. Env.* **35**:4281–4296.

Streets, D.G., and S. Waldhoff. 2000. Present and future emissions of air pollutants in China: SO_2, NO_x, and CO_2. *Atmosph. Env.* **34**:363–374.

Subak, S., M. Hulme, and L. Bohn. 1997. The Implications of FCCC Protocol Proposals for Future Global Temperature: Results Considering Alternative Sulfur Forcing. CSERGE Working Paper GEC-97-19. Norwich: CSERGE, Univ. of East Anglia.

Vitousek, P.M., J.D. Aber, R.W. Howarth et al. 1997. Human alteration of the global nitrogen cycle: Sources and consequences. *Ecol. Appl.* **7**:737–750.

Zabel, G. 2000. Population and Energy, August 2000. http://dieoff.org/

Back: Nebojsa Nakicenovic, Ulrich Cubasch, Billie Turner, Peter Cox, and Will Steffen
Front: Bert Bolin, Liana Talaue-McManus, Paul Crutzen, and Andi Andreae

16

Group Report: Earth System Dynamics in the Anthropocene

W. STEFFEN, Rapporteur

M. O. ANDREAE, B. BOLIN, P. M. COX, P. J. CRUTZEN,
U. CUBASCH, H. HELD, N. NAKICENOVIC,
L. TALAUE-MCMANUS, and B. L. TURNER II

THE EARTH SYSTEM IN THE ANTHROPOCENE

Human-driven changes to many features of the Earth system have become so ubiquitous and significant in magnitude that a new era for the planet — the "Anthropocene" — has been proposed (Crutzen and Stoermer 2001; Clark et al., this volume). Many of these changes are large in magnitude at the planetary scale, sometimes even exceeding natural flows in major aspects of biogeochemical cycling. In addition, anthropogenic changes invariably occur at rates that are much larger than those of natural variability, often by an order of magnitude or more. The magnitudes and rates of these changes, coupled with the fact that changes to a large number of Earth system processes and compartments are occurring simultaneously, has led to the recognition that the Earth is now operating in a "no-analogue state" (Steffen et al. 2004).

The features of this no-analogue state present challenges to human responsiveness, challenges that have not been experienced in coping with any previous environmental changes, which have occurred at local and regional scales. These features include the facts that:

- Large parts of the problem are global in scale, transcending any region, continent, or ocean basin on its own.
- Connectivity between different biophysical processes and geographical areas of the planet is much greater than previously thought, and the connectivity of human activities is increasing at a rapid rate.
- Human-driven changes to Earth system functioning operate on very long timescales, where the consequences of some human actions may be present for decades and centuries, and across very large space scales, where causes are spatially de-linked from consequences.
- The impacts of global change on human–environment systems can no longer be understood by simple cause-effect relationships, but rather in terms of cascading effects that result from multiple, interacting stresses.

- New forcing functions arising from human actions (e.g., synthetic chemicals) and rapid rates of change imply that the natural resilience of ecological systems may not be sufficient to cope with the change.

These features of the Anthropocene are already leading to discernible changes in the functioning of the Earth system and may well lead to accelerating change in many ways throughout this century (IPCC 2001; Steffen et al. 2004). Significant improvements in the ability to observe, understand, and simulate past, contemporary, and potential future change are essential to provide the knowledge base required to achieve global sustainability.

The aim of this report is to examine a few critical areas where human activities are having, or have the potential to have, significant impacts on the functioning of the Earth system. We begin by focusing more closely on the climate component of the Earth system before moving to broader considerations of geography and to a discussion of the potential for abrupt changes in several components of the Earth system. A brief discussion of the challenges for modeling and observation follows. We conclude by examining whether or not technical substitution or fix can address the issues raised earlier in the report.

CLIMATE SENSITIVITY

How does the increasing understanding of the role of aerosols change our understanding of the climate sensitivity to greenhouse gases?

Climate has been changing during the last century. The IPCC (2001) has established that the global mean temperature has increased by $0.6 \pm 0.2°C$. About half of this increase has occurred during the last 40 years or so. It is also clear that this change is unique as compared with the rest of the last millennium (Mann et al. 1999), although there are some uncertainties about the variations of the mean temperature during this period as deduced from the many different indicators that have been used.

The IPCC has also concluded that the change during the twentieth century cannot be explained without including the role of increasing concentrations of greenhouse gases in the atmosphere during this period of time. Although few in the scientific community now contest this conclusion, there are still occasional claims that the recent warming is the result of internal natural variations of the climate system. To accept this as a plausible possibility, the following question must be answered: Why did a major change of the internal variability occur toward the end of the twentieth century? Further, such changes must also be associated with changes of the internal fluxes of energy through the atmosphere. A plausible analysis to support the idea that random internal variations would be the prime reason for the increase of the global mean temperature during the twentieth century has yet to be presented. The conclusion that human activities have led to a significant change of climate, characterized to first order by a global warming, is now beyond reasonable doubt.

The observed increase in the global mean surface temperature is the net result of opposing forcings and effects. Greenhouse gases exert a warming forcing that is modified by feedbacks, most prominently as a result of changes in atmospheric water vapor and clouds. Aerosols have direct, radiative cooling and warming effects as well as a series of indirect effects due to aerosol-induced changes in cloud properties, abundance, and dynamics. Large uncertainty is associated both with the magnitude of the enhancement of greenhouse gas forcing (warming), resulting from cloud feedbacks, and the (net cooling) aerosol effects, including the various aerosol–cloud effects. Consequently, the observed temperature increase could be explained by a large greenhouse gas effect (implying a large greenhouse gas–cloud feedback), which is opposed by a large aerosol–cloud effect, or alternatively by small greenhouse gas–cloud and aerosol–cloud effects. At present, these alternatives yield nearly the same solution for the interpretation of the observed climate history. The magnitude of the greenhouse gas–temperature–cloud feedback is a long-standing uncertainty. The magnitude of the cooling owing to aerosol-related effects is also difficult to determine directly because of the inhomogeneous distribution of aerosols, as well as clouds, and limited knowledge about their optical characteristics.

There is, however, a very important difference in the way these alternatives affect future climate change. "Climate sensitivity" is defined as the amount of climate change per amount of radiative forcing (greenhouse gas, aerosol) added to the atmosphere (usually expressed as degrees of global temperature rise per doubling of CO_2 in an equilibrium climate model run). At present, different climate models predict very different climate sensitivities, mostly because they contain different ways of representing greenhouse gas–cloud feedbacks, and there appears to be no *a priori* way of deciding which of these models gives the "better," more accurate answer regarding climate sensitivity. The net effect of the present-day greenhouse gas and aerosol forcings (including delayed effects due to latency in the system) is simply the observed present-day climate change. Therefore, the potentially large aerosol effects (as summarized by Anderson et al. 2003) would imply that climate sensitivities are more consistent with the high end of the range presented in IPCC (2001).

A quantitative analysis, based on Crutzen and Ramanathan (this volume), of the current situation illustrates the qualitative point made above. The enhanced concentrations of greenhouse gases in the atmosphere have reduced the outgoing long-wave radiation by 2.7 ± 0.5 W m^{-2} (IPCC 2001). This energy flux must be balanced by (a) increases in outgoing radiation due to warming at the Earth's surface, (b) flux of heat into the Earth's surface, primarily the surface ocean, and (c) the sum of aerosol radiative effects and albedo feedbacks among aerosols, greenhouse gases, and clouds. We can deduce the approximate outgoing flux of energy from the observed warming at the surface of the Earth (Ramanathan 1981) to be 1.5 ± 0.5 W m^{-2}. The smaller value of the range (1.0 W m^{-2}) includes the role of feedback mechanisms, primarily due to water vapor, which reduce the long-wave radiation to space (water vapor feedback), whereas the large value (2.0 W m^{-2}) is obtained for an atmosphere without a water vapor feedback. The increase of the temperature in the atmosphere is also driving a flux of heat into the oceans of about $0.32 \pm 0.15 \text{ W m}^{-2}$ (Levitus et al. 2000; Barnett et al. 2001). To achieve a balance of the energy budget, aerosol–greenhouse gas–cloud interaction processes must induce a net flux of radiation back to space, ranging from about 0 W m^{-2} in the case of no water vapor feedbacks, which is known to be unrealistic (IPCC 2001), to $1.9 \pm 0.5 \text{ W m}^{-2}$, corresponding to 60% of the greenhouse gas forcing, if the feedback mechanisms due to water vapor, etc. are considered. In effect, the cooling that results from aerosol-related processes counteracts a significant increase of surface temperature that would have otherwise occurred in their absence. It is important to note that this quantitative analysis is valid for the situation around the year 2000 and does not scale into the future as both greenhouse gases and aerosol loadings in the atmosphere will change with time.

This result has a major consequence for estimates of future climate change. Atmospheric aerosol loadings will not likely increase strongly; they may even decline in the coming decades. Thus, their cooling effect will level off or decrease. On the other hand, greenhouse gases will continue to accumulate in the atmosphere. In the case of high climate sensitivity, this must lead to a considerably sharper increase in global temperatures than has been experienced so far and to temperature increases closer to the upper end of the range given by the IPCC Third Assessment Report (2001), even without extreme emission scenarios.

It is evident that great effort should be invested in resolving these issues. To make progress in this direction, we need (a) to improve the representation of cloud effects in general circulation models (GCMs), (b) to develop parameterizations of aerosol effects on clouds and find ways to incorporate them into GCMs, and (c) to use the analysis of parameters besides temperature (precipitation, heat fluxes, etc.) and of the spatiotemporal distribution of climate change to diagnose the relative contributions of greenhouse gas and cloud effects. In addition, a much more systematic, consistent, and continuous climate observing system is required to test models and to improve understanding of the climate system in general.

EARTH SYSTEM GEOGRAPHY IN THE ANTHROPOCENE

What are the important regions in the Earth system and in what ways? Do midlatitudes really matter compared to the tropics and high latitudes? Can we differentiate geographically between drivers and impacts of global changes?

The Earth's surface is highly heterogeneous, and the distribution of humans and our activities are highly skewed, features that have important implications for the functioning of the Earth system in the Anthropocene. The implications of heterogeneity vary, however, according to the question being asked. For example, very different subglobal patterns of important areas, or *hot spots*, emerge from analyses of the physical climate system and of the socioeconomic sphere of the Earth system, respectively. For any aspect, however, understanding of the Earth system is only as good as its least understood region, implying that the scientific effort must be much better distributed around the globe than in the past.

In terms of *socioeconomic aspects*, an important feature is that the areas that are currently important as drivers of change (the midlatitudes) are not necessarily those that will largely bear the brunt of global change impacts (e.g., Shah 2002) and that maintain critical processes for Earth system functioning (the tropics, e.g., the role of tropical forests in heat and water vapor exchange). Yet understanding, either in a biophysical or a socioeconomic sense, is far less for the tropics than for the midlatitudes, and the disparity in research effort appears to be increasing. The implications of this for the quest for global sustainability are significant, as projections of demographic change suggest that by 2030 about 90% of the population will live in the tropics (UN 2001). The increasing connectivity of the global economy (e.g., through production–consumption chains of key commodities like food) are linking the tropics more tightly to the midlatitudes so that impacts in the tropics will reverberate further in the Earth system.

The impacts of human-driven change in the tropics for the functioning of the Earth system are equally less understood in comparison to the midlatitudes. Much contemporary land-cover change is occurring at local and regional scales in the tropics, usually involving conversion of forest to agriculture and pasture as well as secondary regrowth that is subsequently recut. Nearly all of the rapid change to the structure of the coastal zone ecosystems is also occurring in the tropics. Understanding of the implications of these changes for biogeochemical cycling, biodiversity, and the physical climate system considerably lags behind that for similar changes in the midlatitudes. One exception is the Large-scale Biosphere-Atmosphere Experiment in Amazonia (Nobre et al. 2002), where a decade-long, multinational study involving hundreds of researchers is rapidly building a better understanding of the dynamics and consequences of land-cover change in the Amazon Basin, from the local up to the global scale. Many other examples could be given, all pointing to the need for a significantly enhanced research effort in the tropics in all aspects of global change.

For the *physical climate system*, a few well-known hot spots have received considerable attention by the research community. Examples include El Niño research in the tropical Pacific Ocean and a rapidly increasing number of investigations of the potentially critical branch of the thermohaline circulation (THC) in the North Atlantic Ocean. The Southern Ocean, however, has been much less studied. Issues such as deepwater formation and its role in THC and the relative importance of the Southern Ocean in the marine carbon cycle demand increased attention.

A strong case can also be made for an enhanced effort in the high latitudes, which are now experiencing the most rapid rates of climate change and which also play an important role in the Earth system through the albedo–ice feedback, the taiga–tundra feedback, and the potentially large releases of carbon compounds from the terrestrial biosphere with increased warming. All of these potential feedbacks (e.g., possible disappearance of Arctic sea ice in summer, movement of boreal forests northward) are positive, that is, they enhance the warming that triggered them and could occur within a 50- to 100-year time frame. This suggests that a concerted effort is required to improve the knowledge base of high-latitude systems under rapid warming; the need is particularly acute for northern Eurasia, which is clearly a very important region from the perspective of Earth system dynamics and which currently suffers from decaying scientific infrastructure and a lack of adequate support for the large scientific community that has worked there through much of the previous century.

On a longer time frame (a few millennia), an intriguing question concerns the sensitivity of the area north of 65°, particularly in North America, as it is known to be the site of glacial inception and thus might be the starting point of the next ice age. Although solar output is currently increasing slightly, and thus is a minor contributor to the observed warming, solar insolation will reach a minimum during the next 200 to 500 years as a result of orbital variation. Model experiments suggest that in a 280 ppm CO_2 world, such insolation conditions could lead to a formation of an ice sheet, although this would almost surely require significant additional cooling due to, for example, injection of massive amounts of aerosols into the atmosphere from volcanoes or anthropogenic activities. However, given that the minimum in solar insolation is rather shallow and will not be much different from current insolation (e.g., compared to greenhouse gas forcing), it appears that the projected increased level of CO_2 over the next 200 to 300 years will more than compensate for the insolation minimum. In addition, another study (Loutre and Berger 2000) suggests that changes in orbital forcing alone, without any anthropogenic increase in greenhouse gas concentration in the atmosphere, will lead to a continuation of the present interglacial period for another 30,000 to 50,000 years. Suggestions of glacial inception notwithstanding, it must be clearly stated that the most important issue by far in the high latitudes, in terms of immediacy and rate and magnitude of change, is the current strong warming and the potential for positive feedbacks to the climate system.

"ACHILLES' HEELS" IN THE EARTH SYSTEM

What are the Achilles' heels in the Earth system? Can abrupt changes in the operation of the Earth system be anticipated and predicted? Can those that are most susceptible to triggering by human actions be identified?

Earth's environment shows significant variability on virtually all time and space scales, and thus global change will never be linear or steady under any scenario.

Of particular interest and importance are abrupt changes that can affect large regions of Earth. For example, the paleo-record gives unequivocal evidence of such abrupt climate change in the recent past, such as the Dansgaard/Oeschger (D/O) events that happened in the period 70,000 to 15,000 years ago (Grootes et al. 1993). Although presumably not having been of global scale, the significance of abrupt changes such as D/O events is that (a) they can involve a scale of change, up to 10°C in a decade or so (see Rahmstorf and Sirocko, this volume), which could devastate modern economies should such changes occur in these regions, (b) they have occurred during the time of human occupation of the planet, and (c) they have occurred in regions (western Europe and North America) now heavily populated. Abrupt changes cannot be dismissed as either implausible or irrelevant in terms of spatial or temporal scales.

Furthermore, one abrupt change of a different kind has already occurred. The formation of the ozone hole over Antarctica was the unexpected result of the release of human-made chemicals thought to be environmentally harmless (see Crutzen and Ramanathan, this volume; Schneider et al. 1998). The event was one of chemical instability in the atmosphere rather than an abrupt change in the physical climate system. In addition, it occurred in a far distant part of the planet, well away from origin of the cause. In several ways, humankind was lucky in that the ozone hole could have been global and present through all seasons (Crutzen 1995).

Such evidence of instabilities in the chemical system in the stratosphere and in the THC in the North Atlantic (thought to underlie the D/O and Heinrich events seen in the Greenland ice core records; Ganopolski and Rahmstorf 2001; Clark et al. 2001) gives a warning that human activities could trigger similar or even as-yet unimagined instabilities in the Earth system, in its physical, chemical, or biological components or in coupled human–environment systems.

Abrupt Changes in the Physical Earth System

Initially it may appear an impossible task to anticipate abrupt changes in the Earth system (NRC 2002; cf. Schneider et al. 1998). Abrupt changes, by definition, occupy small regions of a potentially high-dimensional climate phase space, such that it is impractical to search for such changes with a comprehensive model (such as a GCM). However, the special nature of abrupt changes actually makes them amenable to analytical techniques. By "abrupt" we mean changes that occur much more quickly than changes in anthropogenic forcing. In a typical setting, for this to occur requires the existence of multiple equilibria (or "fixed points"), of which the "current" Earth system equilibrium state becomes linearly unstable or even vanishes (such a point in the phase space/forcing diagram is called "bifurcation"). Under these conditions an arbitrarily small perturbation to the formerly stable equilibrium can result in a transition to a different equilibrium state even in the absence of changes in forcing.

The classical example is the ocean's THC, which in its current state transports heat from equator to pole, helping to keep western Europe unusually warm for its latitude. Thermohaline circulation takes warm surface waters to the North Atlantic, where they cool (releasing their heat to the atmosphere), become denser, and sink to depth. Simple models of the THC exhibit both "on" and "off" states with the potential for rapid switching between these states based on the freshwater input to the North Atlantic. The current "on" state can be destabilized by additional freshwater inputs to the North Atlantic, which freshen the surface waters, make them less dense, and inhibit sinking (Rahmstorf 2000). It is hypothesized that such a perturbation arose from the melting of the North American ice sheet, leading to a shutdown of the THC and a cooling of the Northern Hemisphere during the Younger Dryas event 12,000 years ago. Some comprehensive GCMs also suggest that the THC could be similarly shut down by increases in rainfall at high latitudes under greenhouse warming (thereby increasing the freshwater flow in Russian rivers to the Arctic Sea); however, this sensitivity is by no means common to all models.

Thermohaline circulation offers an excellent example of where a possible abrupt change in the Earth system has been anticipated using a combination of models and data. Although the precise timing of a THC shutdown cannot be predicted with any certainty, the topology of the THC phase space is sufficiently well known to inform attempts to monitor for signs of an impending switch to the off state. Furthermore, the transition from one equilibrium state to another (triggered by a bifurcation) is typically preceded by enhanced variability in the THC (Kleinen et al. 2003), offering an additional warning of possible change.

Other aspects of Earth system dynamics are also believed to exhibit multiple equilibrium states and may therefore display abrupt transitions between these equilibria. These include evidence for a transition from a green to an arid Sahara in the mid-Holocene 5,500 years ago (Claussen et al. 1999; deMenocal et al. 2000) and model-derived results which suggest that Greenland can support both ice-covered and ice-free conditions under current CO_2 conditions. These subsystems display "hysteresis" or path-dependence in their response to control variables. Thus, for example, under sustained increases in CO_2 level (equivalent to a 3°C warming over millennia), the Greenland ice sheet is predicted to melt in an irreversible manner (IPCC 2001), such that much lower CO_2 values would be required before it would return.

The generic properties of multiple equilibria, linear instabilities, and bifurcations offer the possibility of cataloguing possible abrupt changes in the Earth system in a much more thorough way than has been achieved to date. In principle, Earth system equilibria can be defined by setting time derivatives to zero within current Earth system models. Linear stability theory requires that only linear terms are kept within the full nonlinear equations, significantly simplifying the analysis. Therefore, the initial cataloguing of possible Earth system instabilities can be based on well-founded analytical and semi-analytical mathematical techniques, potentially providing a map of hot spots in the Earth system where abrupt change is possible. Once an equilibrium has been found in a model, path-continuation numerics (Feudel and Jansen 1992) make it possible to derive automatically a bifurcation diagram. This technique is becoming increasingly feasible even for comprehensive models (Dijkstra 2000).

Another approach is to use the phenomenon of stochastically induced jumps between multiple equilibria. According to Kramer's rule (Gardiner 1994), an increase in noise in a complex system can trigger an abrupt shift from one state to another. The related timescale is determined by the potential well between the equilibrium states and the amplitude of the noise. The interplay between multiple equilibria and noise can amplify an existing periodic forcing ("stochastic resonance"; Gammaitoni et al. 1998) or may trigger an excitable cycle ("coherence resonance"; Pikovsky and Kurths 1997). Stochastic resonance occurs where the period of the forcing matches the time for transitions between alternative equilibrium states of the system. In analogy, coherence resonance occurs where the time for excitation by noise fulfills a certain matching condition with the period of the excited cycle. Stochastic resonance has been suggested as a contributing factor in D/O events (Ganopolski and Rahmstorf 2002).

Instabilities in the Earth system could be explored by subjecting Earth system models to a noise and systematically tuning this noise until a resonance is achieved (defined by a significant amplification of the variability in internal model variables at a characteristic frequency). The resonance would be indicative of multiple equilibrium states, which might yield abrupt changes under anthropogenic forcing, but it would also give insights into the magnitude of the abrupt change and the amount of noise needed to trigger such a state change. Related ideas have already been successfully applied to complex systems (Majda et al. 1999; Fischer et al 2002). In the latter case, the metastable states of a molecular dynamical system were extracted from time series of the stochastically perturbed system. A simlar approach has the potential to yield invaluable insights into abrupt transitions in the Earth system.

Complexity in the Chemistry of the Atmosphere

The stability of chemical systems in the atmosphere is of concern following the discovery of the ozone hole. Tropospheric chemistry is as complex as that in the stratosphere and is of high importance for the health and well-being of humans, as well as the for the functioning of the Earth system. The troposphere is an oxidizing medium, removing compounds emitted naturally by the terrestrial and marine biospheres and pollutants emitted by human activities. It also affects climate in many ways, for example, through the destruction of the potent greenhouse gas methane, CH_4. Without this cleansing ability, a large range of natural and human-made compounds would accumulate in the atmosphere to very high concentrations. The most important of the oxidizing species in the atmosphere is the highly reactive hydroxyl radical, OH.

Because of its short lifetime, the concentration of OH shows large variations in space and time. Models indicate that the regions with the highest abundance of OH are located over the tropics. Therefore, most of the self-cleansing reactions of the troposphere occur in the tropical zone, and this region consequently plays a key role in the regulation of atmospheric composition. In spite of the well-established importance of the OH radical in atmospheric chemistry, measurements of this species are still very sparse. In particular, there are no measurements at all of OH over the tropical continents, where anthropogenic perturbations of the atmospheric oxidant cycle are likely to occur and where they may have the most pronounced effect. Such measurements are urgently needed as a test of our basic understanding of atmospheric photochemistry.

Ecological Complexity and Earth System Functioning

Major anthropogenic activities have manifested their impacts on the global biosphere. Overfishing and eutrophication due to human activities stand out as among the most serious issues threatening the marine biosphere worldwide. Myers and Worm (2003) recently reported that about 90% of the large predatory fish biomass has been removed from the world's oceans, with removal rates being highest with the onset of post-World War II industrial fisheries. Ecosystem impacts include intermediate results of compensation by nontarget fish populations. However, because of accelerated expansion of fishing in the 1980s, fishing pressure exceeded these compensatory mechanisms and has now led to unequivocal evidence of decline in most pelagic and ground fisheries of continental shelves. There is less evidence for oceanic fishing grounds. Given the importance of top-down controls on the dynamics of marine ecosystems, there is the possibility that such overfishing could lead to significant, abrupt changes to marine ecosystems (often called "regime shifts"), which reverberate through to lower trophic levels such as zooplankton, phytoplankton, and bacteria.

Other anthropogenic pressures on the coastal zone have led to abrupt changes (from an Earth system perspective) in the functioning of marine ecosystems. For example, because of its ubiquity, human-dominated waste loading is altering coastal ecosystems on a global scale. This has led to a state of eutrophication, the latter being a biogeochemical response to heavy nutrient loading. Primary producers synthesize organic matter in addition to what is delivered as waste from populations and manufacturing systems. The excess organic matter undergoes oxygen-consuming degradation. From the 1970s to the 1990s, anthropogenic loads of dissolved inorganic nitrogen increased about sixfold to 13.3 Tg (1 Tg = 10^{12} g). Over the same period, dissolved inorganic nitrogen increased fourfold to 1.6 Tg.

There are secondary consequences of eutrophication. Hypoxic zones under certain conditions can release nitrous oxide to the atmosphere during the process of denitrification. This has been documented for the western shelf of India, which obtains dissolved inorganic nitrogen inputs both from seasonal upwelling and from horizontal delivery from land. In the Gulf of Mexico, hypoxia is a major summer feature, but denitrification has not been detected. Competing microbial pathways such as dissimilatory nitrate reduction to ammonium may keep the reactive substrate in the water column.

The ecosystem effects of eutrophication are just beginning to be studied Jackson et al. (2001) argue that historical overfishing, including the removal of suspension feeders because of trawling and other top predators, has resulted in the simplification of trophic and other functional relationships and the microbialization of coastal systems. Phase shifts include the shift from long-lived macrophytes to short-lived epiphytes and the increasing frequency of phytoplankton blooms and cyanobacteria. In sediments, shifts toward heterotrophic microbial processes are evident.

It remains to be seen how overfishing and eutrophication will alter biogeochemical cycles and the resulting global inventories of carbon, nitrogen, phosphorus, and silica. There is, seemingly, consensus that the nearshore estuarine systems most proximal to human populations are carbon sources, being net heterotrophic and microbe dominated. In open shelf and oceanic domains, the systems remain as carbon sinks, being net autotrophic. Despite the apparent capacity of oceanic ecosystems to assimilate the impacts of waste loading and overfishing, governments should consider the imminent collapse of coastal ecosystems as symptoms that demand immediate mitigation.

In contrast to their marine equivalents, terrestrial ecosystems generally lost many of their top predators and underwent trophic pathway simplification several centuries ago. There has not been widespread ecosystem failure as a result. Terrestrial ecologists generally favor a more "bottom-up" view of ecosystem regulation.

There is evidence (Tilman 1999; Loreau et al. 2001) of a relationship between terrestrial biodiversity and aspects of ecosystem functioning, particularly when the biodiversity is expressed in "functional type" terms. However, it appears that quite modest levels of biodiversity are sufficient to maintain processes such as primary production and nutrient cycling at close to maximum levels, and there is no obvious threshold below which loss of ecosystem function or services suddenly occurs.

If such an effect does occur, it is most likely within the radically simplified agricultural systems. Widespread failure of these systems would have dire consequences for human welfare, but not for life on Earth. Agricultural systems not only replace more diverse natural and seminatural systems with a small group of domesticates, they also simplify the landscape when conducted at large scale, and within the agricultural species, the genetic base is progressively narrower.

The argument, largely unsupported by data, is that agricultural systems of low spatial and genetic diversity are more vulnerable to pest outbreaks and environmental change.

Pandemics

Critical breakpoints for the Earth system may also lie in the still very inadequately explored interactions of climate and environmental change, socioeconomic development, and human and animal health. The preeminent feature of the Anthropocene is that human activities have become a geophysical and biogeochemical force that rivals the "natural," nonhuman processes. This implies that major discontinuities in the socioeconomic domain may lead to corresponding disruptions in the biogeochemical/physical domain. An example of such a discontinuity may be the spread of a new disease vector resulting in a pandemic. High population densities in close contact with animal reservoirs of infectious disease make the rapid exchange of genetic material possible, and the resulting infectious agents can spread quickly through a worldwide contiguous, highly mobile human population with few barriers to transmission. Warmer and wetter conditions as a result of climate change may also facilitate the spread of diseases. Malnutrition, poverty, and inadequate public health systems in many developing countries provide large immune-compromised populations with few immunological and institutional defences against the infectious disease. An event similar to the 1918 Spanish Flu pandemic, which is thought to have cost 20 to 40 million lives worldwide at the time, may result in over 100 million deaths worldwide within a single year. Such a catastrophic event, which is not considered to be unlikely by the epidemiological community, might lead to rapid economic collapse in a world economy dependent on fast global exchange of goods and services. In a worst case this might lead to a drastic, and probably long-lasting, change in the way humans affect the Earth system.

Current Knowledge Base on Abrupt Changes

The preceding discussion of the "Achilles' heels" of the Earth system can be summarized as follows:

- It is well established that abrupt changes in major features of Earth system functioning can occur and indeed have occurred. Prominent examples include the D/O events and the formation of the Antarctic ozone hole, which have been regional in scale but may trigger impacts at the global scale.
- It is further known where some of these abrupt changes can occur. In addition to the two examples given above, the switching of northern African vegetation between savannah and desert, the existence or not of Greenland ice cover, and the large regions of permafrost in northern Eurasia are further areas of instability where a part of the Earth system can change relatively rapidly from one well-defined state to another.

- Not all of the potential abrupt changes in all components of the Earth system (climate, chemical, biological, human and their coupling) are known, nor are they likely to be. However, promising techniques exist to identify more of them.
- Beyond knowing that a potential abrupt change might occur, it is more difficult to determine what triggers abrupt changes or how close a system may be to a threshold.
- Both the magnitude and rate of human forcing are important in determining whether an abrupt change is triggered in a system or not. In general, the probability of abrupt changes in complex systems increases with the magnitude and rate of forcing.
- The Earth system as a whole in the late Quaternary appears to exist in two states (glacial and interglacial) with well-defined boundary conditions in atmospheric composition (CO_2, CH_4) and climate (inferred temperature) (Petit et al. 1999). The nature of the controls on the boundary conditions are not known (cf. Watson et al., this volume) nor are the consequences of the present large, ongoing, human-driven excursion beyond these boundaries (e.g., Keeling and Whorf 2000). Model-based exploration of Earth system phase space cannot yet find a third equilibrium state at a warmer, higher CO_2 level than the interglacial (Falkowski et al. 2000).

SYSTEMS OF MODELS AND OBSERVATIONS

What sort of models and data do we need to understand and anticipate Earth system change in the Anthropocene?

The Current State-of-the-Art in Climate Projection

Many critical Earth system characteristics are undergoing rapid change in the Anthropocene, but climate change is the most obvious example of where international research has been organized to address a policy-relevant question. The production of climate change projections for the twenty-first century, as embodied in the assessments of the IPCC, is multidisciplinary (see Figure 16.1). The drivers of climate change (anthropogenic emissions of greenhouse gases and aerosols and land-use change) are derived using socioeconomic models, based on a range of "storylines" regarding population growth, economic development, and technological change. High emissions scenarios assume major technological developments to permit extensive use of nonconventional oil and gas resources. We do not know how plausible such developments might be. The emissions are then used to drive atmospheric chemistry models, which produce corresponding scenarios of changes in the concentrations of greenhouse gases and aerosols for use within climate models. The resulting climate projections are used by impact modelers, who estimate the extent to which the projections will

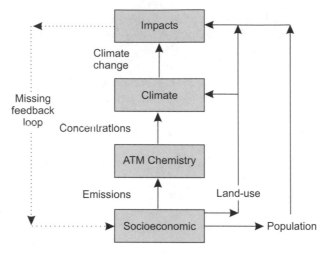

Figure 16.1   Schematic showing the methodology used by the IPCC to produce model projections. Continuous arrows show the inputs and outputs to each model in the chain. Dotted line shows the missing impacts-to-socioeconomic responses loop.

affect humankind (e.g., through climate-driven changes in water and crops, as well as changing demands driven by population growth).

Each stage of this process involves models of some complexity and with widely differing structures. The socioeconomic models operate at large regional scales (e.g., North America, Europe), are not gridded, and are often based on optimization assumptions under equilibrium conditions. By contrast, atmospheric chemistry models and GCMs use a grid (e.g., with boxes of equal size in latitude and longitude) to represent the Earth system and are based on deterministic differential equations. The computational cost of running these models is very high, which limits the resolution they can employ (i.e., the minimum size of the gridboxes) to about 250 km at present. On the other hand, impacts are generally felt at finer scales (e.g., at the scale of a river catchment for hydrology), so it is normally necessary to "downscale" the outputs of the GCM before they can be used to drive impact models, either using statistical techniques or high-resolution regional climate models, which currently operate with gridboxes of about 50 km. The validity of these downscaling models has not yet been well tested.

At each stage of the IPCC modeling process there is a change in the way the Earth system is represented, which leads to difficulties at the interfaces, requiring downscaling, upscaling, or arbitrary definition of the outputs of one model in terms of the inputs to another. In addition, the modeling methodology is "one-way" in the sense that information flows bottom-to-top in Figure 16.1 but not the reverse. This means that the subcomponents of the modeling system do not generally feedback on one another in the way in which the real Earth system operates, which of course is a principal deficiency.

New Tools Required to Guide Policy in the Anthropocene

The methodology outlined in Figure 16.1 has been remarkably successful in co-ordinating different research disciplines to address a key aspect of Earth system change. However, this approach is not capable of answering some of the most critical questions posed by scientists and policy makers. Here we list these questions and suggest the new tools and methodologies that these demand.

1. *How will the coupled Earth system respond to anthropogenic forcing?* As noted previously, the existing climate modeling methodology lacks feedbacks between subsystems of the Earth system. Some recent attempts have been made to include feedbacks between the physical, biological, and chemical parts of the Earth system through a two-way coupling of the various subsystems (e.g., Jones et al. 2003; Johnson et al. 2001). Integrated assessment models also represent the feedbacks between the socioeconomic and natural parts of the Earth system, but they do this at the expense of drastic simplifications in the submodels (e.g., climate may be represented solely by global mean temperature). An intermediate complexity approach is required in which the subcomponents are "traceable" to more comprehensive models but which are sufficiently economical to enable exploration of additional feedbacks.

2. *What are the impacts of climate change at the scale of communities?* This question is difficult to answer because climate models are currently too coarse-grained for regional impacts assessments. Furthermore, impacts are generally determined by climatic extremes (e.g., droughts or floods), and these are not well represented at low resolutions. Higher-resolution climate projections are therefore required, either through embedding regional climate models in GCMs or through basic enhancements in GCM resolution as computer power increases. The latter approach is typified by the Japanese Earth Simulator Centre, which has plans to run global climate projections with a resolution of 10 km, compared to 250 km in current GCMs. Note, however, that such significant increases in resolution may compromise the ability to include the full Earth system feedbacks, to explore abrupt change, and to assess the uncertainty in the projections (see the first and third questions in this section).

3. *What are the uncertainties in the projections?* A key deficiency in climate modeling has been an inability to define "error bars" for projections. Some qualitative measure of uncertainty is given by the spread in results from different GCMs; however, policy makers actually require a more meaningful estimate of the "probability distribution function" (PDF) for future climate. The probability of certain critical thresholds being crossed (e.g., > 2K warming by 2050) is required for risk analysis. Attempts are now underway to define the climate PDF using "physics ensembles," which are made up of structurally identical models each of

which has different plausible sets of internal parameters. Climate projections are then weighted by their ability to reproduce key features of the current climate (with "better models" receiving more weight). This approach is promising but requires many climate model runs (~hundreds) rather than a "one shot" model. There is, therefore, a tension between greater model resolution and sufficient model speed to enable such estimation of uncertainty. A fundamental difficulty remains, however, in that the socioeconomic future of the global society cannot be predicted, since this system is indeed chaotic and in principle unpredictable, except for some overarching features and within some limited period of time. Furthermore, the results of projections of future developments cannot be tested against real data, since in reality there will be only one experiment and we are in the midst of it. In addition, there are still considerable differences in the regional changes of climate as simulated by different models.

4. *Can we reduce uncertainty?* The inclusion of additional feedbacks in the Earth system is likely to increase rather than reduce the spread among model projections, since the additional components provide new ways for the models to differ. However, more complete Earth system models will provide a more realistic (and larger) estimate of the uncertainty in the behavior of the real Earth system. The uncertainty is valuable information in its own right (e.g., for assessing the probability of some abrupt change occurring); however, the fact that it appears to be growing is in danger of being misinterpreted by our paymasters (who may wish to wait for less ambiguous results and conclude that since more money into model development increases uncertainty, less money might have the desired effect!).

Model development alone is unlikely to reduce uncertainty in the foreseeable future, and some uncertainties can never be eliminated since the climate system is chaotic. Still, additional data on changes in the Earth system can constrain models and thereby reduce uncertainty. Thus, there is an urgent need to maintain and develop the monitoring of the Earth system (e.g., through the Global Climate Observing System). A wide spectrum of Earth system quantities needs to be monitored, ranging from the maintenance of historical records (e.g., of riverflow) to the utilization of new satellite data (e.g., CO_2 from space). There is also an urgent need for socioeconomic data of particular interest in the context of climate change (e.g., data on land use and land-use change).

Further Developments in Earth System Modeling

In addition to the developments outlined above, full Earth system models must consider other processes. Dynamics of the biological and human systems of the planet are relevant at the global scale and, through their interactions, must be included in simulations of the dynamics of the whole Earth system. At present, with the focus of many global models on climate change, the primary emphasis in terms of human activities has been on greenhouse gas emissions, land-use change, and aerosol emissions. The influence of biological systems is modeled mainly through their biogeochemical cycles.

There are, however, other important aspects of Earth system dynamics that are not climate related. For example, the growth of the world's population, evolution of technology, transformations of the economy or relevant changes in global lifestyles, and political ambitions occur with or without climate change. Such factors are becoming increasingly global in scale and character. These dynamics in the human part of the Earth system have significant, first-order effects upon the whole system, and future projections of biosphere–sociosphere interactions are undoubtedly crucially important in simulating the future evolution of the Earth system as a whole. Biogeochemically, the material basis of the human economy, which at the core concerns the distribution and redistribution of materials extracted from the physical and biological systems of the world under various constraints, can be treated as an extension of "natural" biogeochemical flows to flows through human systems. As noted in the preceding section, these societal processes are presently included as "given" scenarios external to the model itself and not included in the internal dynamics of the model.

Thus, three types of activities are currently on the near-term agenda to develop more complete Earth system models:

1. Coupling full terrestrial biosphere models (Dynamic Global Vegetation Models, DGVMs) to climate models to capture not just primary feedbacks of the terrestrial biosphere to the climate system but also, in a consistent way, the effects of climate on the terrestrial biosphere (similar models, the so-called "Green Ocean" models, are being developed for putting the marine biosphere into Earth system models).
2. Expansion of DGVMs to incorporate fully human land use, particularly agriculture and water use, including the development of parameters that allow quantification of ecosystem services to society.
3. Coupling of DGVMs through their land-use modules to economic models (including endogenous technology dynamics), themselves perhaps drawing upon models of lifestyle dynamics.

The technical challenges of such model development and coupling are considerable. For example, climate and biosphere models are time-step models, whereas economic models are mostly based on optimization approaches under equilibrium conditions. Economic models are therefore not gridded but rather act upon 10 to 20 world regions, whereas climate and biosphere models are spatially explicit (similar differences occur in the data sets available for parameterizing and driving the models). Coupling requires considerable efforts in downscaling and development of software metastructures for fuzzy information exchange (hard-wired coupling may be less preferable than a "mutual envelope" approach). With respect to the economic system, models of price dynamics have to be interpreted more consciously in terms of material and energy flows, including those that are not currently assigned monetary value (such as use of clean air). In the social sphere, formulating quantified scenarios of lifestyle dynamics seems an urgent task. Progress is being made in all of these fields, but many efforts are still at an early stage. The promise of enhanced understanding is great. For example, such integrated socioeconomic–biophysical models may be used to explore whether gradual changes in the biophysical realm of the Earth system can trigger abrupt changes in the socioeconomic sphere.

Observations designed to monitor the anthroposphere in the context of the Earth system involve the human subsystem as well. Various human dimension and related initiatives have yet to concur on a shortlist of high priority areas that require monitoring, in part because of the large variation in the ways in which different communities perceive the problems inherent in the anthroposphere. Focusing on the immediate or proximate factors that register humanity's demands on the Earth system and resources (e.g., Turner 2002), such a list would include: population variables (e.g., fertility, age structure, rural–urban mix), wealth and changes of behavior associated with changes in wealth (e.g., diet), energy–material consumption and waste emissions by level of economic development location, efficacy of institutional controls on resource–environment issues, and critical land-use/land-cover trajectories. Visionary approaches to building an Earth-observing system focused on socioeconomics are embodied in such projects as the Geoscope (www. sustainability-geoscope.net).

Models of Biodiversity

The biological complexity of the planet also plays a role in the functioning of the Earth system. The capability to predict where, and to what degree, biodiversity is likely to be lost as a result of the combined impact of climate change, land-use change, direct use, and the impact of pollutants is an emerging field. In the climate change field, models have progressed from simple bioclimatic envelope approaches that are applied to whole biomes, to similar models applied to functional types and then individual species (including nonclimatic constraints), to fully dynamic models that track the movement of populations to determine if they can keep up with the rate of change. The next step will be "ecosystem" models, which take into consideration the presence of competitors, mutualisms, food and predator species, as well as habitat structure.

On another track, integrated assessment models aim at expressing the complexity of biodiversity in synthetic index terms (macro-ecological indicator), and then relate changes in this index to various types and intensities of human activities. This allows scenarios to be developed, targets set, and performance to be monitored. Examples include the RIVM natural capital index (ten Brink 2000) and the SafMA biodiversity intactness index (in preparation).

TECHNOLOGICAL FIX AND SUBSTITUTION

How effective will technological substitution be in dealing with increasing impacts of the human–environment relationship on the Earth system?

The preceding sections make a strong case for the necessity of a societal response to global change. A business-as-usual approach to the future will not achieve global sustainability. Prominent among the proposed responses to global change are technological options, ranging from treatment of the fundamental causes of the problem, such as the development of noncarbon-based energy systems, to treatment of the symptoms of the problem through highly controversial geo-engineering approaches.

Throughout history, society has responded in two principal ways to environmental vagaries, flux, hazards, and drawdown, including resource depletion: *move*, either through designed mobility as in pastoral nomadic systems or "forced" relocation owing to environmental or resource degradation as exemplified in the salinization-relocation pattern of irrigation in Mesopotamia (Adams 1965); and *change techno-managerial strategies*, as in the adoption of fossil-fuel energy or genomics (Grübler 1998). The first option has decreased in significance in an evermore crowded and politically controlled world. The second option — to modify or transform biophysical conditions in order to gain a measure of "control" over some portion of the environment or to deliver a substitute for a depleted resource — is not only ancient but has become a defining element of our species (Diamond 1997; Redman 1999; Turner et al. 1990). Such responses are labeled technological fix and substitution. Modern society has raised the bar in pursuit of techno-managerial solutions, with long-standing success in regard to deliveries of food, fuel, and fiber to increasingly larger and highly consumptive populations (Grübler 1998; Kasperson et al. 1995; Kinzig et al., this volume).

This approach to human–environment relations and ensuing problems is the cornerstone of the modern conditions of life, be it the Industrial Revolution or the Green Revolution. Society has become so reliant on ever-increasing advances in technology to overcome the next generation of problems that a disconnect or gap has emerged between the environmental consequences of production and consumption and the public consumer (e.g., Sack 1992). Technological solutions also offer a means to avoid the thorny issues involved in alternative solutions that are often perceived to affect lifestyles.

Science for sustainable development confronts technological fix and substitution in the face of natural and anthropogenic changes in the Earth system, culminating in global environment change. Technology constitutes one of a set of responses to deal with the problems inherent in changes in the Earth system. Indeed, some researchers believe that technological solutions will, in fact, liberate the Earth system of many of its current threats (Ausubel 2000, 2002). As noted above, however, these changes have no known analogues (Steffen et al. 2004), and some of them constitute qualitative shifts in the structure and functioning of the Earth system. These qualities raise a fundamental question and set of sub-questions about the efficacy of technological fix and substitution *alone* to cope with the problems: What evidence exists to indicate that the changes underway in the Earth system constitute a no-analogue situation, not only in the changes themselves but also in the sole use of technological fix and substitution to address these changes?

Furthermore, regarding technological fix and substitution, does the evidence suggest:

- Reduced effectiveness to address changes at the scale of the Earth system?
- Excessive cost to develop and deploy them compared to alternatives (e.g., societal changes or preservation of goods and services of the Earth system)?
- Temporal mismatch between the capacity of the potential fixes to become operative and the increasing environmental problems, with potential abrupt changes?

The antecedents and antiquity of global environmental change notwithstanding, the human–environment condition has entered a new phase that constitutes a qualitative or threshold shift: (a) The capacity of humankind to change directly the biogeochemical cycles that sustain the structure and functioning of ecosystems and the biosphere as a whole. Anthropogenic input into many of these cycles now exceeds nature's input (Steffen et al. 2004; Turner et al. 1990). For example, more nitrogen is now fixed from the atmosphere as a result of human activities than all natural nitrogen-fixing processes in the terrestrial biosphere combined. Technology also introduces new, synthetic compounds into the Earth system. The release of the well-known chlorofluorocarbons (CFCs) is only one of many examples; globally over 100,000 industrial chemicals — many of them unknown in the natural world — are in use today (Raskin et al. 1996). (b) The combination of these emissions has complex, systemic consequences, that is, numerous unforeseen feedbacks with far-ranging consequences and connections invariably leading to "surprises." Perhaps the most dramatic example to date was the formation of the ozone hole over Antarctica.

This qualitative shift in the human impacts on the Earth system generates at least three new conditions to confront technological responses:

1. Earth system changes are global in scale; climate and other environmental changes are taking place worldwide. Whereas these changes vary by region and locale, the historical societal response of moving the location of production and consumption to account for these changes or those of resource depletion or degradation appears to be attenuated. An increasingly occupied and crowded planet reduces new spaces in which to move and fosters more intensive uses of the spaces already occupied.

2. Systemic changes are inherently transboundary, and thus changes in one place affect places far away. For example, the burning of fossil fuels in North America and western Europe probably made a major contribution to drought and subsequent famine and starvation in the Sahel region in the 1970s and 1980s (Rotstayn and Lohmann 2002).
3. Many of the changes currently underway drive processes that operate over long timescales with impacts that will affect the functioning of the Earth system long after the forcing function is relaxed. Examples include the atmospheric emissions of CO_2, whose effects have a lifetime of 50–150 years; the closure of the ozone hole, which despite the reduction in CFC emissions following the Montreal Protocol, is expected to take at least four or five decades, perhaps more, to close fully; and the accumulation of reactive nitrogen compounds in terrestrial and marine ecosystems with consequences that will be played out over century timescales.

To date, humankind has directed technology to environmental problems focused primarily on resource extraction of food, fuel, and fiber, on the reduction in resource stocks (enlarging or changing), or on reducing the consequences of environmental hazards (e.g., drought to floods). The Earth system and the major societal activities affecting it have redirected these characteristics. The impacts of waste from production and consumption, such as CO_2 emissions, are equivalent to or exceed the consequences of resource extraction, including land-cover conversion, and the changes underway have shifted from resource stocks to functioning of ecosystems and the biosphere. It is highly improbable that ecosystems can be significantly altered and their many functions replaced technologically. It is even less probable that technological replacements can be found for the functioning the Earth system as a whole, especially its ability to absorb and process wastes.

The kind of environmental changes underway challenge the historical relationships between technology and environment. Other factors, however, affect this relationship as well.

The temporal dimension of the development and deployment of new technologies varies considerably by case, and the overall process may be accelerating through time (Grübler 1998). The Green Revolution, for example, transpired rapidly; it took no more than thirty years from the founding of research development centers for hybrid crops to dominate the world (Conway and Ruttan 1999). Regardless, changes currently underway in the Earth system are likely to play out over much longer timescales unless technologies of "reversal" are developed (see examples quoted above). In addition, the Earth system could shift in ways that would change the very aims or goals of technological controls. For example, if an abrupt shutdown of the THC in the North Atlantic Ocean leads to no net warming or even cooling, societies in northern Europe would have to abandon plans to change their infrastructure to cope with strong warming.

These characteristics of the Earth system and changes underway indicate that there are few analogues regarding past technological fixes and substitutions. Also, the lock-in of significant growth in human population (Population Reference Bureau 2002) and the near-universal call for increases in per capita consumption within the developing world indicate a world in 2050 that will demand more, not less, from the Earth system (Kates et al. 2001). These conditions require new ways of approaching human–environment problems that deviate from "business as usual" and are capable of provisioning (resources) and conserving (ecosystem–biosphere) more while degrading and changing less (Earth system). The "precautionary principle," uncertainty and surprise, and the no-analogue conditions noted above suggest caution in a solution focused solely on technological fix and substitution and raise consideration of alternatives that address values, institutions, and other societal structures (Kinzig et al., this volume).

These nontechnical solutions need not be necessarily invented anew; various examples exist or are emerging, research on which provides clues for exploration. For example, comparative case study work indicates that sociopolitical structures which facilitate the flow of information among many stakeholders and decision makers tend to encourage learning in such forms as recognition of local and regional threats to environmental systems, a critical step toward any action taken (Kasperson et al. 1995; Social Learning Group 2001). Likewise, structures providing checks and balances on resources and environmental decisions tend to prevent potential threats to extant uses of local and regional ecosystems that might otherwise be inflicted from decisions made from afar. For example, absence of these checks and balances permitted the Soviet government to reduce the Aral Sea ecosystem to near-death conditions, despite local recognition of its demise (Micklin 1988; Kasperson et al. 1995). This observation, however, does not mean that structures promoting strong checks and balances necessarily lead to improved environmental conditions. Finally, it is important to recognize that global structures designed to provide some measures of checks and balances regarding environmental issues constitute a relatively new phenomenon (Young 1999, 2002).

These structures are emerging within a political–economic process labeled globalization in which production, consumption, and information operate in worldwide networks that connect virtually every place. This process is argued by some to attenuate the repercussions of environmental and resource disasters, for example, by marshaling large amounts of food aid to famine areas (Kates and Parris 2003). Alternatively, others claim that it amplifies environmental problems by disconnecting more than ever in human history, the location and impacts of production and consumption, which exacerbates environmental degradation in marginalized locations. The large-scale destruction of the Indonesian forests for the international timber industry (Brookfield et al. 1995; Dauvergne 1997) is a case in point. Less explored is the concept that increasing globalization potentially sets the stage for worldwide collapses of social and environmental systems because the geographical, and in some cases temporal, buffering of subsystems is diminished. In terms of technological substitution, globalization could, in principle, increase the ability of new technologies to diffuse and penetrate more rapidly from their point of development to other regions of the world.

The challenges to technological fix and substitution notwithstanding, technology will constitute part of the solutions directed to environmental problems — global and local — in the future. Indeed, inasmuch as technology is responsible for some of these problems, so can it help to alleviate them. Technological advances promise increasing efficiencies in existing technologies whereas various emerging technologies will likely be critical in the future; these include genomics and biotechnology, nanotechnology and information, as well as "alternative" energy.

RESEARCH CHALLENGES

Significant progress has been made over recent years in understanding the dynamics of the Earth system in the Anthropocene. The complexities of atmospheric composition in influencing the climate system are increasingly well understood; the possibility of abrupt changes in the Earth system is apparent and promising approaches for understanding and anticipating them are being developed; and a suite of Earth system models of varying emphases and complexities is being developed to simulate past, present, and future functioning of the planet. Such progress helps to sharpen the focus of the near-term research effort and leads to a set of research questions to help guide Earth system science over the next five to ten years.

Climate Sensitivity

- What is the quantitative importance of greenhouse gas–aerosol–cloud dynamics in enhancing or counteracting the direct radiative effects of greenhouse gases in the atmosphere?
- What are the radiative and chemical characteristics of aerosol particles, their emission/formation processes, regional and intercontinental dispersion, and deposition on a regional and global basis?
- Can the energy balance at the Earth's surface be closed at the regional and global scales for the Anthropocene? If so, what insight does that give about the climate sensitivity to greenhouse gases?

Earth System Geography

- What strategies are required to achieve a better balance of research and observation effort around the world?

Abrupt Changes

- What is the catalogue of possible abrupt changes in the Earth system resulting from a model-based, systematic exploration of Earth system phase space using equilibrium and stochastic resonance approaches?
- What research approaches can be developed to anticipate abrupt changes in the socioeconomic sphere of complex, coupled human–environment systems?

Models and Observations

- What spectrum of Earth system models is required to examine the wide range of questions associated with Earth system functioning, from exploring critical thresholds and abrupt change to high resolution impacts studies? How can we build a "traceable" spectrum of Earth system models?
- What is the best strategy for developing models that incorporate the human dimension as a fully interactive component of the Earth system?
- How can data-model fusion be developed further to provide a more complete diagnosis of the Earth system? What critical parameters need to be observed routinely to monitor the "vital signs" of Earth system functioning?
- What is the best strategy to test and improve Earth system models in the context of gradually evolving global change punctuated by extreme events in nature and society?

Technological Substitution

- What is the probability that technological change will be able to support the projected global population of 2050 at significantly higher average levels of consumption while reducing the emissions of CO_2, CH_4, and other gases and particles to the atmosphere and slowing down and ultimately stopping the degradation of marine and terrestrial ecosystems?
- Will technological fix and substitution directed to environmental concerns be offset by that directed to other concerns (e.g., economic growth)?
- Which institutional and organizational structures have proven most effective (including public acceptance) in enforcing environmental regulations under different human–environment conditions and different scales of governance (Kinzig et al., this volume)?
- What kinds of programs and policies effectively support the conversion to and maintenance of consumption-production processes (industrial and agricultural) that are more environmentally benign (compared to extant or conventional processes) in both developed and developing countries?

ACKNOWLEDGMENTS

We thank Bob Scholes for his written contributions to this report.

REFERENCES

Adams, R.M. 1965. Land Behind Baghdad: A History of Settlement on the Diyala Plain. Chicago: Univ. of Chicago Press.

Anderson, T.L., R.J. Charlson, S.E. Schwartz et al. 2003. Climate forcing by aerosols — A hazy picture. *Science* **300**:1103–1104.

Ausubel, J. 2000. The great reversal: Nature's chance to restore land and sea. *Technol. Soc.* **22**:289–301.

Ausubel, J. 2002. Maglevs and the vision of St. Hubert, or the great restoration of nature: Why and how. In: Challenges of a Changing Earth: Proc. of the Global Change Open Science Conf., Amsterdam, 10–13 July 2001, ed. W. Steffen, J. Jäger, D. Carson, and C. Bradshaw, pp. 175–182. Berlin: Springer.

Barnett, T.P., D.W. Pierce, and R. Schnur. 2001. Detection of anthropogenic climate change in the world's oceans. *Science* **292**:270–274.

Brookfield, H.C., L. Potter, and Y. Byron. 1995. In Place of the Forest: Environmental and Socioeconomic Transformation in Borneo and the Eastern Malay Peninsula. Tokyo: United Nations Univ. Press.

Clark, P.U., S.J. Marshall, G.K.C. Clarke et al. 2001. Freshwater forcing of abrupt climate change during the last glaciation. *Science* **293**:283–287.

Claussen, M., C. Kubatzki, V. Brovkin et al. 1999. Simulation of an abrupt change in Saharan vegetation at the end of the mid-Holocene. *Geophys. Res. Lett.* **24**:2037–2040.

Conway, G., and V.W. Ruttan. 1999. The Doubly Green Revolution: Food for All in the Twenty-First Century. Ithaca, NY: Comstock Publ.

Crutzen, P. 1995. My life with O_3, NO_x, and other YZO_xs. In: Les Prix Nobel (The Nobel Prizes), pp. 123–157. Stockholm: Almqvist and Wiksell Intl.

Crutzen, P.J., and E. Stoermer. 2001. The "Anthropocene." *Glob. Change Newsl.* **41**: 12–13.

Dauvergne, P. 1997. Shadows in the Forest: Japan and the Politics of Timber in Southeast Asia. Cambridge, MA: MIT Press.

DeMenocal, P.B., J. Ortiz, T. Guilderson et al. 2000. Abrupt onset and termination of the African Humid Period: Rapid climate response to gradual insolation forcing. *Quat. Sci. Rev.* **19**:347–361.

Diamond, J. 1997. Guns, Germs, and Steel: The Fates of Human Societies. New York: Norton.

Dijkstra, H.A. 2000. Nonlinear Physical Oceanography. Atmospheric and Oceanographic Library Series. Dordrecht: Kluwer Academic.

Falkowski, P., R.J. Scholes, E. Boyle et al. 2000 The global carbon cycle: A test of our knowledge of Earth as a system. *Science* **290**:291–296.

Feudel, U., and W. Jansen. 1992. CANDYS/QA-a software system for the qualitative analysis of nonlinear dynamical systems. *Intl. J. Bifurc. & Chaos* **2**:773–794.

Fischer, A., C. Schutte, P. Deuflhard, and F. Cordes. 2002. Hierarchical uncoupling–coupling of metastable conformations. In: Computational Methods for Macromolecules: Challenges and Applications, Proc. 3rd Intl. Workshop on Algorithms for Macromolecular Modeling, ed. T. Schlick and H.H. Gan, Lecture Notes in Computational Science and Engineering Series, vol. 24, pp. 235–259. Berlin: Springer.

Gammaitoni, L., P. Honggi, P. Jung, and F. Marchesoni. 1998. Stochastic resonance. *Rev. Mod. Physics* **70**:223–287.

Ganopolski, A., and S. Rahmstorf. 2001. Rapid changes of glacial climate simulated in a coupled climate model. *Nature* **409**:153–158.

Ganopolski, A., and S. Rahmstorf. 2002. Abrupt glacial climate changes due to stochastic resonance. *Phys. Rev. Lett.* **88**:038501-1–038501-4.

Gardiner, C.W. 1994. Handbook of Stochastic Methods, 2nd ed. Berlin: Springer.

Grootes, P.M., M. Stuiver, J.W.C. White, S. Johnsen, and J. Jouzel. 1993. Comparison of oxygen isotope records from the GISP2 and GRIP Greenland ice cores. *Nature* **366**:552–554.

Grübler, A. 1998. Technology and Global Change. Cambridge: Cambridge Univ. Press.

IPCC (Intergovernmental Panel on Climate Change). 2001. Climate Change 2001: The Scientific Basis. Working Group I Contribution, Third Assessment Report of the IPCC, ed. J.T. Houghton, Y. Ding, D.J. Griggs et al. Cambridge: Cambridge Univ. Press.

Jackson, J.B.C., M.X. Kirby, W.H.Berger et al. 2001. Historical overfishing and the recent collapse of coastal ecosystems. *Science* **293**:629–637.

Johnson, C.E., D.S. Stevenson, W.J. Collins, and R.G. Derwent. 2001. The role of climate feedback on methane and ozone studied with a coupled ocean-atmosphere-chemistry model. *Geophys. Res. Lett.* **28**:1723–1726.

Jones, C.D., P.M. Cox, R.L.H. Essery et al. 2003. Strong carbon cycle feedbacks in a model with interactive CO_2 and sulphate aerosols. *Geophys. Res. Lett.* **30**:1479–1482.

Kasperson, J.X., R.E. Kasperson, and B.L. Turner II, eds. 1995. Regions at Risk: Comparisons of Threatened Environments. Tokyo: United Nations Univ. Press.

Kates, R.W., W.C. Clark, R. Corell et al. 2001. Sustainability science. *Science* **292**: 641–642.

Kates, R.W., and T.M. Parris. 2003. Long-term trends and a sustainability transition. *Proc. Natl. Acad. Sci. USA* **100**:8062–8067.

Keeling, C.D., and T.P. Whorf. 2000. Atmospheric CO_2 records from sites in the SIO air sampling network. In: Trends: A Compendium of Data on Global Change. Carbon Dioxide Information Analysis Center, Oak Ridge Natl. Laboratory. Oak Ridge, TN: U.S. Dept. of Energy.

Kleinen, T., H. Held, and G. Petschel-Held. 2003. The potential role of spectral properties in detecting thresholds in the Earth system: Application to the thermohaline circulation. *Ocean Dyn.* **53**:53–63.

Levitus, S., J.I. Antonov, T.P. Boyer, and C. Stephens. 2000. Warming of the world ocean. *Science* **287**:2225–2229.

Loreau, M., S. Naeem, P. Inchausti et al. 2001. Biodiversity and ecosystem functioning: Current knowledge and future challenges. *Science* **294**:804–808.

Loutre, M.F., and A. Berger. 2000. Future climatic changes: Are we entering an exceptionally long interglacial? *Clim. Change* **46**:61–90.

Majda, A.J., I. Timofeyev, and E. Vanden Eijnden. 1999. Models for stochastic climate prediction. *Proc. Natl. Acad. Sci. USA* **96**:14,687–14,691.

Mann, M.E., R.S. Bradley, and M.K. Hughes. 1999. Northern Hemisphere temperatures during the past millennium: Inferences, uncertainties, and limitations. *Geophys. Res. Lett.* **26**:759–762.

Micklin, P. 1988. Desiccation of the Aral Sea: A water management disaster in the Soviet Union. *Science* **241**:1170–1176.

Myers, R.A., and B.Worm. 2003. Rapid worldwide depletion of predatory fish communities. *Nature* **423**:280–283.

Nobre, C.A., P. Artaxo, M.A.F. Silva Dias et al. 2002. The Amazon Basin and land-cover change: A future in the balance? In: Challenges of a Changing Earth, Proc. of the Global Change Open Science Conf., Amsterdam, NL, 10–13 July 2001, ed. W. Steffen, J. Jäger, D. Carson, and C. Bradshaw, pp. 137–141. Berlin: Springer.

NRC (National Research Council). 2002. Abrupt Climate Change: Inevitable Surprises. Washington, D.C.: Natl. Academy Press.

Petit, J.R., J. Jouzel, D. Raynaud et al. 1999. Climate and atmospheric history of the past 420,000 years from the Vostok ice core, Antarctica. *Nature* **399**:429–436.

Pikovsky, A.S., and J. Kurths. 1997. Coherence resonance in a noise-driven excitable system. *Phys. Rev. Lett.* **78**:775–778.

Population Reference Bureau. 2002. 2002 World Population Data Sheet. http://www.prb.org/pdf/WorldPopulationDS02_Eng.pdf

Rahmstorf, S. 2000. The thermohaline circulation: A system with dangerous thresholds? *Clim. Change* **46**:247–256.

Ramanathan, V. 1981. The role of ocean–atmosphere interactions in the CO_2 climate problem. *J. Atmos. Sci.* **38**:918–930.

Raskin, P., M. Chadwick, T. Jackson, and G. Leach. 1996. The sustainability transition: Beyond conventional development. Polestar Series 1. Stockholm: Stockholm Environment Institute.

Redman, C.L. 1999. Human Impact on Ancient Environments. Tucson: Univ. of Arizona Press.

Rotstayn, L.D., and U. Lohmann. 2002. Tropical rainfall trends and the indirect aerosol effect. *J. Climate* **15**:2103–2116.

Sack, R.D. 1992. Place, Modernity, and the Consumer's World. Baltimore: Johns Hopkins Press.

Schneider, S., B.L. Turner II, and H. Morehouse Garriga. 1998. Imaginable surprise in global change science. *J. Risk Res.* **1**:165–185.

Shah, M. 2002. Food in the 21st century: Global climate of disparities. In: Challenges of a Changing Earth, Proc. of the Global Change Open Science Conf., Amsterdam, 10–13 July 2001, ed. W. Steffen, J. Jäger, D. Carson, and C. Bradshaw, pp. 31–38. Berlin: Springer.

Social Learning Group. 2001. Learning to Manage Global Environmental Risks, 2 vols. Cambridge, MA: MIT Press.

Steffen, W., A. Sanderson, P.D. Tyson et al. 2004. Global Change and the Earth System: A Planet Under Pressure. The IGBP Book Series. Berlin: Springer.

ten Brink, B.J.E. 2000. Biodiversity indicators for the OECD Environmental Outlook and Strategy: A feasibility study. Report 402001014. Bilthoven: RIVM.

Tilman, D. 1999. The ecological consequences of changes in biodiversity: A search for general principles. Robert H. MacArthur Award Lecture. *Ecology* **80**:1455–1474.

Turner, B.L., II. 2002. Toward integrated land-change science: Advances in 1.5 decades of sustained international research on land-use and land-cover change. In: Challenges of a Changing Earth, Proc. of the Global Change Open Science Conf., Amsterdam, 10–13 July 2001, ed. W. Steffen, J. Jäger, D. Carson, and C. Bradshaw, pp. 21–26. Berlin: Springer.

Turner, B.L., II, W.C. Clark, R.W. Kates et al. 1990. The Earth as Transformed by Human Action: Global and Regional Changes in the Biosphere over the Past 300 Years. Cambridge: Cambridge Univ. Press.

UN (United Nations). 2001. World Population Monitoring 2001. Population, Environment and Development. ST/ESA/SER.A/203. New York: UN Publ.

Young, O.R. 1999. Governance in World Affairs. Ithaca: Cornell Univ. Press.

Young, O.R. 2002. Can new institutions solve atmospheric problems? Confronting acid rain, ozone depletion and climate change. In: Challenges of a Changing Earth, Proc. of the Global Change Open Science Conf., Amsterdam, 10–13 July 2001, ed. W. Steffen, J. Jäger, D. Carson, and C. Bradshaw, pp. 87–91. Berlin: Springer.

17

The Mental Component
of the Earth System

W. Lucht[1] and R. K. Pachauri[2]
[1]Potsdam Institute for Climate Impact Research (PIK), 14412 Potsdam, Germany
[2]The Energy and Resources Institute, New Delhi 110003, India

ABSTRACT

This chapter explores how the mental component of the Earth system functions and interacts with the physical and biological systems of the planet. A tetrarchical loop is used to
illustrate the four elements of the mental component:

- GeoScope (the interplay between observation and theory),
- GeoGraphy (the reintroduction of generalized knowledge into social contexts),
- GeoMind (aspects of identity), and
- GeoAction (interplay between governance and representation).

The ability to transfer understanding from local to global levels, e.g., through macroscopes, is a central challenge in managing the future of the Earth system. Management
goals should be identified through a continuous process, based on the awareness that a
multiplicity of global realities and situations exist, that there is partiality of knowledge,
that observation is often influenced by theory, and that an empirical basis is needed for a
science of sustainability.

SUSTAINABILITY

Humankind faces the challenge of managing our planet in a sustainable manner.
To meet this challenge, management goals need to emerge that reflect the practical dimension (How can society and its institutions evolve in such a way so as to
address questions of global sustainability?) as well as the systems theoretical dimension (How can interactions in the Earth system be studied and made conscious so that human actions can be purposefully guided in more sustainable directions?). For a discussion of the political and social sciences dimension, the
reader is referred to Gallopín (1999), ICSU et al. (2002), Juma (2002), Martens
and Rotmans (2002), and Parris and Kates (2003). Here, we focus primarily on
the systems dimension, as it offers a broad framework of mental reference to regional developments. This view is useful when discussing management goals,
since solutions for sustainability will most likely not emerge from simple extrapolations of current practice (Martens and Rotmans 2002). In addition,

understanding the interconnections between different components of the Earth system, including the dimension of human actions, requires novel insights at the systems level. Acknowledging the seminal paper by Schellnhuber (1998), we propose some ideas that will hopefully contribute to the kind of systems understanding needed to formulate management goals for sustainability.

Scientific knowledge about the Earth system has been gained through observations and generalizations, often via formalized theories. For purposes of this discussion, we distinguish three more or less distinct spheres:

1. the physical world, or the world described by chemistry and physics;
2. the biological world, which contains the element of the living;
3. the mental world, which introduces the element of consciousness.

These three spheres (Figure 17.1) are the evolutionary products of the Earth system whereby, through coevolution, the biological sphere emerged from the physical sphere. The recent advent of the mental sphere of human consciousness, characterized by culturally acquired syntactical language and the ability to imagine the future, has enabled human beings to intervene and impact the Earth system on all scales (Crutzen and Steffen 2003). Although this has occurred throughout human history, the past few hundred years have witnessed a dramatic rise in the anthropogenic impact on core elements within the Earth system. This era has been called the Anthropocene (Crutzen and Stoermer 2000).

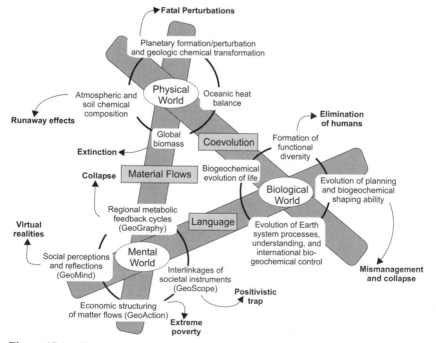

Figure 17.1   The mental component of the Earth system as related to its physical and biological components.

Our awareness that undesired or catastrophic outcomes may result from a continued, uncontrolled coevolution of the mental, physical, and biological spheres has increased over the last decades, substantiated by evidence from the environment and via scientific analysis. Still, we lack information as to the boundary conditions and dynamics through which the Earth system is evolving (Schellnhuber 1998). "Sustainability" refers to the goal of avoiding undesired effects in the Earth system, particularly those states that are deemed unacceptable in terms of human values (Lubchenco 1998; Kates et al. 2001; Tilman et al. 2002; Clark and Dickson 2003). A similar value may also be accorded to preindustrial properties of the biological and physical systems ("nature").

Can the mental sphere evolve quickly and purposefully to a point where the future evolution of the system can be managed consciously toward a state of dynamic sustainability of the whole system? If so, how? Schellnhuber (1999) refers to the evolution of the associated global, but heterogeneous and distributed, consciousness as the *emergence of a global subject*. In this chapter, we discuss non-economic aspects of the workings of the human world and its interactions with the physical and biological spheres. In doing so, we hope to elucidate some of the processes by which goals of global management toward sustainability might be established.

EQUITY

In a largely theoretical discussion aimed at providing ways to address interconnections within the Earth system, it is easy to lose sight of the problems that seriously affect people in many parts of the world. In addition to the legitimate concerns of the Earth system, the goal of achieving sustainability is inseparable from improving the prospects of ordinary people by increasing equity across cultures and societies (Lubchenco 1998). The World Summit on Sustainable Development, held in Johannesburg in 2002, represents a milestone in the international debate on global management. It focused on issues related to the Millennium Goals set by the United Nations (UN 2000), in particular the widespread problem of poverty, which despite efforts by national governments as well as by multilateral and bilateral development assistance organizations remains excessively common and acute.

To highlight one example, the urgent need of meeting the basic requirement of clean water was identified prior to and at the summit. Currently, one billion people do not have access to safe drinking water, and about four million people die annually from drinking contaminated water. Present estimates indicate that 300 million Africans live in water-scarce environments; by 2025, 18 countries in Africa, with 600 million inhabitants, will have shifted to a water-stressed status. In Asia, as well, water demands have increased significantly, and this trend is expected to continue over the next 25 years, putting many areas at risk. Thus, the goal of reducing by half the proportion of people without access to safe drinking water by 2015 represents a significant challenge. Others of similar magnitude exist with respect to living conditions, provision of food, education, gender equity, security, and environmental quality around the world.

Whether the present global system can be sustained should also be questioned on the basis of the enormous disparities that exist between the rich and the poor. Continuing and widening disparities of wealth are likely to create acute frustration and anger in the minds of those living in poverty. It is no longer possible to insulate the poor from images of material wealth and prosperity that exist in the richest parts of the globe. This, in turn, is likely to threaten not only the maintenance of law, order, and peace in the world but may also impact the sustainability of current human and biophysical systems.

In its Third Assessment Report, the Intergovernmental Panel on Climate Change (IPCC 2001) clearly stated that "the impacts of climate change will fall disproportionately upon developing countries and the poor persons within all countries, thereby exacerbating inequities in health status and access to adequate food, clean water and other resources." The inequities in the global system are, therefore, likely to become more acute as a result of climate change.

EARTH SYSTEM KNOWLEDGE AND PREDICTABILITY

Management of the Earth system requires an understanding of how its components function and implies mental images of potential futures. These may be language based, as cultural products or formalized scientific arguments, or may exist in the realm of values and images of the self. Our ability to predict the future varies with respect to the physical, biological, and mental spheres that constitute the Earth system. This variance has implications for predicting the future of interactions between the coevolving components of the Earth system.

The physical system can largely be described by laws identified within the natural sciences. These laws are also applicable to predicting its future; however, limitations can arise in such predictions from (a) processes still unknown, (b) interactions across temporal and spatial scales associated with strongly nonlinear or chaotic behavior of subsystems, (c) unknown values of variables that result from a lack of initialization and validation data, and (d) interactions arising external to the physical system. Although these limitations may be severe in some cases, we can nonetheless expect reasonably well-founded results from a scientific analysis of the physical sphere.

Consider, for example, the science of climate change. Uncertainty in projecting climate change and its impact on various parts of the world has been greatly reduced, as evidenced by IPCC's Third Assessment Report. A large source of uncertainty that remains concerns the future direction of human actions. This uncertainty obscures the exact nature and magnitude of climate change and constrains mitigation efforts. Thus, the current challenge is to address uncertainty in a methodologically sophisticated manner.

In the biological system, the situation is less clear. The biosphere thrives within the conditions of the physical system out of which it evolved. It has partially developed the capability of modifying and adapting the physical conditions within which it exists (e.g., through the buildup of atmospheric oxygen or the formation of soils), leading to a coevolution of both spheres (Vernadsky 1926; Huggett 1999; Samson and Pitt 1999). Some aspects of the physical system are substantially altered through the introduction of pathways in chemical cycles that lead through biological and human systems as they strive to maintain and reproduce.

Although some elements of the nonhuman part of the Earth's biological system may perhaps be described mechanistically, we still lack essential understanding of many important biological processes. Large gaps in available data exist, for example, concerning the reason and role of functional diversity as well as the allometric and metabolic laws governing the division of matter and energy between individuals and between species. We lack a full theory of ecology. Although we have a theory to explain the evolution of species, it is not clear whether this evolution can be described without reference to a mechanism for filtering avenues of change, possibly through as yet undiscovered self-organization at the system level in the form of a system-wide chemical or other capacity for coordination in complex living feedback loops (cf. Lovelock 2003; Kirchner 2002; Samson and Pitt 1999). If discovered, the associated laws gained would be of a most fundamental nature, not only with respect to the evolution of the Earth system. Diversity and the properties of the individual would emerge to be more than simply the product of mutational selection. What was once called "survival of the fittest" would be replaced by a principle that optimizes (as of yet unknown) factors across the system. Whether these would also include the recent rise of mental self-awareness in humans or whether it is an accidental by-product of evolution is impossible to determine at this time; however, even as we engage in speculating on these issues, our actions increasingly intervene in the evolutionary process (Western 2001).

The mental component of the Earth system depends fully on biological reproduction yet emerges as a separate entity. Conscious planning alters established biogeochemical and evolutionary pathways (Vitousek et al. 1997). Today, even simple observations of the physical world, such as temperature readings or the chemical composition of river sediments, can no longer be explained without reference to human interference. The human economy is a system for directing flows of matter and energy through societal infrastructures that encompass our bodies. By endangering biodiversity and considering genetic manipulation, the mental world severely impacts the biological world in which it is physically situated.

An element of unpredictability of the future, fundamental by nature, accompanies the mental sphere as it emerges as a global player to be developed into a reflective agent capable of global management. This unpredictability has its roots in the reflective properties of the feedback cycle between the analytical, speculative, and observational dimensions of the mental system. As the mental world shapes reality through action and reacts to experiences, it alters its own contents. The mental world is a construct that evolves out of its interactions with its own projections, which in turn are reactions to the perceived, possibly altered, physical and biological worlds. It cannot solely be described by its own inherent laws but is dependent on the functions of the entire system, a process that has been investigated in studies of learning and the cultural encoding of experiences. This openness, however, also serves as the basis for hoping that the human mind may develop into a successfully reflective agent within the system.

Even though economics comprise the most formalized part of the social sciences, we still lack a convincing formal theory of socioeconomic evolution that will allow more substantial predictions than are possible today. Theories of financial development, rule, and value evolution exist and are rendered plausible by detailed scientific arguments. Still, they are partially contradictive in their fundamental assumptions and are almost impossible to verify due to a substantial lack of relevant data. Despite considerable progress, this condition also prevails in the political and social sciences. Regardless of our efforts to stabilize the human reproductive systems, our socioeconomic material metabolism, and monetary turnover to avoid dangerous economic crashes, we lack the ability to plan on timescale of decades.

It is impossible to know whether the advent of reflective consciousness in the Earth system is an integral part of the general evolution of a complex system that forms a biosphere. Arising out of the biological system, the feedback loop exhibits properties that are somewhat stable. We call this empirical pattern culture, and it includes political and institutional cultures. Whereas lifestyles and perceptions change over time, underlying principles (e.g., modes of analytical thought, political theory, and cultural heritage) appear to be somewhat more persistent. Distinctions in thought and tradition can be mapped geographically and are not likely to disappear altogether, in spite of increasing effects of cultural globalization. Contemporary historians do not generally believe that there is, or will be, a theory of history; history and the evolution of societies are not perceived to be mechanistic. Experience shows that predictions of the future routinely fail. There is thus only a very narrow formal basis upon which to build theoretical concepts of the mental aspects of global management. Although there is a growing literature that explores this domain and a large body of highly relevant practical experience, our continual inability to predict on a daily basis societal or even market developments is evidence that the mental world is not a deterministic process.

The attempt to unravel the workings of the mental world is an ongoing project of the human mind. Elements of this process reflect the patterns and principles of language, mental images such as mythological prototypes, as well as the relationship between observation, judgment, logic, science, and so on. It is a broad field, not to be covered here, and for our own future, we should be prepared to encounter surprises as well as a host of failed predictions and expectations. Planning for the future is an enterprise that entails the risk of being wrong. Despite all we do know, the future remains fundamentally uncertain.

Given the uncertainty in predicting the physical, biological, and mental domains of the Earth system, managing efforts toward sustainability should be characterized as a process of constant re-evaluation, re-investigation, and re-definition of goals. Since first-order interactions between components are the key feature of the Anthropocene, rational scientific planning in the traditional sense is not possible on the Earth system level. Science is and will remain a central part of the process of managing efforts toward sustainability; however, it clearly cannot be the sole method involved. To identify goals and set courses of action requires the collective input from the whole of human knowledge and experience, beyond the purely scientific domain. The Earth system, which extends beyond the physical into less mechanistic domains of the biological and mental worlds, is far more complex than the possibilities offered by positivistic scientific inquiry.

This, however, should not be surprising: Every day we routinely encounter situations in which we do not know in advance that which may be of importance. Often, society addresses issues even when a strictly formal theory of identifying missing knowledge is unavailable. We do not know, for example, whether cures exist for important diseases, whether the dangers of nuclear technology can be harnessed, what consequences genetic engineering may bring, or what politics, markets, and consumers will prefer in 10 or 20 years. Nonetheless, substantial resources are constantly allocated to advance knowledge and practice in these fields. A similar approach can be taken to identify ways toward sustainability. Decisions must and are being made, often without hesitation, in the face of uncertainties. Actions that appear logical when viewed retrospectively were frequently not at the time they were made (Feyerabend 1975).

This is not to imply that important problems cannot be reasonably analyzed, predicted, and used to base future actions. Not all is uncertain: delivering food to regions where the harvest has failed will prevent starvation; curbing CO_2 emissions will reduce the increase in global temperatures. Failure to act will negatively impact the poorest members of our global society, who neither possess the physical infrastructure, institutional strength, nor early warning systems to withstand or adapt to detrimental developments in the Earth system. The need for coordinated action has been the message of many international meetings over the last two decades. This call should be heeded and action taken without hesitation on many fronts.

FOUR COMPONENTS OF THE MENTAL WORLD

Possible goals of global management toward sustainability and the process by which they are identified are not issues to be considered within the physical or biological systems. They must be approached through human reasoning, that is, within the mental world. Solutions are only possible when the complex interlinkage of observation, analysis, preference, and action is considered. Yet can this be accomplished? If we wish to go beyond a mere optimization of current practice, we should strive to understand the inherent interrelationships at work. By identifying new dimensions of inquiry into the Earth system and its functioning, possible goals can be generated, evaluated, and negotiated in the mental world; this, in turn, will enable a better understanding of the coevolution at work within the Earth system. Such knowledge should be gained in concert with concrete and extremely important practical efforts that are currently underway to develop societal processes in positive directions.

Figure 17.2 depicts the mental world as a circular pattern, subdivided into a tetrarchy of four elements:

- GeoScope refers to the interaction between observation and theoretical formalizations from which a description of a world perceived arises. Observation and theoretical concepts are mutually interdependent.
- GeoGraphy relates to the discourse associated with this knowledge and its localized transformation into patterns of local thought, practice, and images. It is the cultural practice of a science that transforms theories and observations into other, for example, more accessible domains of knowledge. Issues of complexity, scale, and translation between different domains of language are central to this process.
- GeoMind refers to a complex of fundamental questions: Who are we? What do we want to be? These questions are closely linked to issues of place and are deeply enmeshed within the formation and evaluation of identity. The additional question "What do we want?" follows and is a predominant driver in our world that is not normally a result of technical analysis.
- GeoAction refers to measures that are taken which affect the world, e.g., implementing policies or decisions. It is associated with issues of governance, which give order to larger social entities, and representation, which relates to various forms of legitimacy of action. Finally, as action alters the observable and the debatable, the effects of action may be observed through a geoscopic process.

The use of the prefix "geo" signifies a relationship to the Earth system but it does not necessarily imply a global dimension, as it is readily applied to personal and local situations. In the tetrarchical loop, the interplay of all four elements constructs — equally and complementary — the progress of perceived reality in the mental world and its interaction with the physical and biological worlds. Neither science, theory, observation, nor action predominates. Whereas nongovernmental organizations and international institutions may focus on the sphere of GeoAction and science may perceive itself as offering an analytical key to finding solutions, it is the stream of thought arising from such sectoral

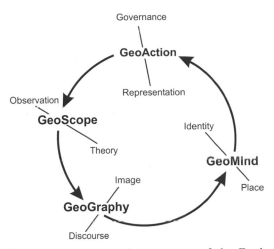

Figure 17.2   Elements of the mental component of the Earth system in the Anthropocene.

contributions and carried into the midst of society that strongly shapes individual identities and creates a GeoMind, which in turn serves as a key source for subsequent decisions. The circle can also be interpreted as consisting of the elements of generalization (GeoScope and GeoGraphy), concretization (GeoGraphy and GeoMind), interpretation (GeoMind and GeoAction), and implementation (GeoAction and GeoScope).

GeoScope: Observation and Theory

Observation can be viewed as an entry point into the functioning of the mental world. When it is not possible to predict the future entirely, observation provides important feedback about developments and consequences of action. Similarly, formalized models require data for their development, initialization, operation, and validation. With respect to a science of sustainability, observation may provide an empirical basis upon which theoretical progress can be made. Observation also entails a fundamental willingness to initiate a mental process by looking, listening, and being open to perceptions. Observation, after all, inherently has the power to surprise us.

Whereas theoretical formalizations may be viewed as universalizations of observations, observation in turn depends on the theoretical concepts and modes of thought of the observer with respect to its nature and meaning. This has been called the theory-ladenness of observation. Observations depend on the social and intellectual context in which they are made. Current images of the globe are produced through a mentally filtered mode and interpreted in culturally laden contexts, be they scientific, economic, political, or otherwise.

One may doubt whether we already possess the means of generating views and concepts of the world suitable to considering sustainability issues (Parris and Kates 2003). From the standpoint of Earth system analysis, viewing the entire system requires a "macroscope" to observe system-order variables and to produce an image of the whole (Schellnhuber 1998, 1999b; Lucht and Jaeger 2001). The principles upon which such a macroscope can be constructed are still under discussion. Large methodological challenges remain to be resolved, particularly with respect to observing social systems.

Despite the immense stream of information generated around the globe each day, we still lack an observational system capable of systematically generating images that can be strictly associated with theoretical concepts of the coevolution of the mental, biological, and physical spheres of the Earth system. Currently, the best systematic global observation instruments available address the physical world and financial aspects of national economies. These include the instrument networks of the natural sciences, which span the globe from Earth-orbiting satellites to ground stations and buoy systems, and the international system of statistical economic reporting, by which quantities such as gross domestic products, inflation rates, or stock market values are observed. Still, these networks provide only a partial view of the world.

Despite international research efforts and the resultant progress on biodiversity and habitat conservation, the Earth's biological systems are not currently under extensive coordinated observation. Considerable gaps in knowledge exist, particularly at the systems level, due primarily to the high degree of complexity and diversity in these systems. A similar situation applies to the mental sphere. Reporting scientifically on the social and political dimensions of the world has increased considerably, especially in areas of education, health care, values, preferences, and political structures. Much of the data, however, are regional, not intercomparable, and unavailable as longer time series. There are notable exceptions, but in comparison to the cost and effort expended in the natural sciences or the statistical offices serving economic planning, observation directly relevant to the many socioeconomic dimensions of sustainability is still poorly developed. Key processes (e.g., the magnitude and distribution of material and energetic exchanges between the environment and human societies or the time and space budgets of people) are not being monitored in a defined and continuous system.

The scientific community faces the challenge of creating a macroscopic instrument suitable to the task (Lucht and Jaeger 2001). If we are to direct the Earth system toward something resembling a sustainable state, a macroscope (i.e., a geoscope in terms of the tetrarchical loop) must consist of a coordinated interplay between relevant, possibly rival, theories and associated select, but deeply meaningful, observations. Such observations must be chosen to allow innovative formulation and evaluation of theories and progress in the discourse between competing ideas by providing a suitable observational reference.

The need for a geoscope is exemplified by the state of debate on sustainability indicators, which are designed to measure progress, or lack thereof, with respect to sustainability, or on the quality of life and the environment (Moldan et al. 1997; Moffat et al. 2001). At present, the theoretical basis and available data upon which such indicators are drawn remains largely unsatisfactory. Despite an increase in their importance, indicators are still based primarily on various degrees of plausibility, rather than on solid theoretical foundations, and are dependent on the available types of data, which were often collected for other purposes.

A suitable observation system is crucial to identify potential goals of sustainability, how they might be achieved, and whether measures advised by our current thinking have the desired effects in this very complex system. We suggest that such a system should initially provide data for comparative regional case studies of concrete measures affecting sustainability. These data can be used to produce time series of key variables that are justified by their relevance to identified theoretical discourse.

It is likely that an initial geoscope will consist of a combination of synoptic-scale observations (e.g., from remote sensing or aggregated statistics) and in-depth, on-the-ground data for selected regions. It could provide spatial realization of information connecting people and pixels across nested scales in domains where such data are not currently available — an effort comparable in time and cost with that of traditional economic statistics or space observation programs. A geoscope will have to be constructed through a cumulative process of learning-by-doing, as it appears impossible to produce a full blueprint at the outset. Elements of what could be called an emerging geoscope are already moving forward as the result of various initiatives. We note that macroscopes may also potentially be used by the security apparatus of global dominant powers. Issues related to the appropriate means of information gathering, ownership, and sharing are therefore highly relevant to the proposition of a geoscope, whose purpose must structurally oppose the implied dangers of total digitization of the globe and the associated potential for political control.

GeoGraphy: Partial and Situated Knowledges

GeoGraphy refers to a scientific practice that goes beyond the acts of observation and formal consideration of theory formulation. It is the process of translating formal knowledge and data into other forms of knowledge by returning them into the complexities of real-world situations (e.g., into localized contexts).

Epistemologically, macroscopic knowledge can either take the form of scientific knowledge (i.e., encoded in a formal structure expressed by rational linkages) or social knowledge (i.e., embedded in social, cultural, and psychological contexts and which may, e.g., be communicated as symbolic mental images expressed visually or through language). When scientific theory and observation meet the multitude of social situations and all of their intricacies, these forms of knowledge are constantly transformed into one another.

Today there is a tendency to move social sciences, such as geography, sociology, or the political sciences, in the formalized direction of the natural sciences. However, the usefulness of the social sciences is most evident when knowledge is translated between systems of reference, structure, and expression. Since mental language varies according to cultural, geographical, and temporal location, this is an ongoing process of exploring adequate expressions for scientific knowledge, expressions that are meaningful in social reference systems. In terms of translating the world into science, generalization and universalization are tasks to be achieved. In terms of translating science into situational meaning, the challenge is to provide a wide array of opportunities for translation. This is the task of GeoGraphy.

Neither a positivistic interpretation (in which the world is but a concrete realization of universal laws) nor a particularistic approach (which negates any relevance of abstract deductions) is productive. Instead, an interplay between generalizing mental analysis and situational complexity, where concepts offered by the mind interact with possibilities of the physical to construct a joint reality, is necessary to describe the dynamics of the Earth system in the Anthropocene as well as the coevolution of the mental, biological, and physical worlds of which it is comprised. In terms of a systems analysis, dynamics of this system are determined by feedback loops between the mind and the physical world.

Central to the process of GeoGraphy is an understanding that the information emerging from observation and formalized into theoretical systems does not have a unique meaning in social reality. Its meaning is tied to the cultural practice of the recipient society. Multiple realizations are possible. Across all scales — from an emerging collective global identity to the particularities of personal opinions — the meaning, interpretation, and importance of specific knowledge varies. Knowledge in all its forms is thus partial and situated (Haraway 1991).

Knowledge is partial in two respects. First, it is not complete in the sense that knowledge gained from within a cultural practice does not necessarily comprise knowledge arising from other cultural practices, nor is it usually possible to unite, without at least some difficulty, knowledge from different cultural practices because of the inherent association of such practices with language, patterns of thought, and identity. Second, knowledge aligns with the cultural preferences of the originator. This does not mean that measurements or results, e.g., of a numerical model simulation are necessarily a construct, rather that the practical meaning of the results is subject to interpretation. This has implications for the interface between science and politics. Science is not a comprehensive prescription for action, although action can react and refer to science.

Knowledge is situated in that it exists under actual conditions that are inseparable from the meaning of that knowledge. Such lines of thought are a product of an increasing realization in our societies that there is a multiplicity of systems of reference with increasing spatial and temporal overlap. Globalization and technology contribute to make this multiplicity of simultaneously existing mind frames more prevalent than in the past.

Understanding the particular generated in such frames in relation to a wider whole is specifically a task of science. The interplay between observation and theory is an instance of relating the particular to the generalized. The associated discourse, firmly anchored in the cultural practice of science, produces knowledge in various forms, both fundamental and as applied to particular problems. Today, however, the search for principles to be generalized is no longer the primary goal, as was the case in classic disciplinary science. Instead, the challenge is to interlink knowledge and its application to highly particular circumstances; most importantly, the task is to return the generalized back into the context of social situations.

GeoGraphy is the cultural practice of a science organizing and creating such partial and situated knowledges of the Earth system. It is likely to be one of the leading sciences of the twenty-first century, taking over the prominent position of physics. The multiplicity of realities as well as the partial and situated natures of knowledge do not mean that the world is merely a sum of many overlapping parts, rather, from a systems point of view, that the world is an internally differentiated whole. Early on, Heraclitus expressed this as the formula *"hen kai pan,"* the one and all, which describes the principle that the world unfolds from one entity and dissolves back into it. Unity then is both the universal whole as well as the sum of the individual parts; that is, it is geoscopic and geographic at the same time. One is limited without the other.

GeoMind: Identity and Place

When we view the lands that surround us, we also include ourselves in this purview and, in doing so, we touch or even transform the lands in one way or another (Cosgrove 1984; von Droste 1995). There is a connection between body, self, and place that is relevant to how we consider ourselves and the world. If observation is one entry point into the mental subsystem of the Earth system, then issues of identity and will are another.

Consider, for example, the extensive transformation of the American continent over the last 500 years: Europeans perceived this expanse of land as being available to them. The South Sea illustrates this attitude as well: it served as an important object of British imperial fantasies about nature in the nineteenth century. Today, when we view the continents from space, what is it that we perceive? What do scientists observe? It is common to think that data flow is simply a digitized stream of photons, but the cultural history of imaging landscapes makes us aware of a substantial body of political, aesthetic, and cultural subtexts. Together, they shape our identity in relation to place (Mitchell 1994). This identity and the willingness to shape the future, which likely includes shaping the land, is a core driver of global transformations.

GeoMind is this contribution of individual and collective identities and wills to making sense of the world. The category of identity is not well recognized in global change science. The dominant assumption is that results of an applied, user-oriented science can feed directly into the process of political decision making. In practice, this is rarely the case. The world of decisions and actions looks toward science for input, but the process is not a formal one. Decisions are made on the basis of intuitive experience and examined in light of available knowledge. Success is measured by outcome. However, it is questionable whether international debate about possible goals of coordinated global management will get far if it does not stem from the self-image of the people participating as well as the people affected; human choice should be accorded a position of departure in the process.

Science neither provides, nor can it provide, answers that encompass all aspects to consider. It also does not, and cannot, serve as the primary motivator for action. Science, however, can play a considerable role when embedded into a larger system of cognition. Knowledge serves to direct decisions into promising directions or to challenge them when made. Scientific analysis has a stimulating and a controlling function, but not usually a creative function per se. Science is a tool that the mind employs to produce analytical challenges to thought patterns.

GeoMind may appear to be an abstract construct, but it is easy to observe in the real world. The United Nations was not founded on a scientific analysis of potential next phases in organizational structures, but as the result of a political will to move in a direction of increased collective checks and balances. Democratic institutions did not emerge from a rational analysis but from the primary desire of people to increase their freedom. Similarly, the goals of sustainable management are unlikely to arise directly from a scientific analysis of the Earth system, but from the mental world, whose questions (who we are and what we want to be) occupy us. For this reason, normative questions about minimal humanitarian standards, the role of equity principles, and intolerable domains have been included into the scientific program (e.g., Schellnhuber's "Hilbertian Programme for Earth System Science"). One question of central importance reads: "What kind of nature do modern societies want?" (Sahagian and Schellnhuber 2002, p. 9).

As we investigate more carefully the subtexts of our images of the world, which co-shape the world despite their often semi-virtual quality, we also have to become more aware of the dangers and possibilities of these new world views that are being created. The self-referential nature of constructed surroundings and cultural settings can either frighten or interest us. At the same time, the real-world consequences of images projected and actions taken are all too real. As Parris and Kates (2002) state: "Defining sustainability is ultimately a social choice about what to develop, what to sustain, and for how long" (p. 8068).

GeoAction: Managing Toward Sustainability

Several vocal parties involved in the debate on action for sustainability are discernible. There are the scientists who analyze the Earth system and consider the dangers of destabilizing the global biological and climatic systems. There are the technologically affluent and optimistic who stress that shifts in associated values are required (e.g., Raskin et al. 2002). There are the materially and ethically oriented critics of the econo-political world of capitalism who champion the perceived aspirations of the poor (e.g., Sachs 1999). There are also the believers in global markets and the technological solutions assumed to spring from them, to which that large part of the development community belongs, which strives to industrialize the as yet unindustrialized parts of the world.

How can we reconcile limits of the environment with the consumption demands of the developed world? How can we balance the justified hopes of poor nations for an increase in material goods with obvious limitations in supply? What principles of equity and minimal humanitarian standards should be applied? These questions remain unresolved.

It is striking not only to study the voices participating in this debate but to consider that large portions of the world population strongly affected by these issues are not currently being heard. This includes members of the technologically and economically less-advanced regions, many indigenous groups, women, and young people. These groups may add other scenarios and mechanisms to the debate beyond markets, technologies, and rationalized values.

The issues under debate are far-reaching in their consequences and are not close to being resolved. For example, does technological innovation have the potential to provide unexpected ways of moving forward without regret? In an extensive study of possible scenarios for the future, Raskin et al. (2002) view such innovation as a key factor in enabling the shifts in values, lifestyles, and modes of organization required to facilitate a "great transition" to sustainability, as opposed to leading the world into less desirable states. Sachs (1999) follows a line with somewhat different implications. He questions whether the imagery of multifaceted global unity, communication, and organization behind some such scenarios is realistic. Observing that "some inhabitants on this Earth of ours are more global than others," Sachs reports (1999, p. ix–x) a "nagging suspicion" that "the Western development model is fundamentally at odds…with sustainability (truly conceived)," which may in the end turn out to be "incompatible with the worldwide rule of economism." Issues of advised action toward sustainability remain contested. On one occasion a western journalist asked Mahatma Gandhi whether he would not want India to reach the same level of affluence as Britain. Gandhi's answer was that it required Britain to conquer half the planet to reach its existing level of affluence. He asked how many planets would India have to conquer to reach the same level.

We only vaguely know the overall shape of possibility space for the Earth system. We have achieved an increased understanding of this space in terms of the stability of natural systems and of resource limitations, but not with respect to ethical and evolutionary questions addressing existing and possible future global and regional structures, goals, and ambitions (Schellnhuber 1998). Not only do we need to list the possible good goals of global management, revise them continually, and discuss their contradictions, we need to focus on the investigations of this process — how goals may be identified and adapted — and position it equally at the center of research (Kates et al. 2001; Martens and Rotmans 2002). Such research necessarily has a strong institutional and political dimension (Juma 2002). Questions about institutions, organizations, process, legitimacy, information, science, participation, and representation immediately emerge (ICSU et al. 2002).

Theories of GeoAction entail social and political theories of action that accept a multitude of realities, the partial and situated natures of knowledge, as well as the theory-ladenness of observation. In such a situation, institutions can rarely have a rationalistic normative character (Finnemore and Sikkink 1998). Rather, they serve as flexible networks of exchange points for knowledge, as providers of frameworks for negotiations, and as stimulators of needed developments. As in GeoGraphy, where formal knowledge is actively brought back into situational contexts to produce applicable social forms of knowledge, GeoAction is charged with producing social forms of action where insights about the evolution of the systems within the Earth system are brought into regional political and economic contexts. Distancing knowledge from politics will not achieve these goals nor will an orthodox normative system of decision making. Only through a living exchange between generalized abstractions and concrete realities, that is, between Earth system considerations and regional necessities, can goals of sustainable management evolve to avoid the pitfalls of adherence to any particular strand of interpretation.

The complexity of the issue prevents the rapid appearance of convincing solutions. A process of learning by doing is unavoidable. A system of adaptive management expressed as GeoAction requires an infrastructure encompassing all facets of the tetrarchical loop (ICSU 2002). The global subject (preferentially to be seen as a web of multiple subjects), which acts consciously in managing the Earth system of the Anthropocene, emerges from a complex process full of contradictions, compromises, and disputes. Indeed, Schellnhuber (1999a, b) describes it as "emerging from the world-wide web of socioeconomic relations and discourses supported by advanced technology" (p. 13), where "one key to its emergence is world-wide communication" (p. C22), and as "a cooperative system generating values, preferences and decisions as crucial commonalities of humanity online" (p. C22). He concludes: "The global subject is real, although immaterial" (p. C22). Nonetheless, the cultural embedding of action is reflected in the remaining, though possibly transformed, place-based geographic nature of global identities.

IPCC is currently involved in one of the most extensive efforts to consider systematically, on a global scale, scientific knowledge upon which action to control global change may be founded. Founded in 1988 by the United Nations Environment Programme (UNEP) and the World Meteorological Organization (WMO), the IPCC has a threefold mandate: to assess the state of existing scientific knowledge on climate change; to examine the environmental, economic, and social impacts of climate change; and to assess response strategies.

IPCC is responsible for producing periodic assessment reports on scientific issues relevant to climate change. Additionally, Special Reports and Technical Papers are produced to address specific scientific needs that arise out of the international negotiations process dealing with climate change and to meet requests received from the United Nations Framework Convention on Climate Change (UNFCCC). The scientific objectivity, transparency, and credibility of these reports are ensured through a rigorous review process, which consists of three steps: the research assessed has already been published in peer-reviewed scientific and technical journals; the resulting draft assessment report is peer-reviewed by a large and diverse group of experts; and finally, the draft report is reviewed by governments.

To produce its reports, the IPCC attracts experts from all over the world. Since its inception, the reports have evolved in their perspective. The IPCC has progressively moved from viewing the problem of climate change, its impacts, and mitigation in isolation to adopting a more holistic approach. In its most recent assessment, the Third Assessment Report (IPCC 2001), the IPCC has presented climate change in the larger context of sustainable development. It recognized the various synergies that exist between climate change and other environmental issues and the opportunities that these synergies present in developing more effective and comprehensive response options. The Fourth Assessment Report (AR4), due out in 2007, will continue exploring these options and delve deeper into the interlinkages. In doing so, it will present an evolving model of systematic global discourse on a scientific, nonprescriptive but politically highly relevant level.

THINKING ABOUT SUSTAINABILITY: A TIMELY TASK

Each generation considers the challenges and opportunities that result from its particular position in history. Questions of Earth system management, formulated in response to a growing scientific basis reflecting a world of increasingly globalized images, are highly characteristic of our particular time.

As we approach these challenges, we must avoid abstract generalizations as well as the anti-analytical preference for only local phenomena. Equally, we should not succumb to the temptations of presuming to possess the perspective of a scientific "sage," with all of our available global imagery, and we must avoid the predominance of segregative systems of organization (e.g., in the form of spatial or temporal segregation of people or a purely disciplinary science).

As Kates et al. (2001, p. 641) write: "The sustainability science that is necessary to address these questions differs to a considerable degree in structure, methods, and content from science as we know it." By integrating observations, theories, places, identities, and actions, and by forging new dialogues across the temporal and spatial dimensions of the mental component of the Earth system, new knowledge can emerge as an epistemologically new type of science

(Gallopín et al. 2001). Through a process such as described by the tetrarchical loop (see Figure 17.2), management goals toward sustainability can be negotiated and continuously revised. This process reflects the current evolutionary state of the Earth system in the Anthropocene.

Ultimately, we are left to reflect upon fundamental questions: Who are we? Who do we want to be? In what kind of world do we want to live? If we attempt to understand the workings of the entire Earth system and not just its physical components, we stand a chance to address these questions. If we continue impacting the Earth in an unsustainable and ill-understood fashion, we (as well as future generations) may well jeopardize the future coevolution of the physical, biological, and mental spheres that comprise it. Harsh realities serve to remind us that our current task is not an academic exercise (Petschel-Held et al. 1999).

Our co-construction of a future, or rather, of futures as perceived in a multitude of perspectives, must be a carefully negotiated enterprise. A common vision should not be assumed in a world of many cultures. Limitations will arise from the workings of the physical and biological systems; dangers will emerge from the association of mental models with very strong political subtexts and fantasies. We must be aware of the many pitfalls present in mental processes and of the uncertainties inherent to projections of the future and its complex interrelated processes. Since we are embarking on an experimental journey — one which may reveal that directions taken are not necessarily those intended — feedbacks from observation into thought and action are essential.

Science is likely to play a decisive role in shaping our understanding while concurrently going through a fundamental reconsideration of its own nature (Lubchenco 1998). One of the most striking outcomes of the Rio Earth Summit of 1992 are local Agenda 21 groups throughout the world. To function, these groups rarely require scientific knowledge to act, as the most pressing problems of a local area and potential courses of action are well known to its inhabitants. Science plays its role when the local action is brought into a global framework of integrated Earth system management. Thus, the practical feasibility of our visions for the future depends on our ability to analyze the workings of a very large and complex system, which evolves while being shaped by our actions. Still, the conclusions drawn will remain ineffective unless they are re-inserted into the complexities of the many real-world situations that make up the planet.

ACKNOWLEDGMENTS

We are indebted to discussions with Carlo Jaeger, Hermann Lotze-Campen, Dörthe Krömker, Leah Goldfarb, Petra Lucht, Marina Fischer-Kowalski, and William Clark. The poems by Li Bai were translated based on English language versions by several translators, among them Xu Yuan Zhong, Yunwei Jia, Binhe Gu, Dafeng Hui, and David Hamilton. The translation of Petrarca from Latin makes use of a translation by James Harvey Robinson, that of von Humboldt is original. Financial support for Lucht's work by the German Climate Research Programme DEKLIM, project CVECA, is gratefully acknowledged.

APPENDIX: FOUR STORIES AS FOOD FOR THOUGHT

Li Bai: Poet

In one of the most famous poems of Chinese literature, the eighth century poet Li Bai speaks of a journey through the Three Gorges of the Yangtze River: "Traveling to Jiangling: Leaving Baidicheng at dawn in clouds, / I've sailed through canyons a thousand li in a day. / The river banks are loud with monkeys. / Passing behind me, a myriad of peaks falls away."

The Three Gorges are known outside China today mainly for the gigantic new dam that is beginning to submerge them. It is a technocratic, industrial-style project in response to China's increasing demand for energy, and its costs are high. The environmental impacts of the dam are thought to be considerable. Losing this landscape cuts deeply into the heart of Chinese culture and mental makeup by eliminating what is at the core of a very old tradition. However, the dam produces neither radioactive waste nor burns greenhouse gas-emitting fossil fuels. To leave industry and homes without additional energy would also impact the outlook and limit the ambitions of millions of people who are justified in expecting increased prosperity and better basic services. Whether the hopes will be fulfilled and the dam will be manageable instead of causing endless problems, as critics suspect, remains to be seen.

Viewed from the perspective of sustainable management, supporters and critics of particular measures often create too easily the impression that advisable directions of action are known and can readily be identified. In reality, things are usually much more complicated. The dam, positioned at the crossroads of economic, environmental, cultural, and political issues, all of which run deep, is a problematic example of the dilemmas encountered when discussing the future as it evolves out of historical conditions.

The same poet Li Bai wrote of a journey going in the opposite direction: "Attempting to go up Three Gorges: The blue sky jammed between the mountains; / we will never reach it with the water rushing down / so strongly. Three mornings we started up the gorge, / at night find we've gone nowhere. For once, / I truly forgot the hair on my brow is turning white." In his life, Li Bai traveled to the region above the Three Gorges also to flee political turmoil further downstream. Returning, he never quite found rest again. Legend has it that he drowned when he fell from his boat while drunk, trying to embrace the moon.

Petrarca (Petrarch) and von Humboldt: Poet and Scientist

On April 26, 1336, the poet Francesco Petrarca ascended to the peak of Mont Ventoux, the highest mountain of the Provence. In the cultural history of Europe this event is regarded iconographically as the beginning of a conscious perception of landscape.

"The highest mountain of this region, which with reason is called Ventosum, 'the windy one,' I have ascended today, led only by the desire to see what so great an elevation has to offer," Petrarca writes the same evening in a letter. After a tortuous ascent, he reaches the summit and is overwhelmed: "At first, moved by the unfamiliar breeze of the air and the whole unhindered view around, I stood as if dazed….I looked back down: Clouds lay at my feet….The Alps themselves, rugged and snow-capped, seemed to rise close by, although they were really at a great distance….I could see with the utmost clearness, off to the right, the mountains of the region about Lyon, and to the left the bay of

Marseilles and the waters that lash the shores of Aigues Mortes, although several day's journey distant. Under our very eyes flowed the Rhone."

But then Petrarca decides to take his eyes off of the landscape and to open a copy of Augustine's Confessions that he was carrying, to whatever he'd happen to find: "I call God as a witness and him who was there, that where I first fixed my eyes, it was written: 'And men go about to wonder at the heights of the mountains, and the mighty waves of the sea, and the wide sweep of rivers, and the circuit of the ocean, and the revolution of the stars, but themselves they forget'." This text gives him cause to admonish himself for having fallen to admiring worldly grandeur when "nothing is admirable except the soul, compared to which nothing is great." And so the story continues: "Then, in truth, I was satisfied that I had seen enough of the mountain. I turned my inward eye upon myself, and from that time not a syllable fell from my lips until we reached the bottom again."

There is justified doubt whether this ascent actually took place as described, though no doubt Petrarca did climb Mont Ventoux at some point in his life. Aside from practical aspects, the whole setup and timing is too ideal and the allegorical nature of the story too obvious to be true exactly as told. It is really a story of attaining the heights of spiritual maturity in the face of obstacles and temptations encountered. As Petrarca ascends the physical mountain, he looses sight of the spiritual mountain, which he returns to remorsefully as he leaves behind him the view of the landscape. However, the fresh exhilaration of his view of the landscape is telling. Emerging from medieval patterns of thought, but barely, Petrarca constructs an opposition of the spiritual realm to the enticement of physical reality in form of the landscape. A beginning was made of entering into a dialogue between these two worlds.

The point here is twofold. First, viewing is elementary. Without gaining a vantage point from which to view the world, the land cannot be perceived as a cohesive whole. Second, observation is closely associated with the mental makeup of the observer. It leads to questions of where one stands and what the important things should be. It causes reflection. There is no reason to believe that before global management toward sustainability can be dealt with as a task, observation of this sort is necessary. Vantage points for viewing the Earth have to be found and the view taken with an open mind, leading to an evaluation of what it is that we should or might do. This vantage point cannot only be the literally highest mountain of modern time, a satellite in low orbit, though it is interesting how many of the very technical personnel that have flown into space have reacted with similarly emotional statements about seeing the Earth. The metaphorical peaks of our time have to permit views of the mental world as well. Thereafter, we should also be ready for some unexpected sights and for some unexpected conclusions as we contemplate, standing at that place, what it is that we really want. At the same time, Petrarca's story may serve to consider the dangers of withdrawing too strongly into the inner perspective, when the view encountered challenges the mind so fundamentally.

In 1802, almost 500 years after Petrarca, the explorer and pioneer of geosciences Alexander von Humboldt ascended a mountain in the South American Andes (Humboldt 1802). He did not come, as had Petrarca, to contemplate the world spiritually, but as a scientist carrying instruments. After much expectation and many disappointments, his party finally reached its goal: "When after many undulations of the ground we had reached on the sheer mountain ridge the highest point of the Alto de Guangamarca, the heaven's dome that had been veiled in haze for so long cleared up….The whole westerly flank of the Cordilleres near Chorillo and Cascas, blanketed by immense blocks of quartz from 12 to 14 feet in length, the plains of Chala and Molinos up to the ocean shores at Truxillo lay, as in miraculous closeness, before our eyes. We saw the Southern Sea now for the first time; we saw it clearly: near the littoral a huge mass of light reflecting back, ascending in its infinity against the barely discernible horizon."

As had Petrarca, von Humboldt then turned to the main purpose of his trip, i.e., the scientific mission, but he confessed: "The joy, which my companions Bonpland and Carlo Montafur vividly shared, let us forget to open the barometer on the Alto de Guangamarca. According to the measurement that we made nearby, but lower than the peak, in an isolated dairy farm in the Hato de Guangamarca, the point at which we first saw the ocean must have been only 8800 to 9000 feet high." As was Petrarca, so the scientist is also affected in his purpose by the impression of the view. His scientific mission is embedded in the whole of his experience as a human being exploring the world.

Buckminster Fuller: Inventor

The American Robert Buckminster Fuller, inventor and mastermind, in his book *Education Automation* (1962) made the following proposal: "The new educational technology will probably provide also an invention of mine called the Geoscope — a large two-hundred-foot diameter (or more) lightweight geodesic sphere hung hoveringly at one hundred feet above mid-campus by approximately invisible cables from three remote masts. This giant sphere is a miniature Earth. Its entire exterior and interior surfaces will be covered with closely-packed electric bulbs, each with variable intensity controls. The lighting of the bulbs is scanningly controlled through an electric computer. The number of the bulbs and their minimum distance of one hundred feet from viewing eyes, either at the center of the sphere or on the ground outside and below the sphere, will produce the visual effect and resolution of a fine-screen halftone cut or that of an excellent television tube picture. The two-hundred-foot geoscope will cost about fifteen million dollars. It will make possible communication of phenomena that are not at present communicable to man's conceptual understanding. There are many motion patterns such as those of the hands of the clock or of the solar system planets or of the molecules of gas in a pneumatic ball or of atoms or the earth's annual weather that cannot be seen or comprehended by the human eye and brain relay and are therefore inadequately comprehended and dealt with by the human mind."

Fuller suggests that "the Geoscope may be illuminated to picture the Earth and the motion of its complete cloud-cover history for years run off on its surface in minutes so that man may comprehend the cyclic patterning and predict. The complete census-by-census of world population history changes could be run off in minutes, giving a clear picture of the demographic patterning and its clear trending. The total history of transportation and of world resource discovery, development, distribution, and redistribution could become comprehendible to the human mind, which would thus be able to forecast and plan in vastly greater magnitude than heretofore. The consequences of various world plans could be computed and projected. All world data would be dynamically viewable and picturable and relayable by radio to all the world, so that common consideration in a most educated manner of all world problems by all world people would become a practical event."

A child of his time, Fuller had great hopes with respect to technological solutions to the world's problems, a vision we are no longer able to share with quite the same enthusiasm. Certainly, his hope that television would usher in a revolution by providing the world with remote access to education has not been fulfilled. He did, however, under-stand that contemplating the future requires observation, comprehensible images as a means of mental access, ways of communicating, and a connection to politics, if science is to be able to contribute. He wrote: "During one-third of a century of experimental work, I have been operating on the philosophic premise that all thoughts and all experi-ences can be translated much farther than just into words and abstract thought patterns. I saw that they can be translated into patterns which may be realized in various physical projections by which we can alter the physical environment itself and thereby induce other men to subconsciously alter their ecological patterning. My own conclusion is that man has been given the capability to alter and accelerate the evolutionary transformation of the *a priori* physical environment, that is to participate objectively, directly, and con-sciously in universal evolution....If he does not do so consciously, events will transpire so that he functions subconsciously in the inexorable evolutionary transformations."

Fuller clearly saw the political dimension of such an undertaking. He says: "But real-ize, at back of the UN Building in New York in the East River is [an island]. And what I wanted to do was build, then, a miniature earth, mounted from those rocks...[It] was go-ing to be 200 feet in diameter [and] would be mounted 200 feet above the water, so...the height would be 400 feet, ...[which is also] the height of the United Nations building. So, it would be a miniature Earth really out confronting the representatives of the world." He summarizes his proposal as follows: "This 200-foot-size Geoscope would make it possi-ble for humans to identify the true scale of themselves and their activities on this planet. Humans could thus comprehend much more readily that their personal survival problems are related intimately to all humanity's survival."

He did build several experimental miniature Earths, or geoscopes, but a full-fledged realization of the idea was never achieved. Today, outgrowths of the basic idea can be found in computer model animation, data visualization, and public displays. However, uniting these capabilities in a single instrument available for everyday use is still lacking.

Mahatma Gandhi: Human Ecologist

One of the major highlights of the 2002 Johannesburg World Summit on Sustainable De-velopment was the reference made to sustainable production and consumption. Attempts to focus on this subject at past meetings and conferences have generally been met with determined opposition from some developed countries, who felt that any discussion of production and consumption patterns would threaten the prosperity they had attained af-ter a century and a half of industrial development. The World Summit agreed to encour-age and promote development of a ten-year framework of programs to accelerate the shift toward sustainable consumption and production. Implied in this decision was ac-ceptance of the fact that current levels of production and consumption were not sustainable.

Seven decades ago a visionary and enlightened leader, Mahatma Gandhi, highlighted these issues, conscious as he was of the persistent exploitation of the poor by the rich, within and across nations (Khoshoo 1995). Gandhi was once asked whether he would like to have the same standard of living for India's many millions as prevailed in England. He responded, "It took Britain half the resources of the planet to achieve this prosperity. How many planets will a country like India require?" He also sounded a warning to the developing countries to chart out a very different path of development when he said,

"God forbid that India should ever take to industrialization after the manner of the West." However, any system that is based on continuing prosperity and high levels of consumption in one set of countries and a substantially lower level of consumption in other nations is not a sustainable solution.

For this reason, commenting on industrial countries, Gandhi said: "The focusing of targets by them for population stabilization in the developing countries must now be backed by their willingly fixing targets for controlling and bringing down resource use in their own countries." Gandhi shunned the idea of developing countries emulating the culture and consumptive lifestyles of the developed countries. The British writer Edward Thompson once mentioned to Gandhi that wildlife was fast disappearing in India, to which Gandhi's response was, "Wildlife is decreasing in the jungles but it is increasing in the towns." He emphasized the fact that a society can be judged by the way it treats its animals. Environmentalists discovered the truth of this saying only toward the end of the twentieth century.

Gandhi's thinking was neither pompous nor based on abstract philosophy. His actions throughout life, including, for example, cleaning the primitive toilets of poor people and inducing them to do the same, emphasized a practical reality. At the Johannesburg Summit, clean water and sanitation were seen as twin challenges necessary to meet if the need for clean water is to be satisfied. Gandhi once said, "Sanitation should occupy the foremost place. There is a fine Latin proverb which says that 'a healthy mind is possible only in a healthy body'." The importance of Gandhi and the articulation of his beliefs underlines the need for leaders who have mass appeal to influence thought and action toward sustainable development. Methods of measuring sustainable development, the deployment of technologies to attain it, and policy initiatives to bring this about will remain incomplete unless human beliefs and values progress in the same direction. Although the Johannesburg Summit developed a ten-year framework, this has to rest on the changing of value systems and a movement away from consumptive lifestyles and a preference for production systems that clearly protect the environment and ensure the efficient use of natural resources.

REFERENCES

Clark, W., and N.M. Dickson. 2003. Sustainability science: The emerging research program. *Proc. Natl. Acad. Sci. USA* **100**:8059–8061.

Cosgrove, D. 1984. Social Formation and Symbolic Landscape. Madison: Univ. of Wisconsin Press.

Crutzen, P.J., and W. Steffen. 2003. How long have we been in the Anthropocene era? *Clim. Change* **61**:251–257.

Crutzen, P.J., and E.F. Stoermer. 2000. The Anthropocene. *IGBP Newsletter* **41**:17–18.

Feyerabend, P. 1975. Against Method. Outline of an Anarchistic Theory of Knowledge. London: New Left Books.

Finnimore, M., and K. Sikkink. 1998. International Norm Dynamics and Political Change. *Intl. Org.* **52**:887.

Fuller, R.B. 1962. Education Automation. Carbondale: Southern Illinois Univ. Press.

Gallopín, G.C. 1999. Generating, sharing and using science to improve and integrate policy. *Intl. J. Sust. Dev.* **2**:397–410.

Gallopín, G.C., S. Funtowicz, M. O'Connor, and J. Ravetz. 2001. Science from the 21st century: From social contract to the scientific core. *Intl. Soc. Sci. J.* **168**:219–229.

Haraway, D. 1991. Situated Knowledges: The Science Question in Feminism and the Privilege of Partial Perspective. In: D. Haraway, Simians, Cyborgs, and Women, pp. 183–201. London: Routledge.

Huggett, R.J. 1999. Ecosphere, biosphere or Gaia? What to call the global ecosystem. *Glob. Ecol. & Biogeogr.* **8**:425–431.

Humboldt, A. von. 1807. Ansichten der Natur. Tübingen: J.G. Cotta.

ICSU(Intl. Council of Sci. Unions). 2002. Science and Technology as a Foundation for Sustainable Development. Dialogue Paper for the Fourth Meeting of the World Summit on Sustainable Development Preparatory Comm. Bali, Indonesia. Paris: ICSU.

ICSU, ISTS, and TWAS. 2002. Science and Technology for Sustainable Development: Consensus Report and Background Document for the Mexico City Synthesis Conf., 20-23 May 2003. Series on Science for Sustainable Development, No. 9. Paris: ICSU.

IPCC (Intergovernmental Panel on Climate Change). 2001. Climate Change 2001. The Scientific Basis. Working Group I Contribution, Third Assessment Report of the IPCC, ed. J.T. Houghton, Y. Ding, D.J. Griggs et al. Cambridge: Cambridge Univ. Press.

Juma, C. 2002. The global sustainability challenge: From agreement to action. *Intl. J. Global Env. Issues* **2**:1–14.

Kates, R.W., W.C. Clark, R. Corell et al. 2001. Sustainability science. *Science* **292**:641–642.

Khoshoo, T.N. 1995. Mahatma Gandhi. An Apostle of Applied Human Ecology. New Delhi: Tata Energy Research Institute (TERI).

Kirchner, J.W. 2002. The Gaia hypothesis: Fact, theory and wishful thinking. *Clim. Change* **52**:391–408.

Li, B. 1996. Selected Poems of Li Po, transl. D. Hinton. New York: New Directions Publ.

Lovelock, J. 2003. The living Earth. *Nature* **426**:769–770.

Lubchenco, J. 1999. Entering the century of the environment: A new social contract for science. *Science* **279**:491–497.

Lucht, W., and C.C. Jaeger. 2001. The Sustainability Geoscope: A Proposal for a Global Observation Instrument for the Anthropocene. In: Contributions to Global Change Research, pp. 134–140. Bonn: German Natl.Committee on Global Change Research.

Martens, P., and J. Rotmans. 2002. Transitions in a globalizing world. Lisse: Swets & Zeitlinger.

Mitchell, W.J.T., ed. 1994. Landscape and Power. Chicago: Univ. of Chicago Press.

Moldan, B., S. Billharz, and R. Matravers. 1997. Sustainability Indicators: Report of the Project on Indicators of Sustainable Development. New York: Wiley.

Moffat, I., N. Hanley, and M.D. Wilson. 2001. Measuring and Modelling Sustainable Development. New York: Parthenon.

Parris, T.M., and R.W. Kates. 2003. Characterizing and measuring sustainable development. *Ann. Rev. Env. Res.* **28**:13–29.

Petrarca, F. 1933. Letter to Francesco Dionigi da Borgo San Sepolcro, April 26, 1336. In: Familiarum Rerum IV. Florence: Ed. V. Rossi.

Petschel-Held, G., A. Block, M. Cassel-Gintz et al. 1999. Syndromes of global change – A qualitative modelling approach to assist global environmental management. *Env. Model. & Assess.* **4**:295–314.

Raskin, P., T. Banuri, G. Gallopin et al. 2002. Great Transitions. The Promise and Lure of Times Ahead. PoleStar Series Report No. 10. Boston: Stockholm Env. Institute.

Sachs, W. 1999. Planet Dialectics: Explorations in Environment and Development. New York: Zed Books.

Sahagian, D., and H.-J. Schellnhuber. 2002. GAIM in 2002 and beyond: A benchmark in the continuing evolution of global change research. *IGBP Newsletter* **50**:7–10.

Samson, P.R., and D. Pitt, eds. Biosphere and Noosphere Reader. London: Routledge.

Schellnhuber, H.-J. 1998. Earth System analysis: The scope of the challenge. In: Earth System Analysis: Integrating Science for Sustainability, ed. H.-J. Schellnhuber and V. Wenzel. Heidelberg: Springer.

Schellnhuber, H.-J. 1999a. Earth system analysis: Integrating the human factor into geophysiology. *Gaia Circular* **2**:12–13.

Schellnhuber, H.-J. 1999b. Earth system analysis and the second Copernican revolution. *Nature* **402**:C19–C23.

Tilman, D., K.G. Cassman, P.A. Matson, R. Naylor, and S. Polasky, 2002. Agricultural sustainability and intensive production practices. *Nature* **418**:671–677.

UN (United Nations). 2000. United Nations Millennium Declaration. Resolution of the General Assembly 55/2. New York: United Nations.

Vernadsky, V.I. 1926. Biosfera. Leningrad: Nauchoe Khimikoteknicheskoe Izdatelstvo.

Vitousek, P.M., H.A. Mooney, J. Lubchenco, and J.M. Melillo. 1997. Human domination of Earth's ecosystems. *Science* **277**:494–499.

Von Droste, B., H. Plachter, and M. Rössler, eds. 1995. Cultural Landscapes of Universal Value. Jena: Gustav Fischer Verlag.

Western, D. 2001. Human-modified ecosystems and future evolution. *Proc. Natl. Acad. Sci. USA* **98**:5458–5465.

18

What Kind of System of Science (and Technology) Is Needed to Support the Quest for Sustainable Development?

G. C. GALLOPÍN

Economic Commission for Latin America and the Caribbean (ECLAC), Santiago, Chile

"The complete truth does not live in one dream but in several dreams..."
— *Pier Paolo Pasolini*

ABSTRACT

This chapter concentrates on the role of science and technology in humankind's efforts to change course toward sustainable development, in the context of deep transformations intensifying as globalization interweaves and clashes with ecological interdependence. Nine fundamental ("nodal") issues, which should be addressed by a scientific and technological system better able to support good global stewardship, are discussed in terms of their strategic implications for global sustainability. These include the basic unit of analysis for Earth system science, integration of research, the criteria of truth, inclusion of qualitative variables, dealing with uncertainty, incorporation of other knowledges, interparadigmatic dialogues, science–policy interface, and stakeholder involvement.

The adaptive management approach, often proposed as a preferred strategy for the use of science and technology for global sustainability, is discussed. Finally, a resulting glimpse of some of the strategic traits of the scientific and technological system for global sustainability is presented, including the possibility of scientific holistic forms of understanding of the Earth system.

INTRODUCTION

The Earth system is currently operating under a no-analogue state (Steffen et al. 2004), with unprecedented magnitudes and rates of change provoked by human interference. Scientific information about the dynamics of the Earth system is accumulating rapidly, but many fundamental uncertainties remain, and, because of imperfect information (and many other reasons unrelated to science), human actions to steward the Earth toward global sustainable development are often ineffectual, tentative, or simply not taken.

The present condition of the Earth system seems to be characterized by massive and deep changes spanning the local to the global scales, in both the anthroposphere and the ecosphere.

The world is now moving through a period of extraordinary turbulence, reflecting the genesis and intensification of deep economic, social, political, and cultural changes associated with the current techno-economic revolution. In addition, the speed and magnitude of global change, the increasing connectedness of the social and natural systems at the planetary level, as well as the growing complexity of societies and of their impacts upon the ecosphere have resulted in a high level of uncertainty and unpredictability. This presents new threats, but also new opportunities, for humankind.

Current trends are seen to be unsustainable, both ecologically and socially (UNEP 2002; UNCSD 1997). The need for a change in direction was officially recognized at the United Nations Conference on Environment and Development (Rio de Janeiro, June, 1992) and reconfirmed at the World Summit on Sustainable Development in Johannesburg in September, 2002.

The complexity of the situation and problems is rapidly increasing (Gallopín et al. 2001; Munn et al. 1999) for a number of reasons, such as the following:

1. *Ontological changes:* Human-induced changes in the nature of the real world are proceeding at unprecedented rates and scales, and are also resulting in growing connectedness and interdependence at many levels. Molecules of carbon dioxide emitted by fossil-fuel burning (mostly in the North) join molecules of carbon dioxide produced by deforestation (mostly in the South) to force global climate change; an economic crisis in Asia reverberates across the global economic system affecting faraway countries.

2. *Epistemological changes:* Changes in our understanding of the world related to the modern scientific awareness of the behavior of complex systems, including the realization that unpredictability and surprise may be built into the fabric of reality, not only at the microscopic level (i.e., the well-established Heisenberg uncertainty principle) but also at the macroscopic level, as described below.

3. *Changes in the nature of decision making:* In many parts of the world, a more participatory style of decision making is gaining in practice, superseding technocratic and authoritarian styles. This, together with the widening acceptance of additional criteria, such as the environment, human rights, gender equality, and others, as well as the emergence of new social actors, such as the nongovernmental organizations and transnational companies, leads to an increase in the number of dimensions used to define issues, problems, and solutions and hence to higher complexity.

At the same time, there is a growing feeling from many quarters that science is not responding adequately to the challenges of our times, and particularly those posed by the quest for sustainable development.

The recognition that a new "Social Contract for Science" is necessary to address the current planetary situation, that "business-as-usual" in science will no longer suffice, that the world today is a fundamentally different world from the one in which the current scientific enterprise developed, has emanated from the mainstream scientific establishment itself (Lubchenco 1997). The challenge to focus on the linkages between the social, political, economic, biological, physical, chemical, and geological systems is seen as a current imperative; dynamic cross-systemic explanations are sought where static and reductionist models once prevailed (as emphasized by the Board of Directors of the AAAS; see Jasanoff et al. 1997).

This theme was also the focus of the World Conference on Science, which met under the rubric "Science for the Twenty-First Century" in Budapest in mid-1999 (ICSU 1999). The documents of the conference emphasize the need for a new relationship between science and society, for a reinforcement of scientific education and cooperation, the need to connect modern scientific knowledge and traditional knowledge, the need for interdisciplinary research, the need to support science in developing countries, the importance of addressing the ethics of the practice of science and the use of scientific knowledge, and other important issues. The conference, however, did not discuss the possibility that science itself may also be in need of change (other than mentioning the need for integration and particularly for interdisciplinary research between natural and social sciences).

Still, the nature of the challenge posed by the management of the Earth system is such that it is highly likely that business-as-usual in science and technology will be as ineffective and as dangerous as business-as-usual in management of the planetary system. This points to the need for changes in the science and technology system itself.

Changes in the methods, criteria, and conduct of science have occurred before; science has been constantly evolving throughout its history. Academic, "curiosity-driven" research gave way after World War II to "industrialized" (Ravetz 1996) or "incorporated" (Rose and Rose 1976) research as the leading form of science and technology system. Its associated form of intellectual property, "public knowledge," is rapidly being driven out of the leading fields (as biotechnology and information sciences) by "corporate know-how."

Those changes in science have not been independent of the unfolding of historical processes in the economic, technological, social, cultural, and environmental domains. The changes reflect, and impinge upon, the social practice and the public image of science and the issue of "quality assurance" of scientific understanding and research. A response to the need for socially relevant criteria for quality assurance has been the proposal of a "post-normal science" (Funtowicz and Ravetz 1992, 1993, 1999). Post-normal science has been developed to deal with complex science-related issues. In these, typically, facts are uncertain, values in dispute, stakes high, and decisions urgent. Science is applied under conditions that are anything but "normal," for the distinction between "hard," objective scientific facts and "soft," subjective value judgments is here inverted. Very often, hard policy decisions must be made where the only scientific inputs available are irremediably soft.

To put the role and potentialities of Earth system science in the proper context, it should be noted that lack of scientific knowledge and understanding are not the only, and not even the major, determinants of the present mismanagement of the planet[1]. In fact, the deep rooted ecological and social unsustainability of world development patterns reflect more the influence of vested interests, political myopia, societal inertia, international and national asymmetries of power, and the overlap of economic, ecological, cultural, political, social, and demographic processes generated by the intersection of globalization with the growing global ecological interdependence, than the limitations of scientific understanding. This means that the success or failure of the attempts to move toward global sustainability will be to a very large degree contingent upon the political processes toward joint action.

Having said that, science and technology can play a crucial role in charting the dangers and opportunities of the road ahead and provide usable knowledge for good global stewardship. The science and technology system affects directly "knowledge and understanding" (one of the ultimate drivers of the sustainability transitions; see Raskin et al. 2002) and also operates indirectly through other ultimate and proximate drivers (Figure 18.1). Throughout the remainder of the chapter, I focus on this particular role.

Many aspects that I mention here are also valid for sustainability science, or science and technology for sustainable development (Kates et al. 2001; ICSU et al. 2002; see also www.sustainabilityscience.org for abundant information), because Earth system science is essentially sustainability science focused on the planetary-level system.

NODAL ISSUES FOR EARTH SYSTEM SCIENCE

The problematic of global sustainability exhibits a number of traits which suggest that changes (or at least serious reexamination) of some fundamental aspects of scientific and technological research will be needed in order to improve the capacity of the science and technology systems to contribute to the sustainability of the Earth. The following are some of these areas that might be called "nodal" (Figure 18.2) in the sense that advances made there would reverberate through many strands of the fabric of scientific knowledge.

---

[1] The clearly insufficient global response to the threat of climate warming, despite the wide scientific agreement and international recognition of the seriousness of the problem, testifies to this.

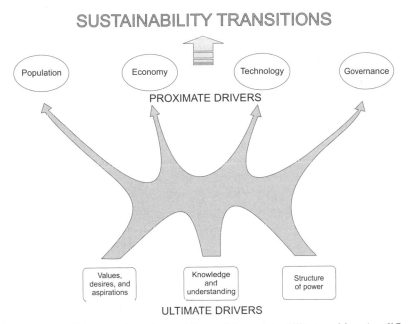

Figure 18.1 Ultimate and proximate drivers of the sustainability transitions (modified from Raskin et al. 2002).

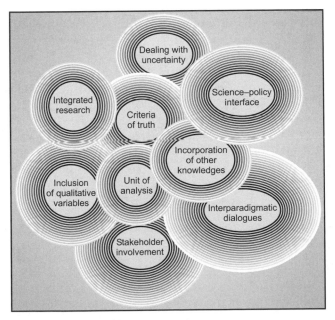

Figure 18.2 Nodal issues for Earth system science.

Unit of Analysis

The ecosphere and the anthroposphere are functionally coupled, with human actions and natural processes intimately interconnected in myriad nonlinear ways. Furthermore, this coupling seems to be intensifying: not only are human actions now reaching the global scale, but new global human processes, establishing new linkages, are now taking place. The current wave of *globalization* (in its trade, economic, cultural, ethical, and demographic dimensions) that is sweeping the planet is an obvious example. The meeting of socioeconomic globalization with ecological interdependence is arguably generating an unprecedented situation in the history of humankind.

The linkages between the anthroposphere and the ecosphere are not always obvious. The events and processes associated with the global economy of rubber linking the Amazon forests, the Malayan rubber plantations, Japanese tire factories, New Jersey marshes, and disease vectors (Kennedy 2001) attest to how subtle these linkages can be.

A clear implication for science and technology for global sustainability regards the appropriate unit of analysis[2] for Earth system science. I have argued (Gallopín et al. 2001), for science and technology for sustainable development in general, that the coupled socioecological system (Gallopín 1991) at different scales represents the fundamental unit of sustainable development and hence the unit of analysis of choice[3]. The global socioecological system has been termed the *Earth system* (Schellnhuber 1998), containing both the "ecosphere" and the "human factor" (anthroposphere plus the "global subject").

Over the last decade, great progress has been made in this respect; however, units of analysis for ecospheric and anthropospheric scientific studies are still largely disconnected (and, in the case of the latter, no generally agreed unit of analysis is yet discernible). A large part of the international global research, such as the International Geosphere–Biosphere Program (IGBP)[4] (including its integration activities, GAIM)[5], the World Climate Research Programme (WCRP)[6], and to some degree, the socially oriented International Human Dimensions Programme on Global Environmental Change (IHDP)[7], is very much biased toward the ecological, or bio–geo–physical, subsystem and only includes the human component either as a box generating perturbations or as the generator of human activities and recipient of environmental impacts. In addition, financial support for IGBP is several times that of IHDP, which perhaps suggests a

---

[2]   The basic entity being analyzed by a study and for which data are collected in the form of variables (standard definition).

[3]   This, of course, does not exclude the use of other analytical units for special purposes and particular studies.

[4]   http://www.igbp.kva.se

[5]   http://gaim.unh.edu

[6]   http://www.wmo.ch/web/wcrp

[7]   http://www.idhp.uni-bonn.de

differential priority allocated to the natural and human components of the Earth system.

This unbalance represents an obvious and hard challenge to the scientific knowledge of the Earth system; filling this gap will certainly not be easy, but it points to a clear research priority. The understanding required for identifying, characterizing, and navigating the future trajectories of the Earth system and defining effective strategies for global sustainability cannot be obtained solely by studying the biophysical subsystems even if "human inputs" are taken into account. The dynamics of the whole system surely depends not only on the interactions between things and things, the geological, biological, atmospheric, or oceanographic components, and the interactions between these and the anthroposphere, but also on the interactions between humans and humans as well as between economic, social, demographic, institutional, and cultural factors. This non-decomposability is beautifully illustrated by a set of quite simple (and local) models of lake-and-managers socioecological systems (Carpenter et al. 1999), extensible to other ecosystems under management (Carpenter et al. 2002). The analysis of the behavior of these coupled models provided various insights of critical strategic importance for the sustainable management of shallow lakes. One of these was the demonstration that unwanted collapse can occur even if the ecosystem dynamics are perfectly known and management has perfect control of the human actors. It was also clear that these insights could not have been obtained by analyzing the lake dynamics and the societal dynamics separately.

One important step toward the integral characterization of the Earth system would be the identification of the minimum set of macro-variables needed to describe the state of the coupled ecosphere–anthroposphere Earth system.

Integrated Research

The fact that the basic unit of observation includes both human and natural subsystems makes Earth system science interdisciplinary by necessity. Integrative research is obviously not just about adding more variables, or broadening the scope to include a larger portion of reality; integration of scientific research in terms of relevance for decision making requires a *holistic approach* (looking at wholes rather than merely at their component parts) and an *interdisciplinary research style*.

Looking at the whole from a scientific viewpoint includes the identification and understanding of the most important causal interlinkages and, more difficult, understanding the dynamics of the system. Nonlinearities and self-organization play a crucial role in the generation of the counterintuitive behavior typical of many complex systems, including the Earth system. This implies that it is necessary to investigate how different components and processes interact functionally to generate system responses and emergent properties, i.e., how the system adapts and transforms itself. This is an area for deep basic and applied research. This understanding is currently much more developed for the biophysical components of the Earth system than for the anthropic ones, and both are more advanced than the understanding of the behavior and dynamics of the coupled socioecological system.

Interdisciplinary research is often required to obtain integration (ICSU 1999; Kates et al. 2001). As with the case of integration, there is a large gap between the rhetoric and the practice of interdisciplinary research. It is not enough to put together a group of researchers from different disciplines to work on a project; it is also necessary to establish a true dialogue between the disciplines, an iterative and interactive process of mutual education and learning. This transformative dialogue is what differentiates interdisciplinary from multidisciplinary research (Thompson 1990). Education and training in how to perform interdisciplinary research is often lacking in most education systems; this is an area in which changes are required.

Criteria of Truth

The criteria used to decide what is "true" (or better, the falsification criteria used to reject scientific hypothesis) and other rules of science need to be reexamined for their adequacy in Earth system science. It is important to ask to what degree (if any) and in which way the existing rules of scientific enquiry, criteria of truth, and practice of science need to be modified in Earth system science. Research frequently focuses on narrow, quantifiable aspects of the problems, thus inadvertently excluding from consideration potential interactions among different components of the complex biological systems of which humans are a part.

Occam's Razor is a good example of a scientific guideline that might be changed in the new context. The rule, as usually stated, is "one should not increase, beyond what is necessary, the number of entities required to explain anything" is still valid when addressing a vastly complex unit of analysis. However, the characterization of "what is necessary" may need drastic broadening to account for the interlinkages between the object of study and other parts of reality, in line with Einstein's aphorism: "Everything should be made as simple as possible, but not simpler."

One example of the differences involved in current science that is applicable to Earth system science is the tension and shifting dominance between the analytical and integrative streams in ecology (Holling 1998). The differences between streams include, among others, basic assumptions on causality, criteria of truth, and epistemological acceptability and evaluation criteria (see Table 18.1).

The *analytical stream* focuses on investigating parts. It emerges from traditions of experimental science, where a narrow enough focus is chosen to pose hypotheses, collect data, and design critical tests to reject invalid hypotheses. Because of its experimental base, the chosen scale typically has to be small in space and short in time.

Table 18.1 Comparing the two streams of the science of ecology (Holling 1998).

Attribute	Analytical	Integrative
Philosophy	• Narrow and targeted • Disproof by experiment • Parsimony the rule	• Broad and exploratory • Multiple lines of converging evidence • Requisite simplicity of the goal
Perceived organization	• Biotic interactions • Fixed environment • Single scale	• Biophysical interactions • Self-organization • Multiple scales with cross-scale interactions
Causation	• Single and separable	• Multiple and only partially separable
Hypotheses	• Single hypotheses and nulls rejection of false hypotheses	• Multiple, competing hypotheses • Separation among competing hypotheses
Uncertainty	• Eliminate uncertainty	• Incorporate uncertainty
Statistics	• Standard statistics • Experimental • Concern with Type I error (in hypothesis testing, rejecting the proposition when it is true)	• Nonstandard statistics • Concern with Type II error (failing to reject the proposition when it is false)
Evaluation goal	• Peer assessment to reach ultimate unanimous agreement	• Peer assessment, judgment to reach a partial consensus
The danger	• Exactly right answer for the wrong question	• Exactly right question but useless answer

The premise of the *integrative stream* is that knowledge of the system is always incomplete. Surprise is inevitable. There will rarely be unanimity of agreement among peers — only an increasingly credible line of tested argument. Not only is the science incomplete, but the system itself is a moving target, evolving because of the impacts of management and the progressive expansion of the scale of human influences on the planet.

This dualism between analytical and integrative approaches is a particular manifestation of the broader differences between the analytical and systemic approaches (Saner 1999; De Rosnay 1975).

The international research programs working on the Earth system belong more to the integrative than to the analytical stream. The challenge of the tight couplings between the natural and social dynamics of the Earth system is to bring this integration to a much higher level than is currently the case.

Inclusion of Qualitative Variables

Nonquantifiable factors are excluded too often from consideration, because the methods used (e.g., classical computer simulation models) cannot incorporate qualitative factors or, worse, sometimes anything not quantitative is simply rejected as nonscientific. The dynamics of the Earth system depend, however, on a large number of complex processes, many of which are not yet quantified, and others (such as cultural processes determining social values) may not be quantifiable even in principle. Still, qualitative factors can be as or more important than the quantitative ones in determining the behavior of the Earth system.

Two comments are relevant here. First, in a number of cases, rigorous (even mathematical) analysis of qualitative factors can be performed (Petschel-Held et al. 1999; Gallopín 1996; Puccia and Levins 1987). Second, even in the cases where a rigorous treatment of qualitative factors cannot be performed, they can be included (at least in narrative form) in the overall conceptualization of the problem or issue, insofar as they are deemed to be causally important. This is often the case with many cultural, social, and political factors, which may even be the dominant element in a problem.

In conclusion, a strong push toward developing rigorous methods and criteria to deal with qualitative information will be required for the science and technology system to be better able to serve the management of the Earth system.

Dealing with Uncertainty

Earth system science confronts many sources of uncertainty; some of them are reducible with more data and additional research, such as uncertainty due to random processes (amenable to statistical or probabilistic analysis) or that due to ignorance (because of lack of data or inappropriate data sets, incompleteness in the definition of the system and its boundaries, incomplete or inadequate understanding of the system). When we consider the complex socioecological systems involved in the Earth system, it is clear that those sources of uncertainty can be insurmountable in practice, even if not in principle. Moreover, fundamental, irreducible uncertainty may arise from nonlinear processes (e.g., chaotic behavior), in the processes of self-organization (e.g., Prigogine showed that the new systemic structure arising from the reorganization of the elements of a dissipative system can be inherently unpredictable even in simple chemical systems), and through the existence of purposeful behavior including different actors or goal-seeking agents. Furthermore, complex "self-aware" (or "reflexive") systems, which include human and institutional subsystems, are able to observe themselves and their own evolution, thereby opening new repertoires of responses and novel interlinkages. In those systems, another source of "hard" uncertainty arises: a sort of "Heisenberg uncertainty effect," where the acts of observation and analysis become part of the activity of the system under study, and so influence it in various ways. This is well known in reflexive social systems, through the phenomena of "moral hazard," self-fulfilling prophecies, and mass panic (Gallopín et al. 2001).

One implication of this situation is that, even in the case of the relatively simpler biogeophysical component of the Earth system, understanding and insight is absolutely not synonymous with capacity to predict. Equally, awareness of risks is not synonymous with capacity to reduce or control the risks. This is even more applicable to the study of the full Earth system including the bidirectional couplings between the ecosphere and the anthroposphere. It is thus open to question whether we are ultimately confronted with a control problem (Schellnhuber 1998, 1999) or with navigating through the phases of a coevolutionary adaptive cycle (Gunderson and Holling 2002).

Therefore, an engineering approach to global sustainability seeking to anticipate all critical situations and building the "perfect model" may not only be doomed to fail, but it could also be exceedingly dangerous for human civilization. The scientific quest for even better understanding and predictive capacity must be complemented by new research and priority-setting strategies that do not merely recognize uncertainty, but even embrace it, becoming part of the process of change as well as probing its transformation possibilities.

Incorporation of Other Knowledges

Reaching a useful and usable understanding of the sustainability, dynamics, vulnerabilities, and resilience of the Earth system will require a strong push to advance focused scientific research, including building up classical disciplinary knowledge from the natural and the social sciences, and an even stronger development of interdisciplinary and transdisciplinary research (Schellnhuber and Wenzel 1998; Kates et al. 2001; ICSU et al. 2002).

The challenge, however, goes beyond scientific knowledge itself. Many discussions and consultations on the role and nature of science and technology for sustainable development emphasize the importance of incorporating knowledge generated endogenously in particular places and contexts of the world, including, for example, empirical knowledge, knowledge incorporated into technologies as well as into cultural traditions (ICSU et al. 2002).

Science for sustainable development creates historic opportunities to use inputs from other forms of knowledge by exploring the practical, political, and epistemological value of traditional/local/empirical/indigenous knowledge. The incorporation of "lay experts" in the processes of public decision making and the research agenda makes good sense in terms of using the expertise that is available, even when it is found in unexpected places.

We lack, however, a comprehensive framework to accommodate the multiplicity of local knowledges that could be used in scientific research but have thus far remained largely unknown to research systems as potential sources of innovation. The key knowledge generated by the lay expert is often contextual, partial, and localized, and has not been easy to translate or integrate into a more scientifically manageable conceptual framework.

In addition to science and technology professionals, the participation of other social actors during the different phases of the scientific and technological research process, and in related decision making, can be crucial for a number of reasons (ECLAC 2002):

1. *Ethical*: The right of the sectors affected to participate in decisions that have a bearing on their well-being (such as the installation of a nuclear or chemical plant in their area) is undeniable.
2. *Political*: It is essential to guarantee society's control over research and development outputs, particularly those that have an impact on health and the environment.
3. *Pragmatic*: In certain cases (e.g., new agricultural technologies, new health treatments), it can be especially important to encourage the social groups who are the intended beneficiaries to have a sense of ownership over the scientific and technological knowledge. For this it may be essential to engage these groups at the R&D phases in order to incorporate their interests and perceptions into the process.
4. *Epistemological*: The complex nature of the sustainable development problematic, in which biogeophysical and social processes usually overlap, often makes it necessary to consider the different perceptions and objectives of the social actors involved. Also, it is increasingly clear that it is important to combine empirical knowledge built up by traditional farmers, other cultures, and ethnic groups with modern scientific and technical knowledge (the constructive combination of diverse types of relevant knowledge).

The need to include other knowledges and perspectives in the science and technology enterprise poses important methodological challenges to science and technology for sustainable development. It requires the adoption of criteria of truth and quality that are broader than those accepted today by the science and technology community, yet not less solid and rigorous (otherwise, the relevance and credibility of science and technology could be gravely damaged).

To what degree, in which situations, what type, and in what form alternative knowledges will need to be incorporated into Earth system science are open questions that need to be addressed.

Interparadigmatic Dialogues

Given the need to foster a sense of common purpose and common understanding among different social actors (e.g., government, business, labor unions, NGOs, community organizations, political parties, minority groups), if sustainable development of the Earth system is to be reached, it will be necessary to move beyond traditional disciplinary thinking, and even beyond interdisciplinarity, toward intercultural, interinstitutional, interjurisdictional, and transdisciplinary exchanges (between scientists and nonscientists, between the modern and the traditional, between the north and the south). This will require constructive communication and cooperation between people who have very diverse mind-sets, world visions, and specific objectives. In short, we need what Mushakoji (1979) called an *interparadigmatic dialogue*.

Interdisciplinary activities in general are defined as involving people from different disciplines, working interactively toward a common purpose. In most cases the disciplines are scientific specializations, or at least professional areas, and therefore the participants share some kind of basic platform of beliefs (e.g., trust in the scientific method). The activities are typically directed to reach a common conceptualization of the issue or problem, and to combine the different knowledges and skills to reach the agreed goal (Thompson 1990).

Interparadigmatic activities (both for research and action) involve a more formidable challenge. A common platform of beliefs cannot be automatically assumed or imposed, and even the sense of common purpose may be lacking, at least initially. In such cases, the issue is not only how to articulate different worldviews, but also different (and legitimate) goals. The reduction of the plurality of viewpoints and interests to a single format (e.g., a mathematical model, a narrative representation) or to a single goal is neither possible nor desirable. The analysis of the objective and subjective conditions and approaches that can generate useful results in those situations, and the experimentation with the approaches, is an important component of the new kind of long-term research that is needed for Earth system science.

Science–Policy Interface

For Earth system science to be used effectively in the quest for global sustainability, the interface between science and policy needs to be better understood. For some scientists, the problem with the utilization of science by policy is that policy makers neither listen to, nor understand, scientists. Conversely, some policy makers view scientists as a closed community unable to act pragmatically or even to agree among themselves. An important requirement for an effective dialogue, for both scientists and policy makers, is to realize that both communities have much to learn from each other in addressing problems involving sustainable development, and that both are required to steward the Earth system to a safe course. The basis for the dialogue must be the recognition of the real differences in criteria and constraints exhibited by the two communities, which make them almost appear as two different subcultures. For instance, scientists (particularly those working in the analytical streams of science) typically dislike making conclusions and offering recommendations until they are satisfied that all necessary data have been collected and alternative hypotheses have been disproved; they also reject subjectivity. By contrast, policy makers are required to act when needed even if scientific knowledge is seriously incomplete; the incorporation of subjective information and value judgments is part of their trade.

Science–policy dialogues are one of the basic loci of integration between understanding and action. Mechanisms to implement the dialogue and to utilize science for policy must include the capacity to make responsible judgment and adequate interpretation of the evidence. The high complexity of the Earth system, as well as its natural and societal subsystems, implies a (often high) degree of irreducible uncertainty that should not lead to policy paralysis. On the other hand, scientific uncertainty should not be interpreted as total ignorance and a license for "anything goes" in the policy realm. Sometimes policy makers, particularly the powerful lobbyists fighting for their interests, are only too happy to reach this conclusion.

We must recognize, however, that in many cases scientific research is not producing the kind of understanding usable by policy makers (Baskerville 1997). Sometimes scientific questions are posed too narrowly, the scales of work are incommensurable with those required, and the policy concerns are not acknowledged.

One way to address this problem is to involve policy makers (personally or, at least, through their technical advisors) from the beginning of a scientific enterprise to identify questions, variables, and indicators usable for policy making. Including them at the outset usually makes it easier to provide policy-relevant knowledge, whereas trying to include them later is usually much more difficult.

Another important reason for early dialogues between science and policy is to ensure that the potential public impact of the research is considered with sufficient anticipation (e.g., by researching risk-avoiding strategies at the same time that risks are investigated). Innovative experiments on how to generate a dialogue and indeed a partnership between science and policy are needed. One of those new attempts is the "Science and Policy Partnership for Sustainability," described online at http://www.consecol.org/Journal/editorial/spps.html.

Stakeholder Involvement

It is clear that the necessary actions to ensure global sustainability cannot be taken by just some countries, powerful as they may be. A collective effort is required involving all regions of the planet.

The potential of the science and technology system to contribute critically to the global sustainability transition is connected to its capacity (and willingness) to incorporate the perspectives and concerns of the major stakeholders around the world, to ensure the relevance of the orientation of Earth system research to collective decision making. This requires the involvement of scientists and technologists in broad processes of international (and perhaps also regional) consultation and dialogue with the relevant stakeholders. One useful model (for the climate dimension) has been the Intergovernmental Panel on Climate Change (IPCC, http://www.ipcc.ch/), involving sustained bi-directional interactions between the science and technology community and the policy community.

The building up of the collective will and the collective institutional limbs of the "global subject" (Schellnhuber 1998, 1999) is essentially a political task — one in which the science and technology system needs to play a facilitating role. One possible direction is the involvement of the Earth system community with policy makers and stakeholders in the construction of alternative scenarios for the Earth system, making use of available and ad hoc simulation models, qualitative analysis, and goal-setting to explore (not predict!) alternative future trajectories of the global system in its ecospheric and anthropospheric dimensions (Schwartz 1991; Gallopín et al. 1997; Cosgrove and Rijsberman 2000). This can be very powerful in making clear uncertainties and irreversibilities (biophysical and social) that are critical for humankind (thus helping to shape the research agenda) as well as the magnitude and complexity of the problem that requires the reconciliation of conflicting and disparate interests.

ADAPTIVE MANAGEMENT

Accepting the fact that there is, and always will be, an element of irreducible uncertainty in the management of the Earth system leads to the recognition not only of the value of the already available information to guide policies, but also of the new information generated by the response of the system to the policy actions implemented.

As put by Holling (1998), successfully managed systems are ever-changing targets because they release the resources for new kinds of human opportunity and expose new classes of human risk. This is the core idea behind *adaptive management*, a structured process of learning by doing. The approach has originated and been tested in context of natural resource management and environmental impact assessment (for an overview, see Holling 1978 and Walters 1986).

Adaptive approaches, explicitly incorporating uncertainty, contrast with *command and control approaches*, which presume knowledge is sufficient and the consequences of policy implementation are predictable. At the heart of adaptive management lies the simultaneous utilization of simulation models to organize knowledge and identify gaps, and large-scale experimental actions to probe into the dynamics of the system.

In practice, it involves conceptualizing the system under consideration through a series of iterations (usually workshops), which involve experts in the different relevant issues, including decision makers. Rather than starting with the favorite scientific questions of each discipline, the process of defining the system starts with the identification of the management goals, a range of alternative objectives, indicators of the accomplishment of the objectives, the management actions to consider, and the bounding of the system in time and space. Only then are the specific variables to be included in the system defined. Computer simulation models are often used as a tool to provide focus and articulation of the issues; however, for some problems, the building of models may not be relevant.

This modeling step is intended to serve three functions: (a) problem clarification and enhanced communication among scientists, managers, and other stakeholders; (b) policy screening to eliminate options that are most likely incapable of doing much good, because of inadequate scale or type of impact; and (c) identification of key knowledge gaps that make model predictions suspect. The modeling step in adaptive management planning allows, at least in principle, to replace management learning by trial and error (an evolutionary process) with learning by careful tests (a process of directed selection) (Walters 1997).

The result of this process is a set of plausible hypotheses about the behavior of the system under various management alternatives. Typically, no action will be superior under all scenarios (if one were, there would be no concern about uncertainty).

The collective understanding of the system (including the uncertainties that are critical in relation to the management goals) is then used to explore different management strategies. An important point is that management actions are evaluated not only in terms of their expected impact on the goals, but also in terms of their role in probing the properties of the system, as sources of future information on the system.

Most often, knowledge gaps involve processes and relationships that have defied traditional methods of scientific investigation for various reasons. It also becomes apparent, in the modeling process, that the quickest, most effective way to fill the gaps would be through focused, large-scale management experiments that directly reveal process impacts at the spatiotemporal scales where future management will actually occur.

The design of management experiments then becomes a key second step in the process of adaptive management, and a whole new set of management issues arises about how to deal with the costs and risks of large-scale experimentation.

Walters (1997) made a critical assessment of the challenges for adaptive assessment using his extensive experience in aquatic natural resources management. According to him, major problems encountered in adaptive management planning included modeling problems due to difficulties in representation of cross-scale effects, lack of data on key processes that are difficult to study, and the confusion of factor effects in validation data. Experimental policies have been seen as too costly or risky, particularly in relation to monitoring costs and risk to sensitive species. Research and management stakeholders have seen adaptive policy development as a threat to existing research programs and management regimes, rather than as an opportunity for improvement. Proposals for experimental management regimes have exposed and highlighted some fundamental conflicts in ecological values, particularly in cases in which endangered species have prospered under historical management and would be threatened by ecosystem restoration efforts. Nevertheless, he concluded that there is much potential for adaptive management in the future, if we can find ways around these barriers.

He discussed four reasons for low success rates in implementing policies of adaptive management; all are institutional. First, modeling for adaptive management planning has often been supplanted by ongoing modeling exercises, based on the presumption that detailed modeling can be substituted for field experimentation to define "best use" policies. There is a further presumption, in such exercises, that best use policies can be corrected in the future by "passively adaptive"[8] use of improved monitoring information. Second, effective experiments in adaptive management often have been seen as excessively expensive and/or ecologically risky. Third, strong opposition to experimental policies by people protecting various self-interests in management bureaucracies has increased. Fourth, there are some very deep value conflicts within the community of ecological and environmental management interests that have become crucial.

Most of the problems with the use of adaptive management identified by Walters apply, with variable importance, to the prospects for applying the approach to the global scale of the Earth system. In particular, some of the problems common to adaptive management become even more important at the Earth system level. One of them is the fact that we have a single system to manage, without replicates. Adaptive management addresses this situation through intimate articulation of modeling and experiment. Another concerns the timescales involved; to detect responses of the Earth system to an experimental intervention may take decades or longer. Finally, one of the most difficult situations is when the proposed management treatments are irreversible (Walters 1997).

The combination of the potential existence of "nasty irreversibilities" lurking in the state space of the Earth system, and the temporal maladjustment between the system responses and the times needed by society to react[9] in the context of irreducible uncertainties, poses very fundamental and unsolved problems to

---

[8] Active adaptive strategies involve the deliberate use of management actions in order to increase information about the system. Passive adaptive strategies involve making use of information as it becomes available, but not modifying the actions with the purpose of gathering information.

[9] Reaction times are affected by the inertia and very slow response of some essential components of the Earth system (such as ocean currents), which make the use of interventions as providers of information on the behavior of the system impossible or irrelevant. They are also affected by the abruptness of some changes once thresholds are crossed, which may exceed the societal reaction time.

adaptive management (or indeed, to any form of management) of the Earth system.

However, the apparently obvious alternative of building increasingly detailed models to bypass real-world experiments by making "virtual experiments" is not appropriate in view of the huge uncertainties involved and the high cost of a failure. Adaptive management seems to remain the best option for global stewardship for sustainability. In this situation, the involvement of the stakeholders becomes essential, and the concept of what is the "socially acceptable" level of risk, a fundamental operational element.

WHAT WOULD A SCIENCE AND TECHNOLOGY SYSTEM FOR GLOBAL SUSTAINABILITY LOOK LIKE?

A science and technology system internalizing the set of challenges discussed above would look quite different than today's dominant model. It would be much more exploratory, receptive to alternative ideas, and visibly more holistic (but not less rigorous) than today. Embracing uncertainty and incorporating the qualitative will lead to an enormous broadening of the universe of solutions (and of questions); new areas of research will be created. Its openness to other forms of knowledge, the interaction with other worldviews, with the problems faced by decision makers, and with stakeholders would result in new, richer ways to set research priorities.

The Earth science and technology system would be exploring, applying, and teaching a constellation of tools and methods rather than relying on a narrow set of models and tools; these tools and methods would be articulated through the search for unifying holistic principles. Flexible international research cooperation networks would be multiplied and strengthened, and interconnected with action-oriented local, regional, and global networks. It would provide indicators of progress toward global sustainability, research results, and capacity-building to policy makers and the civil society, thus supporting the unfolding of the collective global will and capacity to steer the Earth system sustainable development.

The emphasis on interdisciplinary activities and the opening to plural knowledges and perspectives will have immense consequences for the education and training of the new generations of scientists, as well as for the forms of communicating and articulating scientific understanding. Schellnhuber (1999) posits that there are three distinct ways to achieve holistic perceptions of the Earth system: the *"bird's-eye" principle* (observing from space), the *digital-mimicry principle* (constructing computer simulation models), and the *"Lilliput" principle* (building microcosms). The development of Earth system science (and of sustainability science in general) might make possible the growth of a fourth way, the *"direct apprehension" principle*, based on a more direct perception of the operations of wholes, combined with deep understanding of the workings of complexity. This would be a type of pattern recognition, which could be trained in much the same way that people learn to identify statistical regularities in a set of points plotted in a chart. It would be supported by advances in the organization, presentation, and visualization of information; make use of rational and emotional mechanisms of comprehension; and combine cognitive theories with a scientific understanding of the Earth system.

REFERENCES

Baskerville, G.L. 1997. Advocacy, science, policy, and life in the real world. *Conserv. Ecol.* **1**:9.

Carpenter, S., W. Brock, and P. Hanson. 1999. Ecological and social dynamics in simple models of ecosystem management. *Conserv. Ecol.* **3**:4.

Carpenter, S., W. Brock, and D. Lutwig. 2002. Collapse, learning, and renewal. In: Panarchy: Understanding Transformations in Human and Natural Systems, ed. L.H. Gunderson and C.S. Holling, pp. 173–193. Washington, D.C.: Island Press.

Cosgrove, W.J., and F.R. Rijsberman. 2000. World Water Vision: Making Water Everybody's Business. London: Earthscan Publ.

De Rosnay, J. 1975. Le macroscope: Vers une vision globale. Paris: Edition Du Seuil.

ECLAC (Economic Commission for Latin America and the Caribbean). 2002. Science and Technology for Sustainable Development. A Latin American and Caribbean Perspective. Workshop report, Seminars and Conferences Series 24. Santiago: ECLAC.

Funtowicz, S., and J. Ravetz. 1992. Three types of risk assessment and the emergence of post-normal science. In: Social Theories of Risk, ed. S. Krimsky and D. Golding, pp. 251–273. Westport, CT: Greenwood Press.

Funtowicz, S., and J. Ravetz. 1993. Science for the post-normal age. *Futures* **25**:735–755.

Funtowicz, S., and J. Ravetz. 1999. Post-normal science: An insight now maturing. *Futures* **31**:641–646.

Gallopín, G.C. 1991. Human dimensions of global change: Linking the global and the local processes. *Intl. J. Soc. Sci.* **130**:707–718.

Gallopín, G.C. 1996. Environmental and sustainability indicators and the concept of situational indicators: A systems approach. *Env. Model. & Assess.* **1**:101–117.

Gallopín, G.C., S. Funtowicz, M. O'Connor, and J. Ravetz. 2001. Science for the twenty-first century: From social contract to the scientific core. *Intl. J. Soc. Sci.* **168**: 219–229.

Gallopín, G.C., A. Hammond, P. Raskin, and R. Swart. 1997. Branch Points: Global Scenarios and Human Choice. Stockholm: Stockholm Environment Institute.

Gunderson, L.H., and C.S. Holling, eds. 2002. Panarchy. Understanding Transformations in Human and Natural Systems. Washington D.C.: Island Press.

Holling, C.S., editor. 1978. Adaptive Environmental Assessment and Management. Chichester: Wiley.

Holling, C.S. 1998. Two cultures of ecology. *Conserv. Ecol.* **2**:4.

ICSU (Intl. Council of Sci. Unions). 1999. Science International, Special Issue Sept. 1999. Paris: ICSU.

ICSU, ISTS, and TWAS. 2002. Science and Technology for Sustainable Development: Consensus Report and Background Document for the Mexico City Synthesis Conf., 20-23 May 2003. Series on Science for Sustainable Development, No. 9. Paris: ICSU.

Jasanoff, S., R. Colwell, M.S. Dresselhaus et al. 1997. Conversations with the community: AAAS at the millennium. *Science* **278**:2066–2067.

Kates, R.W., W.C. Clark, R. Corell et al. 2001. Sustainability science. *Science* **292**:641–642.

Kennedy, D. 2001. Black carp and sick cow. *Science* **292**:169.

Lubchenco, J. 1997. Entering the century of the environment: A new social contract for science. *Science* **279**:491–497.

Munn, T., A. Whyte, and P. Timmerman. 1999. Emerging environmental issues: A global perspective of SCOPE. *Ambio* **28**:464–471.

Mushakoji, K. 1979. Scientific Revolution and Inter-paradigmatic Dialogues. Manuscript. Tokyo: United Nations Univ. Press.

Petschel-Held, G., A. Block, M. Cassel-Gintz et al. 1999. Syndromes of global change. A qualitative modelling approach to assist global environmental management. *Env. Model. & Assess.* **4**:295–314.

Puccia, C.J., and R. Levins. 1987. Qualitative Modeling of Complex Systems. Cambridge, MA: Harvard Univ. Press.

Raskin, P., T. Banuri, G.C. Gallopín et al. 2002. Great Transition: The Promise and Lure of the Times Ahead. A Report of the Global Scenario Group. SEI PoleStar Series Report 10. Boston, MA: Stockholm Environment Institute.

Ravetz, J. 1996. Scientific Knowledge and Its Social Problems. New Jersey: Transaction.

Rose, H., and S. Rose. 1976. The Political Economy of Science. London: Macmillan.

Saner, M.A. 1999. Two cultures: Not unique to ecology. *Conserv. Ecol.* **3**:r2 [http://www.consecol.org/vol3/iss1/resp2].

Schellnhuber, H.J. 1998. Earth system analysis: The scope of the challenge. In: Earth System Analysis: Integrating Science for Sustainability, ed. H.-J. Schellnhuber and V. Wenzel, pp. 3–195. Heidelberg: Springer.

Schellnhuber, H.J. 1999. Earth system analysis and the second Copernican revolution. *Nature* **402**:C19–C23.

Schellnhuber, H.J., and V. Wenzel, eds. 1998. Earth System Analysis: Integrating Science for Sustainability. Heidelberg: Springer.

Schwartz, P. 1991. The Art of the Long View. New York: Currency Doubleday.

Steffen, W., A. Sanderson, P.D. Tyson et al. 2004. Global Change and the Earth System: A Planet Under Pressure. The IGBP Book Series. Berlin: Springer.

Thompson, K.J. 1990. Interdisciplinarity: History, Theory, and Practice. Detroit: Wayne State Univ. Press.

UNCSD (United Nations Commission on Sustainable Development). 1997. Critical Trends: Global Change and Sustainable Development. New York: United Nations.

UNEP (United Nations Environmental Programme). 2002. Global Environment Outlook 3 (GEO-3). London: Earthscan Publ.

Walters, C. 1986. Adaptive Management of Renewable Resources. New York: MacMillan.

Walters, C. 1997. Challenges in adaptive management of riparian and coastal ecosystems. *Conserv. Ecol.* **1**:1.

19

Institutions, Science, and Technology in the Transition to Sustainability

R. B. MITCHELL[1] and P. ROMERO LANKAO[2]

[1]Department of Political Science, University of Oregon, Eugene OR 97403–1284, U.S.A.
[2]Department of Politics and Culture, Autonomous Metropolitan University of Mexico at Xochimilco, C.P. 04960, D.F. México

ABSTRACT

This chapter explores the implications for a transition to sustainability of understanding the Earth as a complex, interdependent system in which human perturbations produce effects (with corresponding feedbacks to humans) that occur at multiple temporal and spatial timescales. Such an approach poses obstacles but also offers opportunities to better understand how human perturbations influence the Earth system and how to govern those perturbations and our human responses to the corresponding feedbacks. This chapter examines how existing human institutions, and globalization, contribute to environmental impacts on the Earth system and also evaluates efforts of alternative institutions to incorporate science and technology into the policy process in ways that will facilitate a transition to sustainability. Major institutional reforms will be needed for existing institutions to use science and technology effectively in the service of sustainability. They will need, in particular, to improve the integration of science into the policy-making process and the integration of policy concerns into scientific research in ways that help science provide more policy-relevant knowledge to those making economic and policy decisions without undercutting its scientific validity. These are complex tasks that will require many institutions to make dramatic changes in how they operate. Experiences with existing institutions that have been relatively successful at making such changes are used to illustrate the argument.

INTRODUCTION

Making existing patterns of human behavior sustainable poses the most challenging task currently facing humanity. The ability to use science and technology effectively in that enterprise — and to understand the possibilities and limitations of that ability — will be a necessary, though not sufficient, condition for success in moving social, political, and environmental relations toward sustainability. As the rest of this volume clarifies, scientists have begun to develop an understanding of the Earth system as consisting of a set of complex, dynamic, and interdependent processes and components that operate on multiple temporal and spatial timescales. A major insight of that growing understanding has been that humans, that is, the human component of the Earth system (the anthroposphere), are having an increasing impact on the other components of that system (the atmosphere, biosphere, hydrosphere, and lithosphere) and that those impacts are generating an increasing number of feedbacks to the anthroposphere that many humans consider undesirable. Many natural scientists already recognize that viewing the Earth as a system improves our understanding of the state, trends, and dynamics of that system and its component parts. An Earth system approach, however, also offers insight for social science understandings of factors that hinder the social ability to govern human perturbations of, and responses to, the Earth system. Indeed, the obstacles to effective governance can be seen as directly related to two major insights of an Earth system perspective, namely, that the system is characterized by *complex, interdependent processes* within and among system components and that those processes include *multiple temporal and spatial timescales*. As developed in the rest of this chapter, these characteristics *pose obstacles to the creation of policy-useful knowledge* and *inhibit effective governance of human perturbations and responses*. In this chapter we delineate the ways in which existing social systems illustrate these Earth system characteristics among humans and between humans and the environment (i.e., within the anthroposphere and between the anthroposphere and other Earth system components) and explore the successes and failures of social institutions to overcome the obstacles to effective governance posed by the Earth system.

HUMAN IMPACTS ON A COMPLEX, INTERDEPENDENT SYSTEM

Human action has been transforming the Earth for centuries (Turner et al. 1990). In some historical cases, human impacts on the environment have been sufficiently large to cause the demise or relocation of tribes, communities, and whole societies. Generally, however, human impacts have been small, local, and diverse. Increasingly, humans are altering the environment in large, global, and homogenous ways that produce impacts at the Earth system level. Numerous factors undoubtedly contribute to this shift toward human impacts occurring at the Earth system, rather than local environmental, level. Two particularly important contributors are population and globalization, with the former explaining why human impacts are increasingly large and global and the latter explaining why they are increasingly homogenous.

A wide range of global-scale environmental problems now illustrate that human environmental forcings or "signals" have begun to rival or surpass the magnitude of natural variation or "noise." Such global-scale human impacts do not replace but instead overlay an increasing number of local-scale impacts. Natural systems appear increasingly unable to absorb aggregate human perturbations in ways that allow re-equilibration at either prior states of the system or states that humans would consider desirable. If we have not yet reached the carrying capacity of the Earth, we certainly have exceeded the carrying capacity of certain components as evidenced by the decline of aquifers worldwide and the collapse of most top predatory fish species (Postel 1999; Myers and Worm 2003). This shift reflects not only increases in population, affluence, and technology but also deeper drivers and structures such as markets, government policies, and the political contexts that operate at different temporal and spatial scales to encourage and constrain human choices (Ehrlich and Holdren 1972; Commoner 1972; Waggoner and Ausubel 2002).

Globalization has, since the sixteenth century but particularly in its current phase, magnified the impacts that changes in population, affluence, technology, markets, policies, and other drivers of environmental damage have on the Earth system (Chase-Dunn et al. 2000). It has led to profound development changes that, in turn, have produced profound environmental and health impacts (Schaefer 2003). The flow of goods, services, capital, information, ideas, and people has expanded exponentially. Western — particularly American — culture, life-styles, material desires, and perspectives spread with increasing speed through numerous channels. These and other processes captured by the notion of globalization generally have reduced the collective diversity of socio–politico–economic systems. Most countries have power sectors based on fossil fuels, transportation sectors based on automobiles, agriculture sectors based on mono-cropping, and consumer preferences that look increasingly similar despite previously diverse cultures. Without engaging the question of whether globalization's net effects on the environment are positive, current patterns of globalization certainly have many negative environmental impacts. Globalization has propagated consumer culture and generated increased demand for both raw materials and end-use commodities, although at different rates across regions and sectors It has spread new technologies that may decrease the resources used to produce those commodities but does so only in those rare cases in which regulations ensure that prices for environmental resources reflect environmental externalities. In those settings in which globalization has increased affluence, its direct negative effects of increased consumption (and relocation of environmentally intensive activities) have been only partially offset by indirect declines in birth rates or improvements in technology.

The tendency for globalization to homogenize preferences, even while many cannot satisfy those preferences, may contribute most to Earth system stresses by decreasing the diversity of human behavior patterns causing consumption of particular resources or production of particular pollutants to be higher than they would be in a more behaviorally diverse world. A global preference for particular foodstuffs, products, or building materials focuses demand previously distributed over a variety of resources onto a relatively few resources, each of which will be more likely to collapse under the weight of that demand. If human- and livestock-powered transportation systems in China, India, and most African countries transition to fossil fuels (China and India have already begun), the lives of people in those countries will improve even as the burden placed on the Earth's atmosphere grows. Cultural homogenization, for example, by causing increasing consumption of protein from particular types of fish, such as salmon and tuna, transforms what might otherwise be local shortages into global collapses of certain fisheries. The increasing movement of people and goods also converts previously local problems into Earth system problems. Global distribution of particular products (let alone hazardous waste) implies that their disposal introduces any associated pollutants into a wide array of different ecosystems. Ships and airplanes intentionally and inadvertently introduce invasive species into habitats throughout the world, converting local pests into ubiquitous threats. In addition, both human and animal diseases spread around the globe at increasingly rapid rates (e.g., AIDS, SARS, and mad cow disease).

The effects of aggregate human behaviors on the Earth system, as well as the impacts and feedbacks of those effects on humans, prove increasingly difficult to understand or predict because of the complex, interdependent nature of the Earth system. Inputs from the anthroposphere to other components of the Earth system (hydrosphere, atmosphere, biosphere, and lithosphere) enter a system in which neither the impacts and feedbacks nor their causes can be straightforwardly identified. The wide range of inputs from the anthroposphere are themselves complex and interdependent. Human behaviors influencing global climate range from power generation and cement production to rice farming and livestock cultivation producing carbon dioxide, methane, and other chemicals and aerosols. Those influencing the fate of fish stocks involve not only intentional catch but also inadvertent by-catch, municipal and agricultural runoff, oil and chemical spills, and loss of habitat. Even if the Earth system itself were not dynamic, complex, and interdependent, we should not expect such a range of human forcings to produce linear and predictable effects on that system. Yet, even without clear models we can predict that the level of such forcings have become sufficiently large to place the system in a "no-analogue" situation, one likely to include numerous feedbacks that humans consider undesirable.

Impacts of perturbations of a complex, interdependent system can appear at temporal and spatial scales that bear little relationship to that of the initial perturbation:

- Current stratospheric ozone loss above the Antarctic is the result of releases of chlorofluorocarbons at the Earth's surface decades ago. Likewise, if all fossil-fuel use were to cease today, past emissions would have effects on the Earth system for decades if not centuries.

- Although nuclear power plants have been operating for only half a century, storage of uranium will alter environmental conditions for millennia.
- The killing of even a few individuals can push the population of an already threatened species over the brink to permanent extinction.

In these illustrations, the separation of cause from effect lies in the nature of the environmental processes involved, whether the absorption rate of certain chemicals or the recruitment rate of certain species. However, social, political, and economic forces often exacerbate the natural processes that separate causes from effects. Humans "solve" many environmental problems simply by displacing them: moving hazardous waste to distant repositories, dumping sewage into rivers or oceans, or disposing of atmospheric pollutants captured in catalytic converters or smokestack scrubbers in landfills. Globalization fosters a range of other, less self-conscious, distancings of cause and effect by reducing the awareness of both consumers and producers of the environmental effects of their consumption and production decisions: Europeans sitting in teak-paneled boardrooms rarely envision denuded southeast Asian forests, Americans eating salmon rarely see the "deserts" beneath fishpens in Norway, and those drinking their morning coffee or evening tea rarely know what pesticides were used or biodiversity lost to enable their consumption.

FOSTERING A TRANSITION TO SUSTAINABILITY

If the Earth system is characterized by complex, interdependent processes in which causes and effects are often distant temporally and spatially, then fostering a transition to sustainability proves particularly challenging. Success in that endeavor will require the creation of knowledge about the Earth system that is useful for governance and for promoting effective governance of human perturbations and responses of the system.

Creating Useful Knowledge

Complexity and interdependence introduce considerable uncertainty into our understanding of how (and which of) our behaviors as well as production and consumption systems are driving the Earth system to a no-analogue state, what the effects are, and how to respond to any negative feedbacks, either by reducing the levels of our perturbations or adapting to them. At a basic level, such systems prove analytically challenging because much, though not all, of modern "Western" science is based on a model that posits that we can "hold everything else constant" to identify and isolate the influence of one variable on another. Although such a position seems reasonable for understanding some elements of the system, it quickly becomes unreasonable for understanding the Earth system or even major components of it. In such complex, interdependent systems, change in even a single variable or parameter is likely to produce changes in many other variables, making *ceteris paribus* assumptions untenable. This conclusion is reinforced when many perturbations are changing simultaneously, as is the case for current human inputs to the Earth system. If many variables in a system are changing and each has causal links to numerous other variables, then untangling true cause-and-effect relationships becomes impressively difficult. Each cause has multiple effects, and each effect has a multitude of interacting causes. Even if there were perfect descriptive knowledge about the array of variables in the system, uncertainty would arise because of the practical and inherent obstacles to properly understanding and modeling the relationships among those variables. Generating knowledge about such systems that can improve governance of human perturbations of those systems requires changes to how scientific research is conducted as well as to how it is communicated.

Understanding the Earth system requires scientists to adopt more interdisciplinary, synthetic, and holistic approaches. As the Intergovernmental Panel on Climate Change (IPCC) reports have made clear, the net effect of human use of fossil fuels depends not only on the direct effects of introducing carbon dioxide and other greenhouse gases into the atmosphere but also on indirect and interactive effects on such processes as the uptake of carbon in trees and the ocean and increased global albedo due to increased vaporization of water. The global scientific community has begun to develop programs that foster research approaches that can address such problems. Some operate across a range of environmental issues, such as the International Geosphere–Biosphere Programme and the Scientific Committee on Problems of the Environment, whereas others have taken more targeted approaches, such as the Scientific Committee on Oceanic Research and the International Council for the Exploration of the Sea. All these programs recognize that deciphering the complex linkages even within single components of the Earth system requires collaboration across scientific disciplines. Increasingly, these and similar efforts have recognized that a full understanding of the Earth system requires the involvement of social as well as natural scientists, a fact institutionally evident in the form of the International Human Dimensions Programme on Global Environmental Change, projects such as the Global Carbon Project or Global Environmental Change and Food Systems, and the IPCC's inclusion of a wide range of social scientists in its work in all three working groups. Both scientists and funding agencies increasingly realize that understanding the dynamics of complex systems to foster sustainability always requires cooperation among natural and social scientists to integrate understandings of ecosystem functioning and human perturbations, and can often benefit from "place-based" research conducted "in ways particularly relevant to state and local decision makers" (Matson et al. 2003; NOAA 2003).

Complexity poses unique obstacles not only to understanding much Earth system science but also to communicating scientific findings effectively to both government policy makers and individual economic and political decision makers. First, the complexity and interdependence of the system introduce fundamental uncertainties. Making claims about how such systems work, predictions about their future states, or policy recommendations about how humans should interact with them have inherent and inescapable uncertainties that arise simply from the complexity of the systems. This does not imply that scientists cannot make claims, predictions, or recommendations but rather that they will need to do so based on consensus and probabilities more often than on proofs and confidence intervals. Economic and political decisions often cannot be delayed until scientific certainty is achieved; scientists can better inform those decisions by learning to communicate the often-large areas of agreement among scientists rather than highlighting remaining areas of disagreement. For example, the 1995 IPCC report's claim that "the balance of evidence suggests that there is a discernable human influence on global climate" illustrates that careful wording can allow scientists to reflect current science accurately while still raising public awareness more effectively than with other wordings. Groups like Seaweb and the Aldo Leopold Leadership Program help train scientists to communicate more effectively with the public and policy makers (Seawcb 2003; Aldo Leopold Leadership Program 2003).

Making environmental science truly useful to those who must change their behaviors and clarifying the factors driving those behaviors will require, however, deeper changes. It requires more than just "doing good science" and learning how to communicate it effectively. Research must be interdisciplinary because the things being studied require the expertise of various disciplines. If research is to influence policy and behavior, it must also be participatory. Improving the uptake of science and technology into decision making requires increasing stakeholder participation in the scientific process and increasing scientific participation in the policy process. Stakeholders who participate in scientific research tend to be more willing to accept the findings that flow from that science and use it in their decisions (Clark et al. 2002). Stakeholder participation can improve science by providing scientists with access to proprietary corporate data or sophisticated local knowledge regarding trends and causes. Involving stakeholders also increases their capacity to understand scientific findings as well as recommendations and is likely to build their commitment to sustainability as a goal. It also helps scientists to learn from stakeholders. Stakeholder participation makes science more influential by making it more salient, legitimate, and credible to the multiple audiences who must incorporate it into their decisions if a transition to sustainability is to occur (Clark et al. 2002). Broadening participation in science as well as decision making tends to produce decisions that are perceived as more legitimate and in which the problems and risks of both action and inaction are better understood, making successful implementation more likely (Fiorino 1996). Over the long term, "coproduction" of knowledge by scientists, policy makers, environmental managers, and stakeholders can increase the collective commitment to sustainability, to incorporating local concerns into science, and to incorporating science into decision making (Jasanoff 1996). In many countries, a crucial preliminary step will require strengthening civil society in both material and ideological ways so that citizens both can and want to participate meaningfully in scientific and decision-making processes. In too many parts of the world, people still lack the opportunities or the desire to participate in social, economic, and political arenas at a local and national level, let alone at the global level. Of course, successfully increasing participation in science will require avoiding scientific conclusions being dictated by economic or political pressures and will require that psychological processes of "group think" do not lead to the dismissing of important alternatives or blindness to potential nonlinearities and surprises.

Understanding and Reshaping the Science–Policy Interface

Using science and technology to foster a transition toward sustainability also requires building on experience to understand the processes and factors that foster (or inhibit) their incorporation into the policy realm. Contrary to common conceptions that either scientific knowledge is straightforwardly applied to policy problems or that policy makers simply ignore science to pursue political and economic goals, the interface between science and policy often involves a complex interchange reflecting the differing science and policy cultures, including differing relationships to information, institutional constraints, and a fundamental divide between environmental and economic concerns (Keely and Scoones 1999; Jasanoff and Wynne 1998).

Making science useful to policy makers requires bridging the gap that separates their differing cultures. The curiosity many scientists have to answer "basic research" questions is one which demands long-term investments with payoffs in knowledge that are often both uncertain and far off in the future. Government policy makers and economic decision makers, on the other hand, face nearer-term pressures in which both action and inaction may involve costly consequences. In policy making, views are more likely to be determined by political and economic power rather than truth and quality of research methodologies. Political constituencies often want economic or environmental solutions adopted before scientists can confidently say what consequences different policies imply. Economic decision makers — from fishers and farmers to corporate CEOs — often face market decisions about whether to go fishing, what crop to plant, or what power source to install long before anything close to full knowledge is available. Current political and economic contexts in most countries mean pro-environment decisions involve large costs for the decision maker's family, constituency, or stockholders in the short term, regardless of whether they are beneficial to other actors at some point in the future (Behn 1986). Scientists can, and appropriately often do, examine the status and trends of

decisions. Consider two research groups that built competing actor networks to support their divergent scientific positions on biodiversity conservation in Kenya (Cussins cited by Keely and Scoones 1999, pp. 21–22). One group used field experiments to support their claim of a relationship between the elephant concentrations and biodiversity loss, a claim that was contested by the other group. The former validated their claims by reference to the conventions of international scientific practice whereas the other validated theirs by reference to the perceptions of local stakeholders well-informed about local conditions. Although both groups had built strong links with stakeholders, the latter group's view was more widely accepted because it had built a broader network and had developed arguments that were more attractive to key local stakeholders. Such involvement of stakeholders in scientific enterprises engages social, economic, and political sectors in working for sustainability in ways that reverse the disaffection caused by the many cases in which policy makers rely on technical arguments to escape responsibility for politically difficult decisions (Keely and Scoones 1999).

Any successful governance for sustainability will be based on changes in the behavior of billions of people and on changes in the structural factors that constrain or foster certain behaviors such as markets, government policies, as well as production and marketing strategies of corporations. Inducing such changes is likely to be both easiest and most effective if scientific information is not only scientifically credible but is also perceived by lay publics and stakeholders as salient or relevant to their decisions and as having been produced through a fair and legitimate process that took their concerns and knowledges into account (Clark et al. 2002). The acceptance and incorporation of science into the policy process is likely to be fostered by decision-making processes that expect and welcome an active, more equitable and respectful interplay of scientists, key political and economic actors, and stakeholders so that a wide spectrum of knowledges and perspectives can be taken into account.

Effective Institutions for a Transition to Sustainability

Taking sustainable development seriously requires institutions that strive simultaneously toward environmental protection and improvement of human welfare (WCED 1987). This requires surmounting the traditional barriers that separate the governance of economic, social, and environmental affairs. Although what happens in the marketplace influences nature and vice versa, human governance has often been ignorant of or insensitive to these connections. Social, political, and economic relations among humans constitute a complex, interdependent system in its own right, and globalization within that system sets off dynamics in the anthroposphere that are often as dynamic, multi-causal, and poorly understood as those in the natural sphere of the Earth system.

Past institutional efforts at governance have taken three different approaches to environmental protection, the first one focusing primarily on economic growth and trade. To understand how this institutional setting operates, it is useful to examine recent development patterns. Globalization, and particularly trade liberalization, has produced quite varied patterns of participation in global trade, development, and environmental impacts. Global economic growth has produced economic benefits for some sectors, regions, and people but provided fewer benefits to, and often imposed economic and environmental costs on, many others. Developed countries' share of manufacturing exports, for instance, has declined recently while their share of technology-intensive, high-value added exports have increased, allowing these countries to promote technological and institutional innovation and improve their citizens' economic welfare. They have low poverty rates and average incomes almost 40 times those of the 20 poorest countries (World Bank 2000). They have begun to "decouple" economic growth from local environmental degradation through technological, economic, and institutional transformations, even as they continue to contribute to Earth system problems (e.g., CO_2) and displace environmental and social problems associated with raw material extraction, industrial production, and waste to developing countries (Fischer and Amann 2001, p. 28). Newly industrialized countries that developed industrial bases closely integrated into the global trading system in the 1970s and early 1980s saw a drop in absolute poverty levels during the 1990s (Lo 1994; World Bank 2000, p. 3). In many developing countries, by contrast, the increase in manufacturing exports has entailed products that involve intensive exploitation of environmental and natural resources, the use of unskilled labor, and the low-skill assembly stages of transnational production chains (Fischer and Amann 2001; UNCTAD 2002). In these countries, the number of people in poverty rose from 1.2 to 2.8 billion from 1987 to 1998 (World Bank 2003). Rents from depleting natural capital are used to make debt payments, transferred to developed nations through deteriorating terms of trade, or lost to economic inefficiency and political corruption rather than being used to develop technological and human capital.

These changes have often been driven by market and systemic forces that have not been self-consciously managed. To the extent they have been governed, the dominant influences on development have been institutions and organizations focused primarily on economic issues, such as the World Trade Organization (WTO), the Organization for Economic Co-operation and Development (OECD), the World Bank, and the International Monetary Fund (IMF). After World War II, major developed country governments established these institutions to foster development, originally within a welfare paradigm and more recently within a neoliberal paradigm. GATT and WTO efforts have sought to reduce or eliminate tariffs, commodity cartels, subsidies, and regulatory standards and expand bilateral, regional, and global trade agreements to promote international trade and protect patents, copyrights, and trademarks (Schaefer 2003). Pressures from the World Bank, IMF, and developed countries have led many developing countries to adopt a broad range of structural reforms with deep implications for those governments relationships with other governments (trade and financial markets), with domestic markets (privatization and deregulation), and with citizens and workers (reduced health and education expenditures and restructured labor markets) (Gwynne and Kay 2000; Harris 2000; Schaefer 2003). These policies, however, have often also reduced expenditures on environmental protection, weakened environmental regulations, and produced increased pressures on natural resources.

The balancing of economic, human developmental, and environmental goals central to a sustainability transition is only beginning to be engaged seriously. Economic considerations consistently receive higher priority at international, regional, national, and local levels. Although many of the structures and programs of primarily economic institutions at all governance levels now incorporate environmental protection and poverty reduction as goals, these goals are rarely central to these organizations' missions and few have yet found ways to address them comprehensively and coherently. Institutional mandates and incentives generally reward a narrow sectoral focus, while cross-sectoral perspectives that might better identify and manage social and environmental issues often recieve few resources and little support (Wade 1997; Gibbs 2000; Varady et al. 2001). The greater priority given to economic over social and environmental considerations plagues the systemic level as well. Economic institutions are far more powerful and comprehensive in their coverage than are social and environmental institutions. The results of WTO dispute panels exemplify the many ways in which economic concerns receive more attention and resources as well as greater legal status and deference than social and environmental institutions. Trade liberalization has sometimes made natural resource protection more difficult in both developed and developing countries. Harmonization of regulatory standards and labeling sometimes leads to the lowering rather than raising of standards, for example, Codex Alimentarius pesticide residue levels for fruits and vegetables that are below those set by the US EPA (Schaefer 2003). Investment liberalization and deregulation have contributed to altering production location decisions of numerous corporations in ways that reflect new costs, regulations, profit considerations, and terms of market access but rarely reflect environmental impacts. As a result, many resource-intensive industries in which environmental protection involves large cost shares have moved to countries of low environmental standards (Varady et al. 2001; Schaefer 2003).

A second set of institutions has had mixed results when they have sought to address needs crucial to human development without examining obvious and directly related environmental dynamics within a complex Earth system. For example, the International Maize and Wheat Research Center (CIMMYT) was created in the 1960s to promote agricultural productivity and the "Green Revolution" via governmental subsidies and investments in infrastructure and marketing that strongly promoted the use of pesticides, fertilizers, and hybrid seeds. The Green Revolution boosted agricultural productivity that supported growing populations in many developing countries and, together with international trade, benefited consumers worldwide by increasing year-round availability of a variety of products at lower prices. These economic gains were accompanied by social and environmental costs, including (a) erosion and decreased fertility of soils, (b) increased ineffectiveness of pesticides against pests, (c) sterility, pesticide poisoning, and other health risks to farmers, and (d) increased disparity between wealthy farmers and poor peasants (Wright 1986; Simonian 1988). Similarly, efforts by the World Health Organization, the UN Food and Agricultural Organization, and developing country health agencies designed to promote the use of certain pesticides, whether to eradicate mosquitoes or to protect crops, initially produced significant health and economic benefits to local populations. In what has become a recurring pattern, however, these strategies became less effective as increasing pest resistance created an "arms race" between human efforts to eradicate pests and the pests' efforts to survive (Chapin and Wasserstrom 1981). More recently, policy makers and farmers have worked with scientists to develop the alternative of integrated pest management, a strategy that did not require a full understanding of the complex relationships among crops and plants, but the far simpler recognition that those relationships are complex and that taking advantage of, rather than circumventing, those complexities was likely to be more effective.

A third set of institutions have directly and explicitly addressed environmental problems. Transboundary environmental problems of all types have been addressed through hundreds of bilateral, regional, and global institutions that have quite varied success in mitigating these problems (Mitchell 2003). The effects and effectiveness of most environmental agreements have yet to be carefully analyzed. To date, research has identified considerable variation in their effectiveness. Agreements on stratospheric ozone depletion, dumping of wastes in the North Sea, and dumping of radioactive wastes globally are some of those that have been judged as quite influential; those addressing the world's natural and cultural heritage, tropical timber, and many fisheries have usually been judged as less effective (Miles et al. 2001; Victor et al. 1998; Brown Weiss and Jacobson 1998). Such judgments of these and other agreements depend considerably on the criteria used to evaluate effectiveness and on the analyst's skills in estimating what would have happened without the agreement. Considerable research is currently underway to understand the design features of, and conditions under which these international environmental institutions are effective at altering behavior. Yet, research into how well they do at incorporating scientific information, let alone at designing governance structures that respond to the complex, interdependent, and multi-scale nature of Earth system problems, is still in its infancy (Clark et al. 2002).

Designing governance to better control our perturbations of, and guide our responses to, the Earth system is likely to require significant changes from current policy approaches in at least three ways. First, environmental policies have often failed when based on a reductionist rather than system perspective. Effective governance requires policies developed in recognition that, if not precisely how, changes in one behavior and its drivers may initially produce intended and desirable outcomes through direct causal relationships but may subsequently produce unintended and undesirable feedbacks that offset those improvements or create unforeseen problems in arenas previously considered "unrelated." Besides unexpected but undesirable outcomes that have been part of the Green Revolution and pesticide control, consider that chlorofluorocarbons were initially welcomed as a solution to the health hazards of earlier refrigerants (such as ammonia and propane) and only later were proved detrimental to the global ozone layer. Prospectively, the complex set of interactions of pollutants in the atmosphere suggest (as noted by Crutzen and Ramanathan, this volume) that local air pollution policies requiring the reduction of aerosols (such as sulfates and black carbon) may exacerbate the problem of global warming because aerosols reflect sunlight and modify cloud properties in ways that counteract global warming. Like integrated pest management, marine reserves and protected areas are illustrative of new approaches being developed that recognize, in this case at a local level, that the best strategy for protecting complex environmental systems involves eliminating perturbations of some portions of those systems that can serve as buffers for the rest of the system.

The characteristics of human perturbations of the Earth system raise major hindrances to effective governance. The temporal and spatial distance between causes and effects, discussed above with respect to uncertainty, also raises political obstacles to mitigation or adaptation initiatives. First, most existing governance structures do not recognize the need for, nor do they facilitate, policy making in the face of the inherent and fundamental uncertainty of complex systems. We can draw lessons from rare cases of success such as the incorporation of precautionary principle language into institutional mandates at local, national, and global levels. The ability to regulate or ban an activity before the evidence of harm is conclusive, as illustrated in global regulation of ozone-depleting substances and national regulation in Europe of genetically modified foods, becomes particularly crucial in an Earth system in which evidence of harm may not be available until it is too late to take remedial action. Such strategies demand a willingness to incur immediate and clear costs to avert unclear and uncertain risks, a strategy that has yet to become commonplace in most of the world.

Second, the social concomitant of gaps between causes and effects is that those reaping the economic or other benefits of a behavior are not the same as those experiencing the environmental costs of it. Environmental problems are often assumed to involve tragedies of the commons in which all actors benefit from their own engagement in an activity but are harmed if others also do so. Displacement of environmental costs onto future generations or onto people in other regions or countries involves, however, a more malign social problem:

those engaged in an activity have little reason to stop unless they become concerned about the victims of that activity or those victims have the ability to punish or reward them. Spatial displacement of environmental problems generally imposes environmental costs onto the least enfranchised and least powerful of the world's population, whether in poor communities or in developing countries. Temporal displacement moves environmental costs onto future inhabitants of the planet whose voices and concerns can only be expressed through the actions of current inhabitants concerned about those future inhabitants. Global markets can obscure the economic impacts of consumer choices and of location decisions by corporations. Yet, alternative market mechanisms are being developed based on new relationships among consumers, labelers, and certifiers that clarify causal impacts and promote fairer terms of trade and better health and environmental standards for workers and consumers. Certified organic agriculture, silviculture, aquaculture, green manufacturing, and voluntary regulations under the International Organization for Standardization (ISO) offer alternative models that may facilitate the transition to sustainability but reflect several competing and paradoxical aspects.[1]

Demonstrations in Seattle, Genoa, and other cities, the nongovernmental summits that paralleled the 1992 United Nations Conference on Environment and Development and the 2002 World Summit on Sustainable Development, as well as other actions by civil society groups are raising awareness of the negative social and environmental implications of free trade and economic growth. An important institutional step in providing a voice to disenfranchised actors incurring environmental harms has been taken by those institutions that directly involve nongovernmental organizations, citizens groups, and other representatives of stakeholder interests in the policy-making process. Successes with participatory democracy at the domestic level have begun to influence policy making at the international level, as illustrated by the 1998 signing of the Convention on Access to Information, Public Participation in Decision-Making, and Access to Justice in Environmental Matters. Both global and regional trade agreements include provisions and subsidiary bodies mandated to protect the environment, some doing so in relatively transparent ways that facilitate monitoring and accountability by civil society groups. Even where such opportunities do not exist or are small, local communities and grassroots organizations are being successful in pressing governments, corporations, and international organizations to increase their attention to social and environmental concerns (Wilder 2000). As with participation in science, and as is evident in recent

---

[1] Certifiers of organic agriculture are constrained by contractual obligations under EU and ISO certification rules that were initially intended to give voice to national notions of social justice, environmental protection, and health. Producers receive an organic price premium, traded off, however, against production, certification and organization costs, and additional organizational burdens (e.g., new responsibilities, more work; see Muttersbaugh 2001).

experience with the Kyoto Protocol, expanding participation must involve mutual dialogue and understanding that lead to improved decisions rather than simply to compromises among the original participants of stakeholders.

Third, confronting the magnitude, diversity, and multiple scales of human impacts on the Earth system will require concerted effort at all governance levels, from international treaties to national governments to city administrations to individuals. Ensuring that scientific and technical knowledge facilitates such efforts requires long-term efforts to communicate science effectively to policy makers, as outlined above, a less arrogant attitude from scientists, as well as educating stakeholders and the general public about particular environmental problems and, more generally, improving scientific literacy. Designing models that provide resolution at various temporal and spatial scales can help policy makers to use science effectively. In addition, over the long term, processes of involving stakeholders in the coproduction of knowledge can go beyond simply improving understanding of what is known and of uncertainty to enhance the willingness and ability of communities to take action to protect the environment. Fortunately, the range of levels of environmental concern has led to institutional variation and innovation in addressing environmental problems. Despite the reluctance of the United States to commit to mitigating national emissions of greenhouse gases, many American cities and states are taking action to reduce their emissions (International Council for Local Environmental Initiatives 2003). BP (British Petroleum) is the most visible major corporation to take voluntary action to reduce corporate emissions (Browne 2002). Many national governments are going forward with unilateral actions to reduce their impacts on the Earth system rather than waiting until all contributors to the problem are ready to take action. The complexity and uncertainty of the Earth system make innovation, whether within or across institutions, central to our success at managing a transition to sustainability. Institutions must engage in self-conscious trial and error of low-likelihood-of-success but high-payoff experiments, to engage in critical evaluation of their performance against sustainability indicators, and to admit errors and failures when they occur, skills that institutions are notoriously poor at exercising (Social Learning Group 2001). This will require scientific and technological innovation as well as social innovation in more effectively incorporating both science and stakeholders into decision making.

CONCLUSIONS

Managing a transition to sustainability is a decades, indeed centuries, long task that will require human societies at all levels to improve vastly their ability to understand how their behaviors and the subsystems within which they are embedded alter the Earth system, to identify indicators that threaten sustainability, to find windows of opportunity, and to develop, adopt, and implement technologies and policies so that, over time, currently unsustainable development and behavior patterns are transformed into sustainable ones. To achieve sustainability, human institutions collectively should build social consensus regarding sustainability as a goal, find ways to identify threats to sustainability and their sources, prioritize among multiple threats and identify responses to them, and implement those responses effectively. Many existing human institutions are not primarily oriented toward environmental protection or improving the material foundations of citizenship; thus current development trajectories are unlikely to become sustainable if those institutions do not undergo dramatic change in the near- to medium-term future. The major institutional reforms needed for existing institutions to use science and technology effectively in the service of sustainability must do a better job of integrating science into the policy process and policy concerns into science, coordinating institutions across issues and across scales, promoting both scientific and policy innovation, increasing participation in both science and policy processes, and engaging more in processes of self-conscious institutional and social learning. These are large tasks that require, at least for many institutions, dramatic changes in how they operate. They constitute at least part of what is necessary but of course not sufficient if human societies are to succeed in a transition to sustainability. The question is whether human societies are up to the task.

ACKNOWLEDGMENTS

Ronald Mitchell's work on this paper was made possible by generous support from a Sabbatical Fellowship in the Humanities and Social Sciences from the American Philosophical Society. Patricia Romero Lankao's work on this paper was made possible by generous support from a Sabbatical Fellowship in the Center for Latin American Studies of the University of Arizona. We are grateful to Billie L. Turner II, William C. Clark, Oran R. Young, Louis Lebel, Martin Claussen, Tim Lenton, two anonymous reviewers, and the other participants at the 91st Dahlem Workshop for valuable insights, suggestions, and criticisms made on an earlier draft of this chapter.

REFERENCES

Alcock, F. 2001. Embeddedness and influence: A contrast of assessment failure in New England and Newfoundland. Belfer Center for Science and International Affairs Discussion Paper 2001-19 of the Environment and Natural Resources Program, Kennedy School of Government. Harvard University, Cambridge, MA.
Aldo Leopold Leadership Program. 2003. Aldo Leopold Leadership Program. http://www.leopold.orst.edu/. Accessed on: 19 June 2003.
Behn, R.D. 1986. Policy analysis and policy politics. *Policy Sci.* **19**:33–59.
Betsill, M.M., and R.A. Pielke, Jr. 1998. Blurring the boundaries: Domestic and international ozone politics and lessons for climate change. *Intl. Env. Affairs* **10**:147–172.
Botcheva-Andonova, L. 2001. Expertise and international governance: The role of economic assessments in the approximation of EU environmental legislation in eastern Europe. *Global Gov.* **7**:197–224.

Brown Weiss, E., and H.K. Jacobson, eds. 1998. Engaging Countries: Strengthening Compliance with International Environmental Accords. Cambridge, MA: MIT Press.

Browne, L.J. 2002. BP beats greenhouse gas target by eight years and aims to stabilise net future emissions. http://www.bp.com/centres/press/stanford/index.asp. Accessed on: 19 June 2003.

Carter, T.R., M. Hulme, and M. Lal. 1999. Guidelines on the use of scenario data for climate impact and adaptation assessment (Version 1). Geneva: Intergovernmental Panel on Climate Change, Task Group on Scenarios for Climate Impact Assessment.

Chapin, G., and R. Wasserstrom. 1981. Agricultural production and malaria resurgence in Central America and India. *Nature* **293**:181–185.

Chase-Dunn, C., Y. Kawano, and B. Brewer. 2000. Trade globalization since 1795: Waves of integration in the world-system. *Am. Sociol. Rev.* **65**:77–95.

Clark, W.C., R.B. Mitchell, D.W. Cash, and F. Alcock. 2002. Information as influence: How institutions mediate the impact of scientific assessments on global environmental affairs. Faculty Research Working Paper RWP02-044 of the Kennedy School of Government, Harvard University, Cambridge, MA.

Commoner, B. 1972. A Bulletin dialogue on "The closing circle": Response. *Bull. Atomic Sci.* **28**:17, 42–56.

Ehrlich, P.R., and J.P. Holdren. 1972. A Bulletin dialogue on "The closing circle": Critique. *Bull. Atomic Sci.* **28**:16, 18–27.

Fiorino, D.J. 1996. Environmental policy and the participation gap. In: Democracy and the Environment: Problems and Prospects, ed. W.M. Lafferty and J. Meadowcroft, pp. 194–212. Cheltenham: Edward Elgar.

Fischer, M., and C. Amann. 2001. Beyond IPAT and Kuznets curves: Globalization as a vital factor in analysing the environmental impact of socioeconomic metabolism. *Pop. & Envir.* **23**:7–47.

Gibbs, D. 2000. Ecological modernization, regional economic development, and regional development agencies. *Geoforum* **31**:9–19.

Gwynne, R.N., and C. Kay. 2000. Views from the periphery: Futures of neoliberalism in Latin America. *Third World Qtly.* **21**:141–156.

Harris, R.L. 2000. The effects of globalization and neoliberalism in Latin America at the beginning of the millennium. *J. Devel. Soc.* **16**:139–162.

International Council for Local Environmental Initiatives. 2003. Cities for Climate Protection Campaign. http://www.iclei.org/projserv.htm. Accessed on: 19 June 2003.

International Research Institute for Climate Prediction. 2002. Impacts of ENSO: References and links. http://iri.columbia.edu/climate/ENSO/societal/impact/resource/. Accessed on: 17 June 2003.

Jasanoff, S. 1996. Beyond epistemology: Relativism and engagement in the politics of science. *Soc. Stud. Sci.* **26**:393–418.

Jasanoff, S., and B. Wynne. 1998. Science and decisionmaking. In: Human Choice and Climate Change: The Societal Framework, ed. S. Rayner, and E. Malone, pp. 1–87. Columbus, OH: Battelle Press.

Keely, J., and I. Scoones. 1999. Understanding environmental policy processes: A review. IDS Working Paper 89. Falmer, Brighton: University of Sussex.

Lemos, M.C., T. Finan, R. Fox, D. Nelson, and J. Tucker. 2002. The use of seasonal climate forecasting in policymaking: Lessons from Northeast Brazil. *Clim. Change* **55**:479–507.

Lo, F.-C. 1994. The impacts of current global adjustment and shifting techno-economic paradigm on the world city system. In: Mega City Growth and the Future, ed. F.

Roland, E. Brennan, J. Chamie, F.-C. Lo, and J.I. Uitto, pp. 103–130. Tokyo: United Nations Univ. Press.

Matson, P., R. Naylor, and I. Ortiz-Monasterio. 2003. Sustainability in the Yaqui valley. http://yaquivalley.stanford.edu/. Accessed on: 19 June 2003.

Miles, E.L., A. Underdal, S. Andresen et al. eds. 2001. Environmental Regime Effectiveness: Confronting Theory with Evidence. Cambridge, MA: MIT Press.

Mitchell, R.B. 2003. International environmental agreements: A survey of their features, formation, and effects. *Ann. Rev. Env. Resour.* 28. 429 461.

Muttersbaugh, T. 2001. The number is the beast: A political economy of organic-coffee certification and producer unionism. *Env. & Plan. A* 34:1165–1184.

Myers, R.A., and B. Worm. 2003. Rapid worldwide depletion of predatory fish communities. *Nature* 423:280–283.

NOAA (National Oceanic and Atmospheric Administration). 2003. Decision support research: Bridging science and service. http://www.oar.noaa.gov/spotlite/archive/spot_risa.html. Accessed on: 19 June 2003.

O'Neill, T., and G. Hymel. 1995. All politics is local: And other rules of the game. Holbrook, MA: Bob Adams Inc.

Peterson, M.J. 1993. International fisheries management. In: Institutions for the Earth: Sources of Effective International Environmental Protection, ed. P. Haas, R.O. Keohane, and M. Levy, pp. 249–308. Cambridge, MA: MIT Press.

Postel, S. 1999. Pillar of Sand: Can the Irrigation Miracle Last? New York: W.W. Norton.

Schaefer, R.K. 2003. Understanding Globalization. New York: Rowman and Littlefeld.

Seaweb. 2003. What is Seaweb? http://www.seaweb.org/. Accessed on: 19 June 2003.

Shackley, S., and B. Wynne. 1995. Integrating knowledges for climate change: Pyramids, nets and uncertainties. *Global Env. Change* 5:113–126.

Simonian, L. 1988. Pesticide use in Mexico: Decades of abuse. *Ecologist* 18:82–87.

Social Learning Group. 2001. Learning to Manage Global Environmental Risks. Vol 1: A Comparative History of Social Responses to Climate Change, Ozone Depletion, and Acid Rain. Cambridge, MA: MIT Press.

Stanford Fisheries Policy Project. 2000. Stanford fisheries policy project. http://fisheries.stanford.edu/. Accessed on: 19 June 2003.

Turner, B.L. II, W.C. Clark, R.W. Kates et al. 1990. The Earth as Transformed by Human Action: Global and Regional Changes in the Biosphere over the Past 300 Years. Cambridge: Cambridge Univ. Press with Clark Univ.

UNCTAD (United Nations Conference on Trade and Development). 2002. Trade and Development Report, 2002. New York: UNCTAD.

VanDeveer, S. 1998. European politics with a scientific face: Transition countries, international environmental assessment, and long-range transboundary air pollution. Belfer Center for Science and International Affairs Discussion Paper E-98-09 of the Environment and Natural Resources Program, Kennedy School of Government. Harvard University, Cambridge, MA.

Varady, R., P. Romero Lankao, and K. Hankins. 2001. Managing hazardous materials along the U.S.–Mexico border. *Environment* 43:22–37.

Victor, D.G., K. Raustiala, and E.B. Skolnikoff, eds. 1998. The Implementation and Effectiveness of International Environmental Commitments. Cambridge, MA: MIT Press.

Wade, R. 1997. Greening the World Bank: The struggle over the environment, 1970–1985. In: The World Bank: Its First Half-century, ed. D. Kapur, J.P. Lewis, and R. Webb, pp. 611–734. Washington, D.C.: Brookings Institutions Press.

Waggoner, P.E., and J.H. Ausubel. 2002. A framework for sustainability science: A renovated IPAT identity. *Proc. Natl. Acad. Sci. USA* **99**:7860–7865.

Wilder, M. 2000. Border farmers, water contamination and the NAAEC environmental side accord to NAFTA. *Nat. Resour. J.* **40**:873–894.

World Bank. 2000. World Development Report 2000/2001: Attacking Poverty. http://www.worldbank.org/poverty/. Accessed on: 15 February 2003.

World Bank. 2003. Measuring poverty. http://www.worldbank.org/poverty/. Accessed on: 15 February 2003.

World Commission on Environment and Development. 1987. Our Common Future. New York: Oxford Univ. Press.

Wright, A. 1986. Rethinking the circle of poison: The politics of pesticide poisoning among Mexican farmers. *Latin Am. Persp.* **3**:26–59.

Back: Wolfgang Lucht, Bill Clark, Oran Young, and Ron Mitchell
Front: Alison Jolly, Gilberto Gallopin, Patricia Romero Lankao, S. Sreekesh,
       Ann Kinzig, and Crispin Tickell (not shown: Ottmar Edenhofer)

20

Group Report: Sustainability

A. P. KINZIG, RAPPORTEUR

W. C. CLARK, O. EDENHOFER, G. C. GALLOPÍN, W. LUCHT,
R. B. MITCHELL, P. ROMERO LANKAO, S. SREEKESH,
C. TICKELL, and O. R. YOUNG

INTRODUCTION

How can science, technology, and knowledge be harnessed more generally to advance the goals of global sustainability? We approached this ambitious question in two ways: by addressing how a better understanding of Earth system science could help society meet the sustainability challenge, and how a better understanding of the sustainability challenge could help Earth system scientists produce more useful research and development. We emphasize that Earth system science must stress both the biogeophysical and socioeconomic aspects of our world as well as the interactions between them (Steffen et al. 2003; Sahagian and Schellnhuber 2002). As such, science and technology must also be taken to include the full sweep of scholarly activities devoted to understanding this integrated Earth system, from the natural sciences to social sciences, humanities, and engineering.

We based our discussions on the modern view of the Earth system, explored more thoroughly in the other working groups, which views nature and society as a tightly coupled, dynamical system. This tight coupling between nature and society has characterized all of human history, although in the past it was evident predominantly at local or regional scales. We assume that most ancient societies did not wish their own demise, and thus posit a certain level of "self-awareness" or self-direction toward environmental stewardship. Nonetheless, history has seen the decline of many earlier civilizations (Redman 1999). These declines were not always directly related to environmental degradation, but the interplay among environmental, social, and political dynamics often led to their undoing.

Scholars have developed a substantial body of analytically derived and empirically grounded knowledge regarding the determinants of robustness or sustainability in human–environment relationships (Allen et al. 2003; Folke et al. 2002; Gunderson and Holling 2001; Redman and Kinzig 2003; Tainter 1988). Much of this knowledge has arisen from a widely shared sense among many researchers that the "tragedy of the commons" (Hardin 1968) is by no means a universal phenomenon and that many small-scale societies have invented successful methods for "governing the commons" (Berkes and Folke 1998; Ostrom 1990). Major contributions on the part of anthropologists, economists, ecologists, and political scientists have produced important insights regarding the conditions governing the occurrence of the tragedy, and of successful management of common resources as well (Bromley 1992; Burger et al. 2001; McCay and Acheson 1987; for a critical review of the literature, see Ostrom et al. 2002).

Today's situation is more daunting. Humanity is now a global environmental force, altering biological communities, biogeochemical cycles, landforms, and climate on unprecedented spatial scales, with unprecedented rates of change (NRC 1999; Turner et al. 1990; Vitousek et al. 1997). This raises two major scientific challenges. The first centers on the problem of scale: To what extent can we "scale up" findings derived from the study of small-scale systems to shed light on what might occur at larger, even global, scales (Folke et al. 2002; Kates et al. 2001; Ostrom et al. 1999; Young 2002)? Second, humanity operates today in an interdependent world in which global processes affect outcomes at the local level, and many small-scale processes can have global consequences, making the consideration of cross-scale interactions essential (Clark 2000; Gunderson and Holling 2001; Young et al. 1999). Past experience suggests that the world will not fall into sustainability by accident; a certain active reflection on the impacts of our behavior and actions as well as purposeful avoidance or amelioration of threatening consequences are required. This "self-awareness" must now come at a global level, commensurate with the scale of impacts.

One cannot glibly discuss a "global self-awareness" as if the means of achieving it were self-evident or simple. Humans have always fostered group identities that to some extent rested on the notion of "an other" or an enemy. Paralleling the unprecedented changes in the Earth's biogeophysical cycles are equally unprecedented changes in the scope and quality of human social organization. In the last century, for the first time, the world organized under an international banner — first as the League of Nations and now as the United Nations — that lends some precedent to global self-reflection and organization, and provides some insights for which strategies to follow and which to avoid. In the environmental and social realms, this global will contributed to the eradication of smallpox and other diseases, the incipient establishment of a global ethics that includes recognition of universal human rights, international technologies and policies to address the problems of ozone depletion and climate change, and hundreds of treaties protecting endangered species and reducing freshwater and marine pollutants.

What kind of world is humanity trying to create with this "global self-awareness"? Answers to this involve values that properly differ across times and places. Nonetheless, as a point of departure for our discussion, we found useful the goal of "sustainable development," originally crafted by the Brundtland

Commission and subsequently adopted by world leaders in Rio de Janeiro at the UN Conference on Environment and Development. In this view, society wants development "that meets the needs of the present without compromising the ability of future generations to meet their own needs" (WCED 1987). To emphasize the dynamic, open-ended character of nature–society interactions, we followed the broad consensus set forth by the world's scientific community at the 2002 Johannesburg World Summit on Sustainable Development: the goals of sustainability "should be to foster a *transition* toward development paths that meet human needs while preserving the Earth's life-support systems and alleviating hunger and poverty" (ICSU et al. 2002). A simpler formulation we particularly liked was "treating the Earth system as if we intended to stay."

It has long been clear that progress in promoting a transition toward sustainability will require substantial reforms in both the political and economic realms of human activity. Many have argued that it may also require fundamental changes in human values. What has less frequently been recognized is that successful navigation of a transition toward sustainability will necessarily be a knowledge-intensive activity as well. Science and technology have contributed significantly to the vigorous growth of human civilization and associated pressures on the environment that have put at risk "the freedom of future generations to sustain their lives on this planet" (Annan 2002). The question before us now is how the science and technology that unconsciously helped to get us into our present predicament can, through a program of purposeful, self-conscious research and development, best support society's larger effort to sustain our common future. In our discussions we came to three major conclusions:

1. The scientific community should place more emphasis on areas of research that are critical to the prospects for achieving sustainability. These include identification of "safe" and "benign" domains of operation for the Earth system; integrated assessment of production–distribution–consumption systems that provide the basis for a better quality of life; and research on the types of institutions, and institutional change, that will foster a transition to sustainability.

2. We recognize that the scientific community is not adequately organized to contribute its appropriate voice to the global dialogue on the world's future. The scientific community needs to consult regularly with, and learn from, stakeholders in setting the questions and determining priorities for science devoted to sustainability; it needs to use more appropriate techniques in constructing scenarios and analyzing uncertainties; and it must clearly communicate what is and is not known to the users of scientific information.

3. The above conclusions place additional organizational demands on the scientific community. This includes developing a global observation network (with both centralized and distributed elements) to assess the state of key components of the coupled society–environment system and

creating a process with the authority and stature of the Intergovernmental Panel on Climate Change (IPCC), but devoted to scientific assessment of sustainable development and environment on a wider basis. Finally, and perhaps most importantly, we strongly endorse a global effort to develop scientific capacity in the regions of the world where it is currently lacking or weak. Without this capacity, global scientific efforts to inform policy making will be irrelevant at best, and wrong at worst.

CARING FOR THE EARTH SYSTEM

Before turning to our major conclusions, we wish first to review briefly two approaches often put forth as effective ways of coping with human-made hazards in a complex, dynamic, but uncertain Earth system: adaptive management and participatory decision making. We note that there is no hope for humanity to control all of the infinite complexities of the Earth system. However, humans can aspire to manage effectively some of their activities, in ways that enhance social well-being while imposing progressively lower stresses on the Earth system, and correct or mitigate some of the ill effects of their own activities.

Adaptive Management

Adaptive management approaches emerged in the 1970s from a recognition that the complexity and scale of human–environment interactions makes it impossible to predict accurately the ultimate impact of candidate management strategies (Holling 1978; Walters 1986, 1997). Rather than futilely striving for such predictions or hoping to avoid entirely inevitable surprise, "adaptive" approaches sought to treat management and policy as system-scale experiments rather than one-off solutions. Monitoring and evaluation of systems response to management, and institutions capable of learning from error, are central to adaptive management strategies. Because adaptive management requires society's support for its policy experiments, the practitioners of adaptive management also became early advocates of the involvement of stakeholders, alongside scientists and policy makers, in the management process. These participants are brought into an on-going process of scientific and policy experimentation with related feedback systems intended to reduce uncertainties and improve knowledge about the operation of the environmental system, the human social system, and the interaction between the two. In the adaptive management model, the process allows for responses to be frequently evaluated and altered so that managers can be responsive to new information about how these systems work, but also to new information about or changes in the fundamental goals of those most directly affected by and involved in the human–environment system.

Adaptive management approaches have been applied to a wide range of local- and regional-scale environmental problems (Lee 1993; Gunderson et al.

1995; NRC 1996; Dovers and Mobbs 1997). A review of this experience suggests that much good has come from the struggle to implement adaptive approaches to environmental management. The approach nonetheless has yet to fulfill its promise in practice. As scholars have shown, the abilities of managers and other individuals to process information and/or respond to it are modest in comparison with the complexity of the problems and environments confronted (Bourdieu et al. 1983; Simon 1977). In addition, the difficulties encountered, even in relatively small-scale, self-contained settings, will surely be greatly exacerbated in the case of the Earth system transformations that concern us here. We review several of the challenges below, drawing in part on findings of other working groups at this Dahlem Workshop.

- *Uncertainty*: Adaptive management was explicitly developed to deal with uncertainty. The combination of integrated modeling and focused, large-scale management experiments have had some success in addressing uncertainty. However, these approaches have worked best where the uncertainty in question can be formulated in terms of specific probabilities or competing hypotheses. Uncertainties encountered in analysis of the Earth system have generally not been "tamed" to this level, leaving the likelihood that adaptive management approaches can usefully respond to them open to question.
- *Non-decomposability*: Elements that cannot be decomposed are unable to be analyzed piecemeal, one factor at a time, because of functional interlinkages. This is a common feature of ecological systems and has been a fact of life in all adaptive management experiences. Classical factor-isolation experiments are inadequate for analyzing such systems. When integrated models can be constructed for guidance, large-scale management experiments have been useful in addressing non-decomposability. In the case of the Earth system, non-decomposability characterizes the coupled social–ecological system. This presents a challenge for adaptive management, since fully integrated models of the social–ecological system at the level of the Earth system are only beginning to be developed.
- *Nonlinearities*: Nonlinearities are characteristic properties of complex systems, contributing to their unpredictability. Nonlinearities take many forms, such as thresholds, lack of proportion between causes and effects, and chaotic behavior. Some events may be so "nasty," rapid, and irreversible (see below) that they make adaptive management's process of policy formulation–implementation–feedback–assessment–reformulation less effective or impossible.
- *Timescales and response times*: In theory, adaptive management can address processes that occur on a variety of timescales. Two particular types of cases, however, challenge the effectiveness of an adaptive management framework: (a) when changes occur abruptly relative to the capacity of the management system to perceive and respond to the change; and (b) when

processes unfold very slowly or exhibit long time lags between a pressure and response. In both cases, management through experimentation tends to become less effective, as necessary feedback is received long after actions can be taken to avert those consequences. This calls into question the very "adaptive" nature of adaptive management, and this set of problems has not yet been satisfactorily addressed.

- *Singularity*: When the system is unique, standard statistical techniques based on comparisons across a number of replicates cannot be applied. This is common in a number of cases of managed ecosystems or natural resources and is obviously the case for the Earth system.
- *"Nasty" irreversibilities*: Irreversibility in itself is not necessarily a problem for adaptive management. The Earth system could, after all, move into a domain from which there is no return (in practice or principle), but which is generally more attractive than the previous one. Irreversibility is only threatening if the domain that cannot be escaped is a "nasty" one, that is, inhospitable for life and unsustainable for human society. Unfortunately, as detailed elsewhere in this volume, nasty irreversibilities seem to be lurking in many Earth system dynamics. In such cases, large-scale experiments might push the system into a trajectory of nonrecovery, the information thus gained would be irrelevant, and the system would irreparably deteriorate.

An additional challenge is the need to perform management experiments, and the institutional capacity to learn from those experiments. When experimentation risks causing harm to people's livelihoods or well-being, it is likely to be politically difficult and certain to be ethically inappropriate. Unfortunately, such issues are at the heart of sustainable development, calling into question the extent to which managers can experiment under an adaptive management framework. One possibility is to use passive adaptive strategies, which involve making use of information as it becomes available through "natural" experiments as they occur in the policy-making process. Decisions are not intentionally modified for the purpose of probing the system and gathering information on its dynamics.

Adaptive management has been offered by some as a panacea for dealing with uncertainty, obviating the need to invoke precautionary principles. The attributes of the Earth system, including (a) the long timescales involved in some processes, (b) the possibility of abrupt change, (c) the potential of nasty irreversibilities, and (d) the singularity of the Earth system, call this proposition into question. It seems almost certain that a "bright line" does not exist between those cases in which large-scale management experiments offer the best approach and those cases in which the precautionary principle should prevail.

The combination of high uncertainties and high stakes present in the Earth system makes it necessary to apply the best science-based management available, but one bounded by the degree of societal acceptance of the risks involved.

Therefore, the participation of members of society in management is essential for determining what constitutes socially acceptable risk.

Participatory Decision Making

Participatory decision making has been promoted as being capable of resolving many global and regional environmental problems. There are many benefits of such participation — not the least of which is securing people's rights in industrialized societies. However, can we presuppose that such inclusive systems automatically, or even usually, achieve outcomes consistent with fostering the long-term sustainability of the Earth system? There are many reasons to believe, in fact, that such processes are inherently ill-equipped to grapple with the complex dynamics that span large spatial and temporal scales. There may be a tension between "rightness of procedure" and "goodness of outcome" (Sen 1995).

Despite the difficulties, for reasons outlined below, we support participatory decision making whenever possible, without supposing that those processes would usually be democratic in the strictest sense of the word. Participants will assume different roles, assets, strategies, and opportunities for participation. Negotiation of the values society holds or will hold is legitimately within the purview of every stakeholder or citizen. Scientists ought not to have a stronger voice in that negotiation than any other citizen. Scientists can, and should, however, have a stronger voice concerning the likelihood of various future scenarios and their courses and impacts (both beneficial and harmful). Similarly, others with specialized knowledge (lawyers, historians, economists) will have particular roles to play. Final decisions that weigh scientific, economic, political, social, and cultural considerations are ultimately in the hands of legitimately recognized representatives or leaders, when they exist. Many countries, unfortunately, lack such legitimate leadership.

A number of factors are relevant to a consideration of the role of participatory decision making in addressing Earth system issues:

1. *Rationale*: Wide participation on the part of the interested parties may be advocated for normative or instrumental reasons. Normatively, participation may simply be regarded as good in itself. Instrumentally, participation may generate creative input into decision-making processes or increase the willingness of affected parties to implement or comply with commitments made during decision-making processes.
2. *Types of participation*: Different types of participation may be more or less relevant to decision making on Earth system issues. For example, interested or affected parties may be allowed (a) to vote or merely to comment, (b) to participate in agenda setting or final choices among options, or (c) to have equal weight or differential weight in making choices.
3. *Types of decisions*: Participation on the part of interested or affected parties may be more important for some types of decisions than others. For

example, participation is highly important in making basic value decisions (e.g., choices regarding social justice versus economic growth) but
relatively less important in making highly technical decisions (e.g., how
to measure concentrations of greenhouse gases in the Earth's atmosphere). In practice, most decisions or choices are likely to fall somewhere between these extremes. It would be helpful to place different
types of choices on this spectrum and to make decisions about appropriate levels and types of participation accordingly.

A number of additional challenges emerge when applying participatory decision
making to problems of sustainable development. First, it is difficult to include
all interested or affected groups, not the least of which are those members of future generations who will be irrevocably affected by our actions, but whose
voices cannot be strongly or accurately represented. In addition, even for present-day stakeholders, identifying those who should be involved, and enlisting
them, can be daunting, since potential participants extend from individuals to
entire nations. Second, it proves challenging to convey the complex science involved to both those who must negotiate what values should prevail (all citizens) and those who must make decisions. We return to these points below, but
there are no simple answers to this challenge, and it must be recognized that
many, if not most, of the participants making decisions about our complex world
do so with a limited scientific understanding as well as with diverse perceptions,
opinions, and interests regarding what should be done. Third, participatory processes may favor "consensus" solutions that reflect the need for political compromise and incrementalism, rather than reflecting environmental exigencies
that make such compromises environmentally and socially intolerable, even
when the majority of participants wish to avoid such an outcome. Finally, the inherent disparity among interested parties ultimately tends to favor the rich and
powerful, both within a society and between societies. Participatory processes
devoted to questions of sustainable development will have to find a way to
strengthen, and perhaps (given the forces orienting us toward the rich and
powerful) favor, the poor and disenfranchised.

Participatory processes can broaden the legitimacy accorded to environmental decision making and thereby increase the concern and commitment of a
range of actors in society to the goal of sustainability. At the Earth system level,
however, the processes must be designed in ways that ensure that the political
exigencies of participation do not override the environmental exigencies of the
problem being addressed.

WHAT IS NEEDED FROM SCIENCE AND TECHNOLOGY?

Recent efforts have begun to outline the core questions that a mature science of
the Earth system would strive to answer (Carley and Shapens 1998; Kates et al.

2001; Sahagian and Schellnhuber 2002). In our discussions, we focused on three groups of questions directly tied to the reconceptualization of sustainability goals quoted earlier: (a) meeting human needs while putting less pressure on the Earth's life-support systems, (b) identifying relatively "safe" or "benign" domains of operation for the Earth system, and (c) assessing the efficacy of institutions and institutional change.

Meeting Human Needs: Integrated Systems of Production, Consumption, and Distribution

The post-industrialization period witnessed large-scale commodification of natural resources, in many cases without an appropriate consideration of the resultant environmental stresses (Carley and Christie 2000). Emphasis was on overall production and consumption, with little attention paid to equitable distribution. An essential need is to explore how alternative systems of production, consumption, and distribution can be configured to provide greater levels of human prosperity while producing significantly lower levels of environmental stress, and while simultaneously accounting for regional and sectoral differences in such systems. Work has already been conducted on the production side, with efficiency improvements, "green design" principles, and pollution recovery or sequestration measures proposed to reduce environmental pressures per unit of goods produced. In addition, the science community has begun to understand the sources of variance of human consumption patterns beyond mere income and opportunity (Princen et al. 2000; Heap and Kent 2000). As part of the research agenda on economic globalization, attention has begun to focus on ways in which the increasing physical separation of production and consumption activities can lead to substantial additional environmental burdens because of extensions to distribution systems (Chisholm 1990). Nonetheless, only in the areas of energy needs and global product chains does the scientific community have even the beginnings of an integrated understanding of the environmental pressures imposed by alternative production–consumption–distribution systems (ICSU et al. 2002; von Moltke et al. 1998). Such integrated analyses of full "systems" options for advancing human well-being are badly needed as management tools to avoid the technical pitfalls inherent in focusing policy on only one or another dimension of the production–consumption–distribution chain. There are also political and equity imperatives for developing such integrated views due to the increasing tendency for rich regions' consumption to be produced in poor regions, which thereby incurs a disproportionate share of the resultant environmental burden (Gwinne 1999). Without an objective and verifiable understanding of such inevitable asymmetries, the prospects for rational management of a sustainability transition will be severely constrained. Investigation of alternative integrated systems of production, consumption, and distribution are needed in at least each of the basic needs identified by the UN Secretary General in his "WEHAB" agenda (i.e., water, energy, health, agriculture, and conservation of biodiversity; Annan 2002). A good case can be made for such analyses in the area of human habitation as well.

Protecting Lifestyles and Livelihood: Delineating "Safe" Regions, Trajectories, and Their Boundaries

One of the major challenges in achieving a transition to sustainability is the capacity of the Earth system (or parts of it) to experience sudden and irreversible shifts to undesirable or even catastrophic domains (e.g., global collapse of fisheries, cessation of thermohaline circulation, inundation of low-lying areas by sea-level rise). These "precipices" can be particularly difficult to avoid when early-warning signals of impending disaster are weak, unnoticed, or tardy relative to the time when mitigative or remedial action must be taken. Unfortunately, science has progressed very little in its capacity to identify relatively "safe" or "benign" domains in which the Earth system can operate, or safe trajectories that provide a reasonable expectation of remaining in these domains. Part of the difficulty lies in the definitions of safe and benign. At their most conservative, the words might be taken to mean that excursions across "nasty and irreversible" state changes have little chance of occurring, or can be avoided. The words as applied to the socioeconomic system may be more problematic. The history of complex civilizations suggests that life is rarely safe or benign for those at the lower levels of society. Defining safe and benign domains appropriately may require identifying "minimal" criteria that avoid increasing the share of society currently at risk due to environmental degradation, more desirable criteria in which environmental risk is actually decreased, and a minimum acceptable quality of life for all people. Whether this latter state would be considered benign by those in the lowest social strata is a matter of speculation.

Making progress in this area first requires that scientists be able to identify the principal "state variables," both socioeconomic and biogeophysical, that describe the Earth system. These state variables could then be used to identify the existence, nature, and location (in time and space) of thresholds and the more desirable domains of attraction. Since nasty and irreversible thresholds should be avoided both on regional and global scales, a more sophisticated understanding of cross-scale interactions and dynamics is warranted. Early-warning signals of change should be identified whenever possible and used to develop indicators of sustainability that can be updated and checked on a regular basis.

This will not be easy. Addressing the issue of thresholds and domains of attraction alone may take the better part of a generation. However, it must be done.

Improving the Capacity to Cope with Environmental Change: Institutions and Institutional Change

It is widely recognized that scientific information is not the only, and may not even be the primary, limitation to achieving a transition to sustainability. Lack of knowledge is frequently not the impediment to action. Global and national institutions will therefore have to evolve in many ways if humanity is to cope more effectively with global environmental problems. These changes will include, among other things: (a) recognition that Earth system maintenance is a necessary step in the pursuit of social good; (b) more effective means of handling long time horizons, including construction and negotiation of future scenarios; (c) recognition that the global marketplace can only provide some of the desired social goods (primarily those associated with the allocation of resources); and (d) capacity to recognize the possibilities and limitations of extending regionally tested institutions (e.g., emissions trading or participatory processes) to global scales. These changes must ultimately be driven by the public and governmental sectors; however, science has a role to play in analyzing and distilling past successes and failures, and providing guidelines for future institution-building efforts. We briefly elaborate on some of these research challenges below.

Scenario Building

Model development and scenario building involve the social components of the Earth system. At the simplest (although by no means simple) level, social drivers must be included in various models of the system. This, in turn, requires collecting, analyzing, and including certain social variables that have frequently been ignored in such exercises. Scenarios are not predictions, but rather explorations of the future. Strictly speaking they are not science, but careful exercises of imagination informed by science and other sources (Schwartz 1991). Scenarios can, however, be useful in understanding the choices that may face humanity in elucidating the long-term consequences of our actions and in providing a focus for science–policy dialogues. As such, they can provide useful perspectives for defining scientific priorities and exploring the implications of scientific findings in terms of sustainable development (Gallopín 2002). Scenario building, in general, needs to incorporate more quantitative elements, criteria of plausibility, and new understandings generated by Earth system science.

Effective scenario building also requires identifying the indicators or parameters of interest to the stakeholders whom the scenario-building exercise is hoping to inform. Success at such efforts requires involving stakeholders in defining the output parameters of the models or scenarios that are most relevant to the decisions they will need to be making. This process will most likely be an iterative one in which stakeholders identify initial parameters of interest, but then those parameters are revised in response and reaction to the scenario-building exercises undertaken. As with initial efforts in the natural sciences, first efforts from the social sciences are likely to be simplistic and unsophisticated, even to those creating them. However, over time, efforts in this direction are likely to pay large dividends in terms of the ability to develop models that more accurately reflect likely scenarios for the future of the Earth system than are currently being constructed.

Investments

Technological change is driven by a suite of drivers, but among them are investments. The determinants of investment decisions are, however, poorly understood by economists. Many models used in climate economics (e.g., the Ramsey model) implicitly assume that there is a complete set of future markets that guarantees intertemporal efficiency. However, recent developments have shown that capital markets are by no means intertemporally efficient. If this is true, the formation of expectations of investors as a process of social learning becomes an important issue. This formation could be influenced by global environmental management: quotas for renewable energy, tradeable permits, or carbon sequestration bonds can help to enforce investments that have a long time horizon. Until now, this insight has been widely neglected by global environmental policy. A challenging research area would be to assess institutional designs that have the potential to stabilize expectations and redirect capital flows to make an economy more sustainable.

Time Horizons and Discount Rates

Solving environmental problems frequently requires a long time horizon in decision making. Therefore, the discounting of damages of environmental change and the costs of mitigation are a highly debated issue, not least because policy advice depends heavily on the assumed discount rate. Many economists argue for the use of a discount rate that can be observed in capital markets. This argument becomes less convincing, however, in light of the increasing discount rate in many OECD countries over the last three decades, which suggests that the time horizon of investors and politicians has been reduced. Little is known about the determinants of discount-rate variation. Some social scientists have argued that both capital markets and democratic institutions tend to reduce time horizons and increase discount rates. If discount rates are susceptible to such influences, we may want to investigate how institutions can use policies to self-consciously lengthen the time horizons and decrease the discount rate, of politicians and investors, since that can increase attention to long-term environmental effects.

Extending Successful Institutions to Larger Scales

In many cases, there are local institutions that work quite effectively at resolving tragedies of the commons and other forms of environmental problems (Kaul 1996; Keohane and Ostrom 1995; Ostrom 1990; Ostrom et al. 1999). Currently an important question is whether and how lessons from these local institutions can be generalized to other localities and "scaled up" to national, regional, or global scales. Efforts in this direction require research into not only why particular local institutions were effective, but also into identifying both analogues to the sources of success in the case analyzed as well as factors in the local context that might facilitate or inhibit such strategies from working in other local settings or at higher levels of social aggregation. There is considerable room for research in identifying how lessons from useful but non-analogous situations have been and could be applied to institutions at other levels and other contexts.

Institutional Renewal

Organizations tend to grow toward large and inflexible bureaucracies aimed more toward their self-perpetuation and enhancement of power than toward the original goals they were designed to pursue. This is particularly true for organizations and agencies working outside the marketplace (where such inflexibility tends to be self-correcting). This hazard faces many natural-resource management agencies and organizations.

There is a healthy and growing literature on the kinds of institutional structures and rules that allow continued "renewal" (flexibility, adaptation) and avoid excessive, and ultimately crippling, bureaucratization. These include a fostering of small-scale "creative destruction" cycles that allow contained questioning of current conditions and creation of new strategies and approaches that can come to permeate the larger organization. More work is needed in understanding how the continued renewal and adaptation is best achieved.

A SCIENCE–POLICY DIALOGUE

If scientists are to realize their full potential in helping to pilot a transition to sustainability, they need to understand the hybrid character of a science that serves policy (Jasanoff and Wynne 1998) and the context-specific character of the interface. In particular, both scientists and policy makers must be constantly aware that they are different actors bound by diverse sets of goals and tools, with different abilities to perceive and digest uncertainty (Weiss 1978). Moreover, there are constraints to the use of scientific decision makers by policy makers, and to the use of public information by scientists, that arise from the beliefs, paradigms, and cultures surrounding the various participants in the dialogue (March and Olsen 1989). We elaborate on some of this below.

Setting Priorities for Science and Technology

Traditionally, scientists have viewed themselves as the best arbiters of appropriate questions and areas of inquiry. Negotiation with non-scientists concerning these questions is often viewed with skepticism because the resulting science is perceived as being tainted. More recently there have been calls for a "new contract" of science with society, formalizing the responsibility of scientists to be responsive to society's articulated needs (Lubchenco 1998). Similar relationships already exist in the areas of health and industrial and corporate development. There is therefore nothing fundamentally new about the notion that certain branches of science should be devoted to meeting human needs and can operate most effectively when listening to societal expressions of which outcomes are and are not desirable. This sort of social contract has not, however, been strongly present in the area of the environment. Its time has come. The science community must facilitate communication with the public and policy makers in ways that allow legitimate participation by a wide variety of stakeholders in determining the types of questions that will be asked by the scientific community and the appropriate level of resources devoted to these questions. Such advances will not be easy; in politically charged policy-making environments, for instance, scientific information is often used to insulate policy making from public accountability (Lemos et al. 2002).

Scientists need not merely wait for non-scientists to cast their votes. They can and should have strong input into the process in two ways. First, they can expand the agenda considered by the public by revealing hazards or compelling areas of inquiry of which the public may be unaware. Second, they can help in ordering the priorities, from those that are most pressing or most ripe, to those where achievement may rest on other advances, and thus be more distant.

Finally, a note of caution. It should not be assumed that humanity will have "smooth sailing" if only it could get past the current environmental problems. Were the world to solve these, others would appear. Society cannot begin to anticipate the kinds of knowledge or information we might need to solve these (unknowable) future challenges. There will always be a place for curiosity-driven science to expand the body of knowledge from which the world can make use. This pushing of intellectual frontiers is also what ultimately makes us human. We are thus not suggesting that all science be subjected to public negotiations concerning immediate needs. However, the challenges today are pressing enough that a greater proportion of our efforts need to be placed in a responsive science devoted to questions of sustainable development.

Scientific Analysis and Inquiry

The modes of objective analysis developed and refined as part of the scientific process have served the community well, and there is little room for negotiation with the public and policy makers concerning these approaches. Nonetheless, we highlight three areas in which it is critical for scientists to make advances in analysis and inquiry.

Lay Knowledge

Several studies have noted the tremendous benefits that can be derived by tapping the sophisticated understanding of coupled human–environment interactions held by many in society who have no formal advanced scientific training but do have extensive firsthand experience with the consequences of those interactions, sometimes referred to as the "lay public." These benefits include the capacity to gain access to and even test predictions immediately against data about the past already available in the minds and memories of the lay public, without having to wait for additional long-term experimentation or observation; the ability to use several different independent information sources in analysis; and an increased transparency and legitimacy of the scientific process in the eyes of the ultimate users of scientific information.

In spite of these benefits, the use of lay knowledge is limited both by the abilities of scientists to identify and sift such knowledge, and by a scientific culture that places a high value on data derived in standard scientific fashion. The scientific community must overcome both, which will place certain demands on our training program. How is lay knowledge effectively and respectfully elicited? Who are the holders of lay knowledge, and how does one know when one has enough? How are data of several different types   from narratives to lists of numbers — to be reconciled and compared?

The scientific community currently engaged in gathering lay knowledge has, however, developed certain guidelines for its incorporation. For example:

1. Make no judgments about the adequacy of conceptual models or methods.
2. Apply the same broad evidentiary principles to all sources of knowledge:
   (a) Traceability: The source of knowledge must be known and shared.
   (b) Repeatability/testability: There must be sufficient internal consistency in the knowledge system that different practitioners within that system provide compatible information and the same practitioner provides compatible information at different times (allowing for learning).
   (c) (Un)certainty: A statement of confidence should accompany key findings, along with an explicit statement about the limits of knowledge (e.g., under which circumstances is it valid, and for which period and area).

Evidentiary Standards

The scientific process is built on the goal of advancing knowledge — penetrating and reducing the reaches of what is not known. Each advance is built on knowledge acquired earlier. The cost of incorrect knowledge is therefore quite high, affecting not only that building block in the foundation, but those that follow. Science has thus evolved procedures whose primary goals are to protect against incorrect generalizations. This includes the use of relatively strict evidentiary standards designed to assure the generalizability of results.

This emphasis on strict evidentiary standards selects for fairly reductionist approaches to studying phenomena whose drivers can be tightly controlled and manipulated. Until recently, this meant largely avoiding study of precisely the kinds of environmental systems that society is influencing most profoundly: global climate systems, entire ecosystems, or landscapes. Scientists must also advance their capacity to convey what is known and what is unknown in these (sometimes highly) uncertain systems. Scientists will have to enhance their capacity to perform synthesis and assessment (drawing in large part on the past experiences with such efforts), to use Bayesian and other approaches for revising our measures of uncertainty in the light of new knowledge, and employ alternative forms of conveying uncertainty (confidence limits, consensus statements on what is and is not known). This places many demands on the training of graduate students, and an overhaul of graduate curriculums may be necessary.

Communicating the Results of Science

Much has been written about the need to improve communication among scientists, the public, and policy makers. This dialogue is crucial not only for introducing needed scientific information into decision-making processes, but for defining the appropriate areas for scientific inquiry, as we have already noted. The scientific community could improve this dialogue in several ways.

The first is to recognize that policy makers are not the only, and may not even be the most important, recipients of scientific information. Changes in policy frequently come as the result of a rising awareness in civil society about a class of problems that are inadequately addressed in the policy arena. Such dynamics characterize, at least in part, the politics of climate change; civil society also played a crucial role in spawning the environmental movement of the 1960s, which led to many new national and international environmental policies. Scientists must spend more time distilling scientific information for the public, thus enhancing their ability to understand science. This means more public lectures, more articles for magazines, and greater efforts at web-based or video-based lectures.

Perhaps the thing that most hampers effective scientific communication, however, is the arrogance that scientists often bring to the table. Scientists have been trained to hone their analytic abilities and frequently believe, or at least behave as if they believe, that their abilities are superior to those of anyone else. This is revealed in their assumptions that policy makers are unable to deal with complexity or uncertainty, and that the scientists themselves know best about how to choose among alternative courses, even when choosing among these courses ultimately involves a (subjective) choice among competing values. Scientists have a unique role to play in society, but it is by no means a role superior to others being taken.

"Communication" must therefore be a conversation, not a lecture. Scientists must expend more effort listening to and conversing with multiple publics in an effort to understand better the concerns and questions as well as models and evidence that drive broader social views of human–environment interactions.

WHAT DOES THIS MEAN FOR THE ORGANIZATION OF SCIENCE AND TECHNOLOGY?

Achieving the above, in terms of extending science to new areas of inquiry and facilitating a dialogue between the scientific community, civil society, and policy-making bodies, will mean introducing some new structures and organizing principles to the scientific community. Here we focus on three new endeavors: (a) building a global observation network for sustainability, (b) implementing global-level institutions for sustainability assessment, and (c) enhancing scientific capacity in all regions of the world.

Observation Networks: A "Macroscope"

Management that seeks to span scales in space ("think locally, act globally"), time ("take care of our grandchildren"), and topical categories ("combine human welfare with environmental stability") requires observational data to support analysis, argument, and action across these scales. Given that the departure point of experience and observation is almost completely local, creating generalized views on coarser scales is the main challenge. To build such "macroscopes" (instruments that do not view the far, such as telescopes, or the small, such as microscopes, but rather bring into view interconnections of first order between major subsystems of a complex world; see Lucht and Pachauri, this volume) requires a close coevolution of observation and corresponding formalization.

Observational data are necessary to ground thinking about the Earth system in the reality of what that system actually does, rather than in guesses about, or metaphors of, what it might do. Today, and with respect to a science of sustainability, the world lacks adequate global-scale systems for the selected observation of ecosphere–biosphere interactions. Such macroscope systems of observation and reporting, built to complement the existing systems of environmental, economic, and social measurement and accounting, are required to provide an empirical foundation, a system of empirical reference, for sustainability science.

Building macroscopic observation and reporting systems that span the whole of the Earth system, particularly the coupling of the physical, biological, and human components, is a central challenge to Earth system analysis. Properly executed, such systems would facilitate bidirectional switching — between local realities and macroscopic views of the planet — that is a prerequisite for the occurrence of a collective (though spatially and temporally heterogeneous) global intent and action. It would also support the communication between different actors at different scales that will be necessary for the emergence of a democratic and participatory form of global will. The knowledge generated through such macroscope systems cannot be the privileged possession of only some actors in the global system, but will have to be widely distributed and accessible to all.

Global Organizations for Earth System Analysis

Taking care of the Earth system and ensuring ultimate sustainability require arrangements that far transcend the scope of local communities, regions, and sovereign states. Without the full support and participation of greater entities, little can be realistically achieved.

There are a few broad methods for achieving such support. The first is to make better use of existing institutions. For the United Nations, there exists a multiplicity of topical programs ranging from the UN Environmental Program to the UN Development Program, plus such decision-making and operational groups as the General Assembly, the Economic and Social Council, and the Commission on Sustainable Development. On the scientific side, there is the International Council for Science, with its existing international research programs and its new commitment to sustainability science (ICSU 2003); the International Social Science Council (ISSC); and the Third World Academy of Science and new Inter-Academy Council, which are bringing together the world's main scientific academies on sustainability issues. There is also a range of ongoing international scientific assessments (ICSU et al. 2002). On the development side, there are countless organizations at all levels involved in sustainability issues (see, e.g., the web site of the International Institute for Sustainable Development at www.iisd.org). These groups have produced a variety of ongoing management programs as well as several hundred multilateral environmental agreements with different degrees of status and authority (Mitchell 2003).

There are some advantages to having the diverse set of approaches and perspectives represented in these organizations, not the least of which is the development of knowledge and expertise relevant to particular local or regional problems as well as a breadth of "natural experimentation" in trying to find compelling solutions to environmental problems. Nonetheless, their sheer multiplicity can mean that, collectively, they lack coherence and may undermine one another's effectiveness. No doubt more could and should be done to avoid overlap and make the work of existing programs more accountable and better coordinated. However, as things are, lack of coordination can blur the messages. Governments and business corporations can usually find means to ignore or sidestep their recommendations. Groups that want quality scientific information may not know where to turn. Moreover there is the constant danger that conflicts of interest and obligation will arise, particularly with the World Trade Organization.

A second method of proceeding is to create a new international organization that would coordinate existing institutions and devote itself to building bridges between science, technology, and the environment, on one hand, and their practical application for sustainability, on the other. The idea of a World Environment Organization has been promoted by a wide range of actors — including the

German Chancellor, the French President, and the outgoing Director General of the World Trade Organization as a way of achieving a balance between scien tific, environmental, and trade considerations on a basis of broad equality, with the necessary arrangements for judging and settling any disputes or conflicts between them. Proponents have argued that such an organization would give sustainability, in all its complexity, a single and powerful focus at a global level. Opponents cite the dubious track record of "super" organizations trying to encompass environment and development at the international or national level. In our discussions, the desirability of a single world *environmental* organization to address issues of global *sustainability* remained very much a subject of debate.

A proposition favored by some, but not all, in our group is the creation of an international process for environment and sustainability assessments that would have a stature, authority, and scientific integrity at least the equivalent of the IPCC. There would be a single international coordinating body, but with regional forums on all the populated continents of the world. These regional forums would be devoted to *assessment* of the scientific knowledge base for sustainable development; the international coordinating body would facilitate dialogue among the regional forums, as well as serve as a clearinghouse for information from other ongoing regional and global assessments. As in the IPCC, the scientific analysis itself would not be subject to revision by policy makers, although the decisions concerning which areas of analysis are most useful and how they should be reported would be subject to negotiation. Other ongoing assessments should be studied to determine what does and does not work in fostering dialogue between scientists, politicians, and stakeholders, while still maintaining scientific independence and integrity in assessing the state of the world. Work done to date suggests that regional forums would need to play a much larger role in such sustainability efforts than they do currently in most international efforts (Clark et al. 2002; Farrell et al. 2001). These regional forums are of critical importance in ascertaining the different challenges that visit the major regions of the world, particularly the different sustainability challenges faced by the richer and poorer nations. To maximize learning, a fluid exchange of practitioners and information among these regional forums is necessary. Regional forums could also coordinate the crucial acquisition of lay knowledge and public priorities for assessment. Coordination in a single body holds the promise of raising the international profile of sustainable development, which is certainly as deserving of attention as global climate change.

Increasing Regional Scientific Capacity

There is an immense need to increase the capacity for research and development in many of the low- and middle-income regions of the world. Without this increased capacity, the world has little hope of solving the challenges of sustainable development: our knowledge of regional- and place-based dynamics would be incomplete, and our understanding of how cultural persuasion and social history influence perceptions of futures and efficacy of policies would be inadequate. Substantial new resources, far beyond those currently available for such capacity building, will be needed. The richer countries of the world need to see both the ethical and practical mandates for supplying such resources.

At the same time, it needs to be recognized that there are many developing countries that already have significant scientific capacity. The scientific community needs to draw on this capacity whenever possible. Human–environment interactions likely have different dynamics in different regions of the world; failure to test the insights gained in one region against another means both risking the opportunity to identify valid generalizations and missing the opportunity to probe local findings with relevant perspectives from elsewhere.

Individual

The capacity to train individual scientists from nations currently lacking in adequate scientific capacity must be increased. For many of these students, the best graduate education and training can still be found in the richer nations of the world; however, these programs need to encourage students to return to their home regions for dissertation research. It is also crucial that training and support not end with completion of a graduate degree. Modest postgraduate resources should be provided, including small "start-up" funds to begin in-country research, funds for a few return trips to the graduate institution and/or for travel to scientific meetings, as well as funds for journal subscriptions and crucial textbooks. It is essential that university programs come to consider this an indispensible part of the degree-granting process. Otherwise, individual scientists, returning to countries with weak support for science, risk failure or isolation, which seriously reduces the value of the degree.

Collaborative Research Teams

North–South and South–South research collaborations, and their equivalents, are a valuable way to enhance regional scientific capacity. The focus of much sustainability research should be place-based, either comparatively across sites located in several countries or at sites in less-developed nations. It is essential that these collaborative efforts be built on the assumption and practice of equity among all the participants, even though the funding contributions will rarely be equal. The design meetings for the research projects must, whenever possible, be held in the less-developed country to enhance the visibility of local scientific participants and to emphasize the local importance of the research being conducted. Similarly, outreach programs in which all scientific members participate, including "open days" and policy briefings, are crucial for visibility and future enhancement of capacity building. These outreach programs also provide a forum for obtaining societal feedback concerning the appropriateness of the research and the applicability of the findings. Data archives and networks must be maintained locally or regionally, in recognition of the "digital divide" that prevents many developing country scientists and citizens from accessing large data files stored in other locations. Information technology, therefore, needs to be an additional target of capacity building. All scientific findings should be made available locally and in local languages, either through direct translation of papers appearing in professional journals or through the production of reports containing the critical findings.

Institutions and Networks

The establishment of regional synthesis centers devoted to sustainability–environment issues can promote in-region scientific capability, both by supplying employment opportunities for local scientists and through networking with other institutions that can leverage resources and information. Regional synthesis centers should be established with clear connections to existing centers and organizations and, when possible, close physical proximity to some or all of those partner agencies. South–South networking is as crucial as North–South networking, and partner organizations in other countries in the region should be identified whenever possible. These regional synthesis centers could also serve as crucial links in the global to regional to local science–policy–public dialogue that we envision as a crucial component of the transition to sustainability. These centers could also facilitate South–South networking of individual scientists or research teams through workshops, short courses, and annual meetings.

The scientific community must make a particular effort to build capacity in regions that are isolated, or isolate themselves, from the global community as a result of political disagreements or dynamics. Long-term isolation from the global community has often resulted in serious environmental degradation, crushing economic problems, and tragic social dislocation. Recovery from this state of affairs will require scientific and technical capacity, and having even a modest foundation from which to start can alleviate significant human suffering.

CONCLUSIONS

We emerged from this Dahlem Workshop with recommendations in three areas: scientific inquiry, scientific communication, and organization of the science community (see Box 20.1). These recommendations necessitate changes in the culture of science (what scientists see as important and how they view their role in the environmental dialogue), in scientific training (producing students with a grasp of synthesis, improved abilities for broad dialogues, and the tools to analyze and convey complexity and uncertainty), and the resources devoted to science (new research programs, establishment of an international assessment body, and regional capacity building).

Box 20.1 Recommendations emerging from the 91st Dahlem Workshop.

1. Promote new Earth system sustainability research in the areas of:
   - Meeting human needs by designing integrated systems of production, consumption, and distribution that radically reduce environmental impact.
   - Protecting life and livelihood support systems by delineating relatively "safe" or "benign" domains in the Earth system, including with it the identification of the principal biogeophysical and socioeconomic variables that define the state of the Earth system.
   - Improving human capacity to cope with environmental problems by improving the efficacy of institutions in linking the global and the local.
2. Improve scientific communication by:
   - Including the public and policy makers in identifying key questions and priority research areas in Earth system studies.
   - Incorporating lay knowledge in scientific analysis and assessments.
   - Employing a broader suite of tools for analyzing and conveying complexity and uncertainty.
3. Enhance the capacity of the science and technology community to provide scientific information on sustainability by:
   - Developing global observing networks for sustainability with an integrative regional focus.
   - Promoting internationally coordinated scientific assessment on the implications of different development paths for sustainability.

Given the plethora of books, reports, and articles written on sustainability, sustainable development, and the need for reform in the scientific community, we suspect there is little in this document that is new. Nonetheless, there is ample evidence that progress is made not only by saying something new, but by saying something, old and true, repeatedly. Thus, we say it again. The current state of affairs — including the extant approaches to scientific research and the capacity to sustain a reasonable dialogue among scientists, policymakers, and the public — is simply incommensurate with the challenges the world faces.

History has shown that humanity will not discover a pathway toward sustainability by accident. A conscious awareness of civilization's present and possible future trajectories as well as a conscious effort toward continuous learning and course correction is necessary. Moreover, humanity must improve upon the abilities of previous, and now absent, complex civilizations for such self-reflection. Technological, scientific, economic, and political advances give rise to an optimism that we may develop an adequate "global will" for sustainable development. The immensity of the challenge — including its global nature, potential for irreversible excursions into undesirable or catastrophic states, and long time dynamics — is more sobering.

A FINAL NOTE

There was a broad sweep of topics at the 91st Dahlem Workshop on Earth System Analysis for Sustainability: from a retrospective assessment of the emergence of life over the 4-billion-year history of the planet (see Chapter 6, this volume) to a forward-looking assessment of the prospects for a transition toward sustainability in our group. What connects these dialogues in a single conference? A guiding principle for some, but not all, of the participants is Gaia — a notion that life creates the conditions for its own persistence. If humanity is to achieve a transition to sustainability, it will likely require a fundamental shift in the prevailing view of the world: from linear, compartmentalizable, mechanical to complex, interconnected, living. In this, Gaia may provide some hope and some answers. We resist, however, the suggestions of some that Gaia may also provide a blueprint for humanity's transition to sustainability. A principle that may explain the emergence and persistence of life broadly over an ancient and archaic sweep of time seems to have little in common with the efforts of a single species to not just maintain life in general but to enhance it for all its members in the space of a century or so. Humans are a unique species, with language, foresight, memory, and dreams. Those dreams tell us that the goal of humanity is not merely to persist, but to thrive.

ACKNOWLEDGMENTS

We have benefited from fruitful conversations with all of the participants at this Dahlem Workshop. We wish to thank particularly Bert Bolin, Alison Jolly, Bob Scholes, Will Steffen, Nebojsa Nakicenovic, and Liana Talaue-McManus for thoughtful comments that contributed directly to the discussions and conclusions contained in this report.

REFERENCES

Allen, T.F.H., J.A. Tainter, and T.W. Hoekstra. 2003. Supply-side Sustainability. New York: Columbia Univ. Press.

Annan, K. 2002. Toward a sustainable future. *Environment* 44:10–15.

Berkes, F., and C. Folke, eds. 1998. Linking Social and Ecological Systems: Management Practices and Social Mechanisms for Building Resilience. Cambridge: Cambridge Univ. Press.

Bourdieu, P., J.-C. Passeron, and C. Chamboredon, eds. 1983 (1968). Le métier de sociologue: Préalables épistémologiques. Paris: Mouton.

Bromley, D., ed. 1992. Making the Commons Work: Theory, Practice, and Policy. San Francisco: ICS Press.

Burger, J., E. Ostrom, R.B. Norgaard, D. Policansky, and B.D. Goldstein, eds. 2001. Protecting the Commons: A Framework for Resource Management in the Americas. Washington, D.C.: Island Press.

Carley, M., and I. Christie. 2000. Managing Sustainable Development. London: Earthscan Publ.

Carley, M., and P. Shapens. 1998. Sharing the World: Sustainable Living and Global Equity in the 21st Century. London: Earthscan Publ.

Chisholm, M. 1990. The increasing separation of production and consumption. In: The Earth as Transformed by Human Action: Global and Regional Changes in the Biosphere over the Past 300 Years, ed. B.L. Turner II, W.C. Clark, R.W. Kates et al., pp. 87–101. Cambridge: Cambridge Univ. Press.

Clark, W.C. 2000. Environmental globalization. In: Governance in a Globalizing World, ed. J.J. Nye and J. Donahue, pp. 86–108. Washington, D.C.: Brookings Press.

Clark, W.C., R. Mitchell, D.W. Cash, and F. Alcock. 2002. Information as influence: How institutions mediate the impact of scientific assessment on global environmental affairs. Working Paper RWP02–04. Cambridge, MA: Kennedy School of Govt.

Dovers, S., and C. Mobbs. 1997. An alluring prospect? Ecology and the requirements of adaptive management. In: Frontiers in Ecology: Building the Links, ed. N. Klomp and I. Lunt, pp. 39–52. London: Elsevier.

Farrell, A., J. Jäger, and S. VanDeveer. 2001. Environmental assessments: Four under-appreciated elements of design. *Global Env. Change* **11**:311–333.

Folke, C., S. Carpenter, T. Elmqvist et al. 2002. Resilience and Sustainable Development: Building Adaptive Capacity in a World of Transformations. Series on Science for Sustainable Development, No. 3. Paris: ICSU (Intl. Council of Sci. Unions).

Gallopín, G.C. 2002. Planning for resilience: Scenarios, surprises, and branch points. In: Panarchy: Understanding Transformations in Human and Natural Systems, ed. L. Gunderson and C.S. Holling, pp. 361–392. Washington, D.C.: Island Press.

Gunderson, L., and C.S. Holling, eds. 2001. Panarchy: Understanding Transformations in Human and Natural Systems. Washington, D.C.: Island Press.

Gunderson, L., C.S. Holling and S.S. Light, eds. 1995. Bridges and Barriers to the Renewal of Ecosystems and Institutions. New York: Columbia Univ. Press.

Gwinne, R. 1999. Globalization, commodity chains, and fruit exporting regions in Chile. *Tidschrift voor Economische en Sociale* **90**:211–226.

Hardin, G. 1968. The tragedy of the commons. *Science* **162**:1343–1348.

Heap, B., and J. Kent, eds. 2000. Towards Sustainable Consumption: A European Perspective. London: The Royal Society.

Holling, C.S., ed. 1978. Adaptive Environmental Assessment and Management. London: Wiley.

ICSU, ISTS and TWAS. 2002. Science and Technology for Sustainable Development: Consensus Report and Background Document for the Mexico City Synthesis Conf., 20-23 May 2003. Series on Science for Sustainable Development, No. 9. Paris: ICSU.

ICSU. 2003. Report of the CSPR Assessment Panel on Environment and its Relation to Sustainable Development, ed. R. Watson, A. Buttimer, A. Cropper et al. Paris: ICSU.

Jasanoff, S., and B. Wynne. 1998. Science and decision-making. In: Human Choice and Climate Change: The Societal Framework, ed. S. Rayner and E. Malone. Columbus: Battelle Press.

Kates, R.W., W.C. Clark, R. Corell et al. 2001. Sustainability science. *Science* **292**:641–642.

Kaul, M.C. 1996. Common Lands and Customary Law: Institutional Change in North India over the past Two Centuries. Delhi: Oxford Univ. Press.

Keohane, R.O., and E. Ostrom, eds. 1995. Local Commons and Global Interdependence. Thousand Oaks, CA: Sage Publ.

Lee, K. 1993. Compass and Gyroscope: Integrating Science and Politics for the Environment. Washington, D.C.: Island Press.

Lemos, M.C., T.J. Finan, R.W. Fox, D.R. Nelson, and J. Tucker. 2002. The use of seasonal climate forecasting in policymaking: Lessons from Northeast Brazil. *Climatic Change* **55**.479–507.

Lubchenco, J. 1998. Entering the century of the environment: A new social contract for science. *Science* **279**:491–497.

March, J.G., and J.P. Olsen. 1989. Rediscovering Institutions. New York: Free Press.

McCay, B.M., and J.M. Acheson, eds. 1987. The Question of the Commons: The Culture and Ecology of Communal Resources. Tucson: Univ. Arizona Press.

Mitchell, R.B. 2003. International environmental agreements: A survey of their features, formation, and effects. *Ann. Rev. Env. Resour.* **28**: 429–461.

NRC (National Research Council). 1996. Upstream: Salmon and Society in the Pacific Northwest, ed. Board on Environmental Studies and Toxicology, Committee on Protection and Management of Pacific Northwest Anadromous Salmonids, p. 472 ff. Washington, D.C.: Natl. Acad. Press.

NRC 1999. Our Common Journey: A Transition Toward Sustainability. Washington, D.C.: Natl. Acad. Press.

Ostrom, E. 1990. Governing the Commons: The Evolution of Institutions for Collective Action. Cambridge: Cambridge Univ. Press.

Ostrom, E., J. Burger, C.B. Field, R.B. Norgaard, and D. Policansky. 1999. Revisiting the commons: Local lessons, global challenges. *Science* **284**:278–284.

Ostrom, E., T. Dietz, N. Dolsak et al., eds. 2002. The Drama of the Commons. Washington, D.C.: Natl. Acad. Press.

Princen, T., M. Maniates, and K. Conca, eds. 2000. Confronting Consumption. Cambridge, MA: MIT Press.

Redman, C.L. 1999. Human Impact on Ancient Environments. Tucson: Univ. Arizona Press.

Redman, C.L., and A.P. Kinzig. 2003. Resilience of past landscapes: Resilience theory, society, and the longue durée. *Conserv. Ecol.* **7**:14[online]. http://www.consecol.org/vol7/iss1/art14/index.html

Sahagian, D., and H.-J. Schellnhuber. 2002. GAIM in 2002 and beyond: A benchmark in the continuing evolution of global change research. *IGBP Newsl.* **50**:7–10.

Schwartz, P. 1991. The Art of the Long View. New York: Doubleday.

Simon, H. 1977. Models of Discovery and Other Topics in the Methods of Science. Boston: Reidel Publ. Co.

Steffen, W., J. Jäger, D.J. Carson, and C. Bradshaw, eds. 2003. Challenge of a Changing Earth: Proc. Global Change Open Science Conference, Amsterdam, The Netherlands, July 10–13, 2001. Newbury: CPL Press.

Tainter, J.A. 1988. The Collapse of Complex Societies. Cambridge: Cambridge Univ. Press.

Turner, B.L. II, W.C. Clark, R.W. Kates et al., eds. 1990. The Earth as Transformed by Human Action: Global and Regional Changes in the Biosphere over the Past 300 Years. Amsterdam: Cambridge Univ. Press.

Vitousek, P.M., H.A. Mooney, J. Lubchenco, and J.M. Melillo. 1997. Human domination of the Earth's ecosystems. *Science* **277**:494–499.

von Moltke, K., O.J. Kuik, N.M. van der Grijp et al. 1998. Global Product Chains: Northern Consumers, Southern Producers, and Sustainability. UNEP Environment and Trade Series 15. Geneva: UNEP.

Walters, C. 1986. Adaptive Management of Renewable Resources. New York: MacMillan.

Walters, C. 1997. Conservation ecology: Challenges in adaptive management of riparian and coastal ecosystems. *Conserv. Ecol.* **1**:1 [online] http://www.consecol.org/vol1/iss2/art1/

WCED (World Commission on Environment and Development). 1987. Our Common Future. Oxford: Oxford Univ. Press.

Weiss, C.H. 1978. Improving the linkage between social research and public policy. In: Knowledge and Policy: The Uncertain Connection, ed. L.E. Lynn, pp. 23–81. Washington, D.C.: Natl. Acad. Press.

Young, O.R. 2002. The Institutional Dimensions of Environmental Change: Fit, Interplay, and Scale. Cambridge, MA: MIT Press.

Young, O.R., A. Agrawal, L.A. King et al. 1999. Institutional Dimensions of Global Environmental Change (IDGEC) Science Plan. Hanover, NH: IDGEC Intl. Project Office. [online] http://fiesta.bren.ucsb.edu/~idgec/html/publications/publications.html

Name Index

Subject Index